REMOTE SENSING AND GLOBAL
ENVIRONMENTAL CHANGE

REMOTE SENSING AND GLOBAL ENVIRONMENTAL CHANGE

SAMUEL PURKIS
AND
VICTOR KLEMAS

⟨W⟩WILEY-BLACKWELL

A John Wiley & Sons, Ltd., Publication

Registered Office
John Wiley & Sons Ltd, The Atrium, Southern Gate, Chichester, West Sussex, PO19 8SQ, UK

Editorial Offices
9600 Garsington Road, Oxford, OX4 2DQ, UK
The Atrium, Southern Gate, Chichester, West Sussex, PO19 8SQ, UK
111 River Street, Hoboken, NJ 07030-5774, USA

For details of our global editorial offices, for customer services and for information about how to apply for permission to reuse the copyright material in this book please see our website at www.wiley.com/wiley-blackwell

Library of Congress Cataloguing-in-Publication Data 1006409294

Purkis, Samuel J.
  Remote sensing and global environmental change / Samuel Purkis and Victor Klemas.
    p.   cm.
  Includes index.
  ISBN 978-1-4443-3935-2 (cloth) – ISBN 978-1-4051-8225-6 (pbk.)
 1. Global environmental change–Remote sensing.   2. Environmental monitoring–Remote sensing.
I. Klemas, V.   II. Title.
  GE149.P87 2011
  550.28′4–dc22

                                                2010043279

A catalogue record for this book is available from the British Library.

This book is published in the following electronic formats: eBook 9781444340266; Wiley Online Library 9781444340280; ePub 9781444340273

Set in 10/12.5pt Minion by SPi Publisher Services, Pondicherry, India
Printed and bound in Malaysia by Vivar Printing Sdn Bhd

1   2011

Contents

This book has a companion website: www.wiley.com/go/purkis/remote

Preface

This book is intended to provide the reader with a broad grounding in the science of Earth observation (EO) of our changing planet. It contains a comprehensive sequenced discussion covering the significant themes of global change, their causes and how they can be monitored through time. In doing so, it represents a good source of basic information while providing a general overview of the status of remote sensing technology. The text will serve as an invaluable reference for managers and researchers, regardless of their specialty, while also appealing to students of all ages.

The scope of the work yields a reference book that presents the science of EO through a series of pertinent real-world environmental case studies. It offers excellent background material for a course curriculum, being tuned in terms of its length and material covered to fit a variety of teaching scenarios.

The book has been written with students from both bachelors and masters degree programs in mind. For the former group, it contains sufficient material to support a full-semester course; for the latter, the work is intended to serve as a user-friendly introductory text for newcomers to a master's program in environmental or climate change where the role of EO is considered. The book is thus aimed at a broad audience concerned with the application of remote sensing in Earth science, biological science, physical geography, oceanography and environmental science programs. It is designed for a reader who does not require an in-depth knowledge of the technology of EO, but who needs to understand how it is used as a key tool for mapping and monitoring global change. As such, the book is also intended to serve as an easily approachable text for professionals already working in fields such as ecology, environmental science and engineering, land use planning, environmental regulation, etc., who come across remote sensing in their work and would benefit from learning more about its practical uses, but who are disinclined to take another master's course or delve into the mathematical tomes on the subject.

The first four chapters of the book introduce the fundamentals of EO, available platforms and the basic concepts of image processing. This will provide an introductory treatment of remote sensing technology to readers who have not been previously exposed to the field. Chapters 5 through 12 each present an important environmental application that:

   i is relevant to global change or the status of the biosphere; and
  ii lends itself to remote monitoring. Each case study offers insights into new EO techniques.

The work presents the fundamental mechanisms of environmental change and the technology underpinning each sensor relevant for its detection and measurement. An in-depth mathematical treatment of the science is purposely avoided, which should make the text especially appealing to many students and professionals with a non-numerate science background. The final chapter provides a look into the near and more distant future of EO as a tool for monitoring global change, before closing with a sombre but pragmatic look at how climate change has become the defining issue of our time.

The book's framework is based on using examples from the recent literature, via a gap-bridging cross-disciplinary approach, rather than presenting a classical 'textbook' that simply projects the author's understanding or perception of the use of remote sensing for the study of global environmental change. This inevitably leads to the citation of numerous references that may be of considerable value to those readers desiring to pursue a topic in more detail, and also ensures that these sources will be duly credited and that readers have direct access to them.

Given its emphasis on transmitting concepts rather than techniques, the book does not include problems or exercises. At the end of each chapter, however, a series of 'key concepts' are presented that summarize the most crucial points covered. These provide appropriate material that can be developed into challenging exercises, even for advanced courses, if the instructor chooses to elaborate on the themes with a more mathematical underpinning of the technologies illustrated.

Samuel J. Purkis
National Coral Reef Institute
Nova Southeastern University
Dania Beach, Florida, USA

Victor V. Klemas
College of Earth, Ocean and Environment
University of Delaware
Newark, Delaware, USA

Acknowledgements

We wish to thank Dr. Ian Francis for the opportunity to undertake this project and his guidance of the text from conception to birth. Invaluable comments, for which we are extremely grateful, were provided by two anonymous reviewers. Throughout this endeavour, Sam Purkis was supported by the National Coral Reef Institute and Nova Southeastern University's Oceanographic Center. Similarly, support was provided to Vic Klemas by the College of Earth, Ocean and Environment, University of Delaware. We are indebted to Chris Purkis for his patient and unwavering assistance with the artwork. The text draws upon large numbers of images and illustrations from the work of others, and we appreciate the generosity of the many individuals and publishers who have made available source materials and gave permission for their use.

Our children and grandchildren, Isis, Grace, Andy, Paul, Tom, John Paul and Asta, provided an oft-needed distraction and perspective on the priorities of life. Writing this book demanded considerable time and energy, a burden that was shared as much by family as by the authors. We thank our respective wives, Lotte Purkis and Vida Klemas, for their support, which stretched far beyond the call of duty.

It is our hope that the publication of this book will provide stimulation to a new generation of students and researchers to perform in-depth work and analysis of our changing Earth using remote sensing.

Introduction

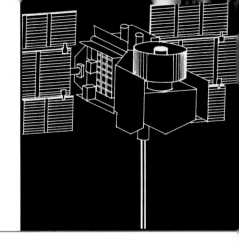

The Earth's climate is now clearly changing and getting warmer. Many components of the climate system are changing at rates and in patterns that are not natural and are best explained by the increased atmospheric abundances of greenhouse gases and aerosols generated by human activity during the 20th century. These changes include temperatures of the atmosphere, land and ocean; the extent of sea ice and mountain glaciers; the sea level; the distribution of precipitation; and the length of the seasons (AGU, 2008).

The Intergovernmental Panel on Climate Change (IPPC), made up of hundreds of scientists from 113 countries, reached a consensus in 2007 that, based on new research concluded in the last decade, it is 90 per cent certain that human-generated greenhouse gases account for most of the global rise in temperatures. The IPPC was very specific, predicting that, even under the most conservative scenario, the global increase of temperature will be between 1.1 °C and 6.4 °C by 2100, and the sea level will rise between 18 cm and 58 cm during that same time period (Collins *et al.*, 2007; IPCC, WMO/UNEP, 2007).

The Earth has warmed consistently and unusually over the past few decades in a manner that can be explained only when a greenhouse process is overlaid on orbital variation, solar variation, volcanic eruptions and other natural disturbances. Observational evidence, complex modelling and simple physics all confirm this. Whatever the proportion of human-induced rise in global temperature versus natural rise, there is no doubt that the temperature and the sea level *are* rising, the Greenland and Antarctic ice sheets *are* disintegrating, and major weather patterns and ocean currents *are* shifting (Figs. 1.1 and 1.2). The Earth's warming is already causing severe droughts and flooding, major vegetation transformations in deserts and forests, massive tundra methane releases and the degradation of the Amazon rainforest and Saharan vegetation (Gates, 1993). It is also starting to impact the Indian Ocean Monsoon, the Atlantic Conveyor Belt and El Niño weather patterns. The economic impacts of droughts in the USA alone cause $6–8 billion in losses per year (Chagnon, 2000).

Remote Sensing and Global Environmental Change, First Edition. Samuel Purkis and Victor Klemas.
© 2011 Samuel Purkis and Victor Klemas. Published 2011 by Blackwell Publishing Ltd.

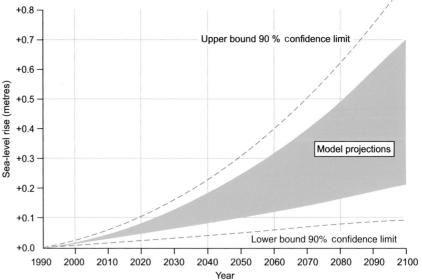

Figure 1.1 The projected range of global averaged sea level rise, re-plotted from the IPCC 2001 Third Assessment Report for the period 1990 to 2100.

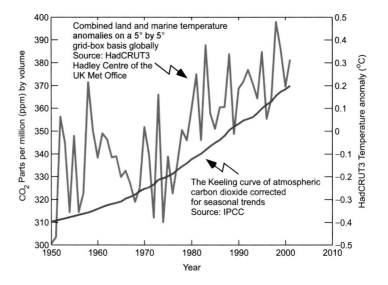

Figure 1.2 The atmospheric concentration of carbon dioxide versus the occurrence of temperature anomalies from 1950 to 2001 (IPCC, 2007). An anomaly, in this case, is expressed as the difference between the observed annual global land-surface air temperature and the 1961 to 1990 mean.

Decision-makers and scientists need reliable science-based information to make informed judgements regarding policy and actions in order to be able to address the risks of such changes and variability in climate and related systems. To have any hope of mitigating or adapting to these mostly undesirable changes, we must be able to monitor them continuously and over large global regions. Ship and field observations have provided important data on these phenomena in the past and will do so in the future. However, to be able to observe environmental changes globally, it is necessary to use remote sensors on satellites and aircraft in order to extend local measurements to large areas.

For example, without satellite remote sensing we could not have mapped accurately the changes in the Antarctic ozone hole or the disintegration of ice sheets in Greenland and other areas. In fact, it was the ability to view these global changes with remote sensors from satellite altitudes that brought the enormity and severity of these environmental changes to the attention of scientists, politicians and the general public. Many such important datasets are now available in near-real time at no cost, through web portals such as *Google Earth* and *Google Ocean*, allowing 'citizen' scientists to accomplish research objectives and promoting public engagement with science in general.

Remote sensing is now a mature enough technology to answer some of the fundamental questions in global environmental change science, namely:

1 How and at what pace is the Earth system changing and what are the forces causing these changes?
2 How does the Earth system respond to natural and human-induced changes?
3 How well can we predict future perturbations to the Earth system and what are the consequences of change to human civilization?

Ice sheets, ocean currents and temperatures, deserts and tropical forests each have somewhat different remote sensing requirements. For instance, ocean temperatures are measured by thermal infrared sensors, while ocean currents, winds, waves, and sea level require various types of radar instruments on satellites. Most ocean features are large and require spatial resolutions of kilometres, while observations of desert or forest changes may require resolutions of tens of metres and many bands within the visible and near-infrared region of the electromagnetic spectrum. Monitoring of coral reefs demands even finer spatial resolution and multiple bands pooled in the short-wavelength visible spectrum.

Fortunately, by the turn of this century, most of these requirements had been met by NASA (the National Aeronautics & Space Administration) and NOAA (the National Oceanic & Atmospheric Administration) satellites and aircraft, the European Space Agency (ESA) and the private sector (Jensen, 2007). Furthermore, new satellites are being launched, carrying imagers with fine spatial (0.6–4 m) and spectral (200 narrow bands) resolutions, as well as other environmental sensors. These provide a capability to detect changes in both the local and the global environment even more accurately. For the first time, constellations of satellites are being launched with the sole aim of quantifying aspects of the Earth's climate synergistically. With such technology available, governments are no longer alone in being able to monitor the extent of tropical forests and coral reefs, the spread of disease and the destruction caused by war.

Advances in the application of Geographical Information Systems (GIS) and the Global Positioning System (GPS) help to incorporate geo-coded ancillary data layers in order to improve the accuracy of satellite image analysis. When these techniques for generating, organizing, sorting and analyzing spatial information are combined with mathematical climate and ecological models, scientists and managers can improve their ability to assess and predict the impact of global environmental changes and trends (Lunetta & Elvidge, 1998).

To handle the vast quantities of information being generated by today's Earth observation programmes, there have been significant advances made in the use of the Internet to store and disseminate geospatial data to scientists and the public. The Internet is set to play an even greater role in the handling of products delivered by future missions.

This book is intended to provide the reader with a broad grounding in the science of Earth observation of our changing planet. It contains a comprehensive sequenced discussion that covers the significant themes of global change, their cause, and how they can be monitored through time. In doing so, it represents a good source of basic information and a general overview of the status of remote sensing technology.

The text will serve as an invaluable reference for managers and researchers, regardless of their specialty, while also appealing to students of all ages. The scope of the work yields a reference book that presents the science of remote sensing through a series of pertinent real-world environmental case studies. It offers excellent background material for a course curriculum, being tuned in terms of its length, and the material covered, to fit a variety of teaching scenarios. Each chapter presents an important environmental phenomenon that:

    i is relevant to global change or the status of the biosphere; and
   ii lends itself to remote monitoring.

Each case study offers insights into new remote sensing techniques. The work presents the fundamentals of the technology underpinning each sensor type and delivers sufficient detail for the reader to grasp the mode of operation of the instrument and how it can be used to detect and measure the environmental parameters at hand. Thus, the book is aimed at a broad audience concerned with the application of remote sensing in Earth science, biological science, physical geography, oceanography and environmental science programmes. It is designed for a reader who does not require an in-depth knowledge of the technology of remote sensing, but who needs to understand how it is used as a key tool for mapping and monitoring global change.

1.1　Key concepts

1 The Earth's climate is getting warmer and the patterns of the weather and ocean currents are changing. Severe droughts and flooding are becoming more prevalent and the ice sheets of Greenland and at the poles are disintegrating.

2 The global sea level is rising by about 2 to 3 mm per year, threatening to inundate many coastal areas by the end of this century.

3 Remote sensors on satellites offer an effective way for monitoring environmental trends on a global scale. They can detect physical and biological changes in the atmosphere, in the oceans and on land. Satellite systems have become the defining technology in our ability to quantify global change.

4 The accuracy and applicability of satellite imagery is constantly improving due to technological advances, such as finer spectral/spatial resolution, more powerful computers, the Global Positioning System (GPS) and Geographical Information Systems (GIS).

5 When these techniques for generating, organizing and analyzing spatial information are combined with mathematical and environmental models, scientists and managers have a means for assessing and predicting the impact of global environmental changes.

2

Remote sensing basics

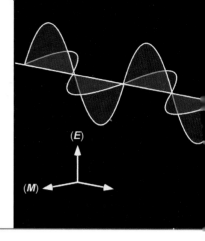

Remote sensing is primarily concerned with collecting and interpreting information about an object or landscape from a remote vantage point. The platform can be anywhere, ranging from a balloon just above the surface of the Earth to a satellite hundreds of kilometres away in space (Figure 2.1). Examples of remote sensing include aerial photography, satellite imagery, radar altimetry and laser bathymetry. Coupled with ground measurements, remote sensing can provide valuable information about the surface of the land, the oceans and the atmosphere.

Techniques for acquiring aerial photographs were already developed in the 1860s; however, colour and colour-infrared (CIR) aerial photographs were not widely used until the 1940s and 1950s. The 1950s and 1960s marked the appearance of remote sensing applications for airborne radar and video technologies.

A significant event in terms of land remote sensing was the 1972 launch of the first Landsat satellite, originally called ERTS-1. The satellite was designed to provide frequent broad-scale observations of the Earth's land surface. Since 1972, additional Landsat satellites have been put into orbit. Other countries, including France, Japan, Israel, India, Iran, Russia, Brazil, China, and perhaps North Korea, have also launched satellites whose onboard sensors provide digital imagery on a continuous basis.

2.1   Electromagnetic waves

Electromagnetic (EM) energy refers to all energy that moves with the velocity of light in a harmonic wave pattern. A harmonic pattern consists of waves that occur at equal intervals in time. EM waves can be described in terms of their velocity, wavelength and frequency. All EM waves travel at the same velocity (c). This velocity is commonly referred to as the *speed of light*, since light is one form of EM energy. For EM waves moving through a vacuum, $c = 299,793 \, \text{km sec}^{-1}$ or, for practical purposes, $c = 3 \times 10^8 \, \text{m sec}^{-1}$.

Remote Sensing and Global Environmental Change, First Edition. Samuel Purkis and Victor Klemas.
© 2011 Samuel Purkis and Victor Klemas. Published 2011 by Blackwell Publishing Ltd.

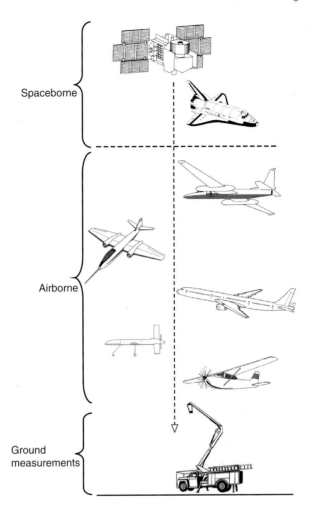

Figure 2.1 Remote sensing platforms.

The *wavelength* (λ) of electromagnetic waves is the distance from any point on one cycle or wave to the same position on the next cycle or wave (Figure 2.2). The units of measurement usually used in conjunction with the electromagnetic spectrum are the micrometre (μm), which equals 1×10^{-6} metres, or the nanometre (nm), which equals 1×10^{-9} metres. This book adopts the nanometre as its unit of wavelength.

Unlike velocity and wavelength, which change as EM energy is propagated through media of different densities, frequency remains constant and is therefore a more fundamental property. Electronic engineers use frequency nomenclature for designating radio and radar energy regions, but this book uses wavelength rather than frequency in order to simplify comparisons among all portions of the EM spectrum.

Velocity (c), wavelength (λ), and frequency (f) are related by $c = \lambda f$. The direction of the E field vector determines the polarization of the wave. Thus the EM wave in Figure 2.2 is vertically polarized.

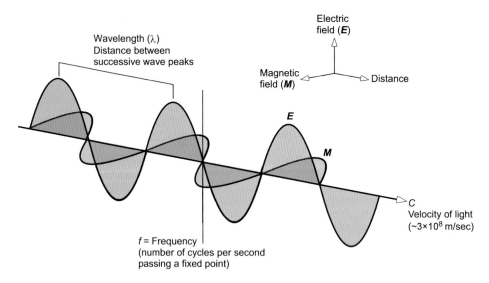

Figure 2.2 An electromagnetic wave, including a sinusoidal electric wave E and a similar magnetic wave M at right angles, both being perpendicular to the direction of propagation.

2.2 The electromagnetic spectrum

The *electromagnetic spectrum* is the continuum of energy that ranges from nanometres to metres in wavelength, travels at the speed of light and propagates through a vacuum such as outer space. All matter radiates a range of electromagnetic energy such that the peak intensity shifts toward progressively shorter wavelengths as the temperature of the matter increases. Figure 2.3 shows the EM spectrum, which is divided on the basis of wavelength into *regions*. The EM spectrum ranges from the very short wavelengths of the gamma-ray region (measured in fractions of nanometres) to the long wavelengths of the radio region (measured in metres).

Remote sensors have been refined over the past several decades to cover the visible, reflected infrared, thermal infrared and microwave (radar) regions of the EM spectrum (Figure 2.3). Besides photographic and video cameras, digital satellite sensors are used for mapping vegetation, general land cover, and ocean features. As we will see in Chapter 3, some of these sensors are passive, like infrared scanners and microwave radiometers, while others are active, such as radar and laser altimeters.

2.3 Reflectance and radiance

Electromagnetic waves that encounter matter, whether solid, liquid, or gas, are called *incident* radiation. Interactions with matter can change the following properties of the incident radiation: intensity; direction; wavelength; polarization; and

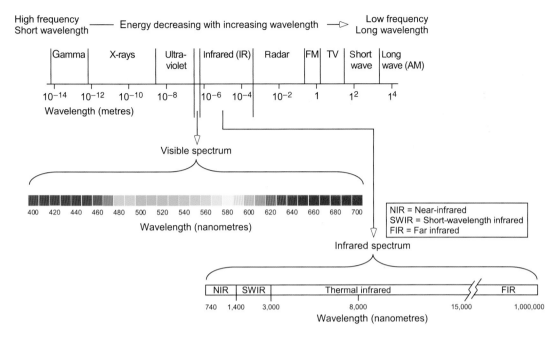

Figure 2.3 The electromagnetic spectrum.

phase. Remote sensors detect and record these changes. We then interpret the resulting images and data to determine the characteristics of the matter that interacted with the incident electromagnetic energy (Sabins, 2007).

During interactions between electromagnetic radiation and matter, mass and energy are conserved according to basic physical principles. Therefore, the incident radiation can only be:

- *transmitted*, i.e. passed through the substance. Transmission of energy through media of different densities, such as from air into water, causes a change in the velocity of electromagnetic radiation.
- *absorbed*, giving up its energy largely to heating the matter.
- *emitted* by the substance, usually at longer wavelengths, as a function of its structure and temperature.
- *scattered*, that is, deflected in all directions. Relatively rough surfaces which have topographical features of a size comparable to the wavelength of the incident radiation produce scattering. Light waves are scattered by molecules and particles in the atmosphere whose sizes are similar to the wavelengths of the incident light.
- *reflected*, that is, returned from the surface of a material with an angle of reflection equal and opposite to the angle of incidence. Reflection is caused by surfaces that are smooth relative to the wavelength of incident energy. As shown in Figure 2.4, the reflectance of different land covers varies significantly as a function of wavelength over a wide range of values.

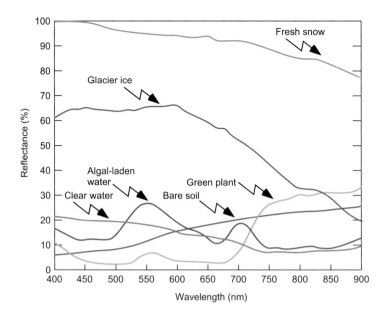

Figure 2.4 Spectral reflectance of various land cover types in the visible spectrum.

Emission, scattering, and reflection are called *surface phenomena* because these interactions are determined primarily by properties of the surface, such as colour and roughness. Transmission and absorption are called *volume phenomena* because they are determined by the internal characteristics of matter, such as density and conductivity. The particular combination of surface and volume interactions with any particular material depend on both the wavelength of the electromagnetic radiation and the specific properties of that material. These interactions between matter and electromagnetic waves are recorded on remote sensing images, from which one may interpret the characteristics of matter (Sabins, 2007).

Remotely-sensed imagery is usually digitized so that it can be stored, analyzed, and displayed in digital form on a computer. Although imagery derived from camera and video sensors can be digitized, modern satellite sensors produce data that is already in digital form. Like the images on a television screen, a digital image is composed of an array of picture elements, or 'pixels', which represent the smallest part of a digital image. Each pixel contains reflectance information about the features being imaged. Reflectance is what allows us to distinguish the features in an image. It is a measure of the amount and type of energy that an object reflects (rather than absorbs or transmits).

Reflectance is given the notation R and is unitless, meaning that it is represented by a scale between 0 and 1 or 0 and 100 per cent. Reflected wave intensity (power) modified by the atmosphere between the ground and the sensor is called radiance (L). Radiance is what the sensor measures, and it may be very different from the ground reflected intensity because of haze and other substances which scatter light (Lachowski *et al.*, 1995). Radiance has specific units and is typically quoted in

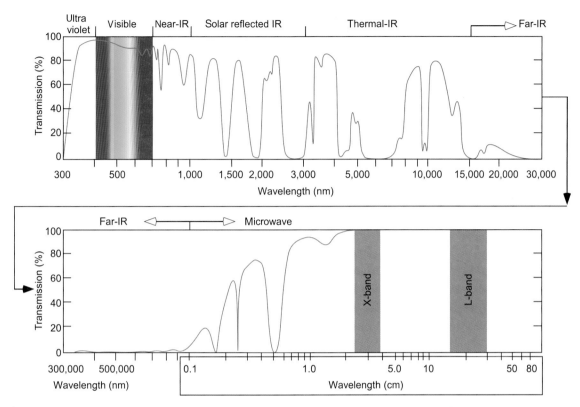

Figure 2.5 Atmospheric absorption and transmission bands. Note change of wavelength units from nanometres (nm) to centimetres in the microwave spectrum. $1\,nm = 1.0 \times 10^{-7}\,cm$.

watts per steradian per square metre per nanometre ($W\,sr^{-1}\,m^{-2}\,nm^{-1}$). These terms will be revisited in more detail in Chapter 8, where we consider the flux of photons in the shallow waters atop coral reefs.

2.4 Atmospheric effects

To an observer, the atmosphere seems to be essentially transparent to light, and we tend to assume that this condition exists for all electromagnetic energy. In fact, the gases of the atmosphere absorb electromagnetic energy at specific wavelength intervals called *absorption bands*. Figure 2.5 shows these absorption bands, together with the gases in the atmosphere responsible for the absorption.

The most efficient absorbers of solar radiation are water vapour, carbon dioxide and ozone. Wavelengths shorter than 300 nm are completely absorbed by the ozone (O_3) layer in the upper atmosphere (Figure 2.5). This absorption is essential to life on Earth, because prolonged exposure to the intense energy of these short wavelengths destroys living tissue. Clouds consist of aerosol-sized

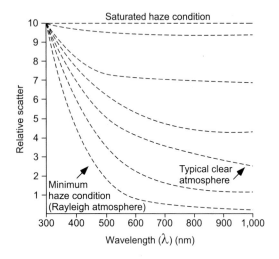

Figure 2.6 Relative scatter as a function of wavelength for various levels of atmospheric haze.

particles of liquid water that absorb and scatter electromagnetic radiation at wavelengths less than about 0.1 cm. Only radiation of microwave and longer wavelengths is capable of penetrating clouds without being scattered, reflected or absorbed (Sabins, 2007).

Atmospheric scattering is caused by particles in the atmosphere deflecting EM waves in all directions and can be of three types: Rayleigh scatter, Mie scatter and non-selective scatter.

Rayleigh scatter dominates when radiation interacts with atmospheric molecules and other tiny particles which have a much smaller diameter than the wavelength of the EM radiation. Rayleigh scatter is inversely proportional to the fourth power of the EM wavelength; therefore short wavelengths are scattered much more than long wavelengths. The blue sky colour is caused by Rayleigh scatter, since it scatters blue light more than the longer wavelength colours such as green and red. Rayleigh scatter is one of the primary causes of haze in imagery. Haze can be minimized by using camera filters that do not transmit the short wavelengths (Figure 2.6).

Mie scatter dominates when the atmospheric particle diameters are approximately equal to the radiation wavelength. Water vapour, dust and various aerosols are the main causes of Mie scatter. Mie scatter affects all EM wavelengths, including long ones.

Non-selective scatter occurs when the diameter of the scattering particles is much larger than the EM wavelength. Thus, water droplets, having diameters from 50 to 1000 nm, scatter visible, near-IR and mid-IR wavelengths nearly equally.

To minimize image degradation by the various types of scatterers in the atmosphere, a wide range of atmospheric correction techniques have been developed, including 'dark-object subtraction' and aerosol modelling. This is particularly important for remote sensing from satellites and high-altitude aircraft (Jensen, 2007; Lillesand & Kiefer, 1994).

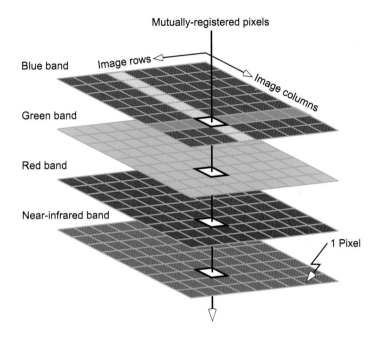

Mutually-registered pixels

Figure 2.7 The anatomy of a multispectral QuickBird image containing unique information in four spectral bands.

2.5 Multispectral feature recognition

Two objects that may be indistinguishable in one portion of the electromagnetic spectrum, because their reflectance is similar in this portion, may be highly separable in another, where their reflectance differs more markedly. To help capture the spectral uniqueness of ground objects, many sensors are designed to collect information in specific regions of the electromagnetic spectrum (Figure 2.3).

Figure 2.7 shows the four spectral bands of a QuickBird satellite image, including their mutually-registered pixels. A pixel or picture element is the smallest part of a digital image, and in most cases the pixel size represents the spatial resolution of the image. Most digital sensors collect information this way, and the acquired multispectral data are stored in spectral bands, where each pixel has a unique intensity value expressed as a digital number (DN). The DN represented by each pixel is an averaged reflectance value for the corresponding objects on the ground.

Typically, changes in land cover (e.g., vegetation) will correspond to changes in the digital numbers of pixels in an image. If the image is multispectral (i.e. it contains fewer than ten or so broad spectral bands), the changes occurring in each band can help differentiate one land cover type from another. Figure 2.8 shows typical reflectance curves for some common ground covers. Since remote sensing platforms, such as satellites and aircraft, are moving rapidly and their sensors are scanning a wide swath of ground, there is inadequate time for a sensor to also scan the entire spectrum for each point on Earth. As

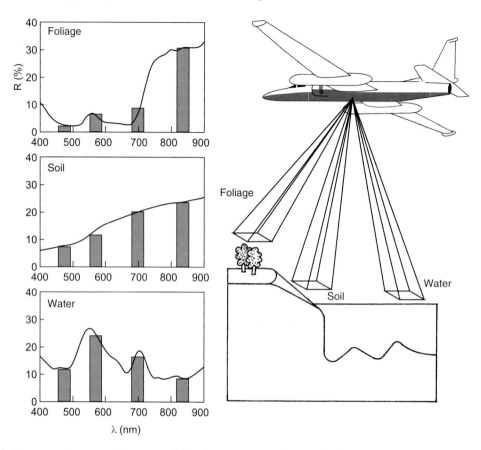

Figure 2.8 Representative spectra for a terrestrial landscape as acquired using the hyperspectral AVIRIS sensor aboard the ER-2 platform (black lines), compared to the crude spectral capability of a 4-band multispectral satellite (gray bars). In both cases, the three land cover types are separable. Most airborne sensors point continuously downward along the local vertical.

a result, most multispectral scanners have a limited number of spectral bands resulting in discontinuous spectral curves, as shown in Figure 2.8 for a 4-band system. The 4-band spectrum in Figure 2.8 is a poor representation of the continuous spectrum, yet quite adequate for discriminating between major types of land cover.

2.6 Resolution requirements

With the wide variety of remote sensing systems available, choosing the proper data source for observing land cover, the oceans or the atmosphere can be challenging. Characteristics often used to describe and compare these analogue and digital systems are grouped into four different types of resolution: spatial; spectral; radiometric; and temporal. Resolution is commonly attributed to an image and the sensor that provides the image data.

Spatial resolution is a measure of sharpness or fineness of spatial detail. It determines the smallest object that can be resolved by the sensor, or the area on the ground represented by each picture element (pixel). For digital imagery, spatial resolution corresponds to the pixel size. Spatial resolution is often represented in terms of distance (e.g. 30 metres, 1 km, etc.) and describes the side length of a single pixel. Thus, the smaller the distance, the higher the spatial resolution (the finer the 'grain') of the image (Lachowski *et al.*, 1995). For instance, some weather satellites have resolutions of 1 km, with each pixel representing the average brightness over an area that is 1 km × 1 km on the ground.

Spectral resolution is a measure of the specific wavelength intervals that a sensor can record. For example, while normal colour photographs show differences in the visible region of the electromagnetic spectrum, colour infrared photographs and the majority of digital sensors can provide information from both visible and infrared (IR) regions of the spectrum. For digital images, spectral resolution corresponds to the number and location of spectral bands, their width, and the range of sensitivity within each band (Jensen, 2007).

Radiometric resolution is a measure of a sensor's ability to distinguish between two objects of similar reflectance. Radiometric resolution can be thought of as the sensor's ability to make fine or 'subtle' distinctions between reflectance values. For example, while the Landsat Thematic Mapper (TM) has a radiometric resolution of 256, the Moderate Resolution Imaging Spectrometer (MODIS) has a radiometric resolution of 4,096. This means TM can identify 256 different levels of reflectance in each band, while MODIS can differentiate 4,096. Thus, MODIS imagery can potentially show more and finer distinctions between objects of similar reflectance (Campbell 2007). One can also say that TM has 8-bit resolution (8 bit = 2 raised to the power of 8 = 256) and MODIS has 12-bit resolution (12 bit = 2 raised to the power of 12 = 4,096).

Temporal resolution is a measure of how often the same area is visited by the sensor. Unlike the three types of resolution discussed above, temporal resolution does not describe a single image, but rather a series of images that are captured by the same sensor over time. While the temporal resolution of satellite imagery depends on the satellite's orbit characteristics, aerial photography obviously requires special flight planning for each acquisition. Temporal resolution for satellite imagery is represented in terms of the amount of time between satellite 'visits' to the same area (e.g. two days for SeaWiFS, 16 days for Landsat TM, 26 days for SPOT, 35 days for IKONOS).

The temporal and spatial resolution requirements for detecting, mapping and monitoring selected terrestrial, oceanic, and atmospheric features and processes are shown in Figure 2.9. For instance, urban infrastructure and emergency response both require high spatial resolution, yet urban infrastructure does not change rapidly and needs to be observed much less frequently than emergency situations such as when hurricanes make landfall. On the other hand, monitoring of the

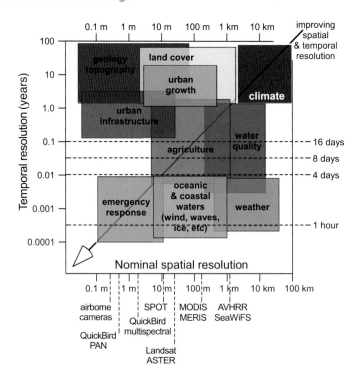

Figure 2.9 The temporal and spatial resolution requirements vary widely for observing terrestrial, oceanic, and atmospheric features and processes. Modified after Jensen (2007) and Phinn *et al.* (2010).

weather and climate both require less spatial resolution, yet climate changes slowly whereas weather conditions can change very rapidly.

As shown in Figure 2.9 and will be seen in later chapters, each application of remote sensors has its own unique resolution requirements (Bissette *et al.*, 2004; Mumby & Edwards, 2002). Since none of the existing remote sensing systems will perfectly meet all resolution requirements, one must always make trade-offs between spatial resolution and coverage, spectral bands and signal-to-noise ratios, etc.

Demanding the highest spatial resolution available may not always result in better image classification accuracy (Wang, 2010). Factors such as computer processing time must also be considered when choosing types of imagery; an increase in resolution will increase the processing time required for interpreting the information. Also, smaller pixels require more storage space on computers for a given area. For example, an image with 10-metre pixels would require significantly more storage space and time to process than a lower-resolution 20-metre pixel image, since the amount of data bits to be processed will increase at least four-fold.

2.7 Key concepts

1 Remote sensors on satellites and aircraft operating at various altitudes use electromagnetic (EM) waves effectively to detect and map features and changes on the surface of the land, sea and in the atmosphere.

2 Remote sensing systems can provide digitized data in the visible, reflected infrared, thermal infrared and microwave regions of the electromagnetic spectrum.

3 Differences in reflectance allow remote sensors to distinguish between objects and features on the ground.

4 The atmosphere absorbs and scatters EM waves. Remote sensors are designed to avoid absorption bands and they use filters or atmospheric correction techniques to diminish the effects of scattering.

5 When planning a remote sensing campaign, one must define the spatial, spectral, radiometric and temporal resolution requirements for the project.

3

Remote sensors and systems

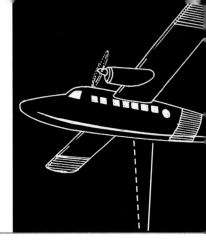

3.1 Introduction

Aerial photography started approximately in 1858 when the famous photographer, Gaspard Tournachon, obtained the first aerial photographs from a balloon near Paris. Since then, aerial photography has advanced, primarily during wartime, first to include colour infrared films (for camouflage detection) and later to use sophisticated digital cameras. Aerial photography and other remote sensing techniques are now used successfully in agriculture, forestry, land use planning, fire detection, mapping wetlands and beach erosion, and many other applications. For example, in agriculture it has been used for land use inventories, soil surveys, crop condition estimates, yield forecasts, acreage estimates, crop insect/pest/disease detection, irrigation management and, more recently, precision agriculture (Jensen, 2007).

Since the 1960s, 'remote sensing' has been used to describe a new field of information collection that includes aircraft and satellite platforms carrying electro-optical and antenna sensor systems (Campbell, 2007). Up to that time, camera systems dominated image collection and conventional photographic media dominated the storage of the spatially varying visible (VIS) and near-infrared (NIR) radiation intensities reflected from the Earth.

Beginning in the 1960s, electronic sensor systems were increasingly used for collection and storage of the Earth's reflected radiation and satellites were developed as an alternative to aircraft platforms. Advances in electronic sensors and satellite platforms were accompanied by an increased interest and use of electro-magnetic radiant energy, not only from the VIS and NIR wavelength regions, but also from the thermal infrared (TIR) and microwave regions. For instance, the thermal infrared region is used for mapping sea surface temperatures and microwaves (radar) are used for measuring sea surface height, currents, waves and winds on a global scale (Martin, 2004).

Remote Sensing and Global Environmental Change, First Edition. Samuel Purkis and Victor Klemas.
© 2011 Samuel Purkis and Victor Klemas. Published 2011 by Blackwell Publishing Ltd.

Table 3.1 Classification of remote sensors.

Classification by application	Classification by wavelength	Classification by mode
Imagers (mappers)	Visible	Active
Photographic (film)	(array or film)	LiDAR
Multispectral (array)		Radar
Radar	Near infrared	†Sonar
†Side-scan sonar	(reflected)	
		Passive
Radiometers	Thermal infrared	Visible
	(emitted)	Infrared
Spectrometers		Microwave
	Microwave	
Profilers (rangers)		
LiDAR, Radar	†Sound waves	
	†Seismic waves	

† Not electromagnetic waves

3.2 Remote sensors

As shown in Table 3.1, remote sensors can be classified by application, wavelength or active/passive mode. Under applications we have imagers, which produce two-dimensional images and can be used for map-making; radiometers measure the radiant energy in a few specific bands very accurately; while spectrometers provide the energy distribution across a spectral continuum or many spectral bands.

Profilers, such as radar and LiDAR, measure the distance to features, allowing us to determine the topography or bathymetry of an area. Radar and LiDAR are primarily 'active' devices, since they provide their own pulse of energy. Most other sensors are 'passive', because they use electromagnetic energy provided by the sun or the Earth.

3.2.1 Multispectral satellite sensors

The multispectral sensors operate in three major wavelength regions: the visible (400–700 nm), the reflected infrared (700–3,000 nm) and the thermal infrared (3,000–14,000 nm) regions. The reflected infrared can be further subdivided into the near-infrared (700–1,300 nm) and mid-infrared (1,300–3,000 nm) regions (Figure 2.3). Thermal infrared sensors use primarily the 10,000 nm atmospheric window (Figure 2.5).

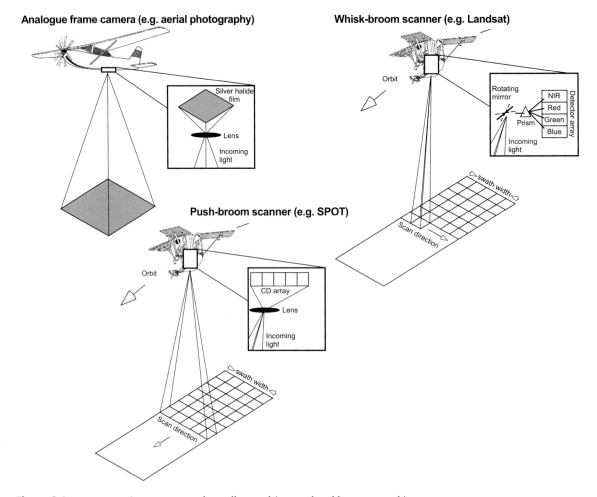

Figure 3.1 Remote sensing systems used to collect multispectral and hyperspectral imagery.

In electro-optical multispectral sensors, the visible region is divided into many bands, whereas aerial photography uses blue, green, and red bands plus one reflected band in the near-infrared. Figure 3.1 illustrates the operation of an analogue frame camera and two types of multispectral scanners – the cross-track or whisk-broom and the along-track or push-broom types. After being focused by a mirror system, the radiation from each image pixel is broken down into spectral bands, with one image being produced in each spectral band (Figure 3.2).

3.2.2   Digital aerial cameras

A major advance in aerial remote sensing has been the development of digital aerial cameras (Al-Tahir *et al*., 2006). Digital photography is capable of delivering photogrammetric accuracy and coverage as well as multispectral data at any

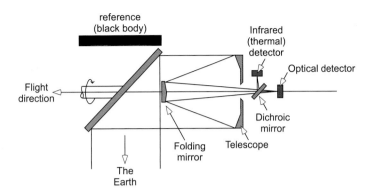

Figure 3.2 Components of a multispectral scanner.

user-defined resolution down to 0.1 m ground sampling distance, thus providing photogrammetric positional accuracy with multispectral capabilities for image analysis and interpretation. As no chemical film processing is needed, the direct digital acquisition can provide image data in just a few hours, compared to weeks using a traditional film-based camera. Another advantage over traditional film is the ability to assess the quality of data taken in real-time during the flight. The data can be used directly in a Geographical Information System (GIS) and has proven to be valuable in interpreting satellite imagery.

Digital aerial cameras typically have Charge Coupled Device (CCD) arrays that produce images containing about $3,000 \times 2,000$ to $7,000 \times 5,000$ lines (pixels). Most record an 8-bit to 12-bit black and white image (256–4,096 gray levels). The shutter can be mechanical or electronic. The exposure time and aperture setting is adjusted before the overflight, depending on conditions and the brightness of the mapped features. Each frame is instantaneously recorded so, unlike multispectral line scanners, there is minimal geometric distortion due to aircraft motion during acquisition. The aircraft altitude and the focal length of the lens system determine the ground resolution or ground sampling distance (GSD). Typical GSD values range from 15 cm to 3 m.

For a typical camera system, pixels with 0.5 m side-length can be obtained from altitudes of about 2,000 m. The area imaged at 0.5 m GSD with an $3,000 \times 2,000$ array would be about 1.5 km × 1 km. The digital image would contain 6,000,000 pixels and would be 6 megabytes (Mb) in size, assuming an 8-bit camera. Such a relatively small file size makes this technology suitable for rapid turn-around and PC-based image processing (Ellis & Dodd, 2000).

Most digital cameras are capable of recording reflected visible to near-infrared light. A filter is placed over the lens that transmits only selected portions of the wavelength spectrum. For a single camera operation, a filter is chosen that generates natural colour (blue-green-red wavelengths) or colour-infrared (green-red-nearIR wavelengths) imagery. For multiple camera operation, filters that transmit narrower bands are chosen. For example, a four-camera system may be configured so that each camera filter passes a band matching a specific satellite imaging band, e.g. blue, green, red and near-infrared bands, matching the bands of the IKONOS satellite multispectral sensor.

Two examples of digital mapping cameras are the Emerge Digital Sensor System (DSS) from Leica Geosystems and the Digital Modular Camera (DMC) from Z/I Imaging, which were developed to address requirements for extensive coverage, high geometric and radiometric resolution and accuracy, multispectral imagery and stereo capability. These two cameras, and similar ones from other companies, are digital frame cameras whose design is based on charge-coupled-device (CCD) array technology. The ground resolution depends not only on the number of pixels in the digital camera's area array, but also on the flight altitude. Thus, by choosing the appropriate flight altitude, resolutions ranging from a few centimetres to metres can be obtained.

Leica's DDS uses a digital camera area array that acquires imagery containing $4{,}092 \times 4{,}079$ pixels. Users can specify colour (blue, green, red) multiband imagery or colour-infrared (green, red, near-infrared) imagery in the spectral region from 400 to 900 nm. In the near-infrared mode, the camera has a spectral response similar to that of Kodak colour-infrared film, but with higher dynamic range. The data may be recorded with 8 or 16 bits per pixel. The DSS collects real-time differentially corrected Global Position System (GPS) data about each digital frame of imagery. These data can be used to mosaic and orthorectify the imagery using photogrammetric techniques (Jensen, 2007; Light, 2001; Leica, 2002).

The Z/I DMC uses four $7{,}000 \times 4{,}000$ CCD area arrays to obtain one complete frame of panchromatic imagery. All CCDs have their own optics and function as stand-alone digital cameras that are synchronized to collect data at the exact same instant. Four additional CCDs with $3{,}000 \times 2{,}000$ detectors are used to obtain blue, green, red and near-infrared multispectral data at the same instant. Therefore, the multispectral bands have reduced ground resolution compared to the panchromatic data. Individual frames of imagery are obtained just like a traditional frame camera, with user-specified end-overlap between frames (Jensen, 2007; Hinz *et al.*, 2001).

3.2.3   Thermal infrared sensors

While most geologists, geographers, and foresters are familiar with aerial photography techniques (Sabins, 1978; Avery & Berlin, 1992), relatively few scientists have had the opportunity to use thermal infrared, radar and LiDAR data. Since the thermal infrared (TIR) radiance depends on both the temperature and emissivity of a surface, it is difficult to measure land surface temperatures, since the emissivity will vary as the land cover changes. On the other hand, over water the emissivity is known and nearly constant (98 per cent), approaching the behaviour of a perfect black body radiator (Ikeda & Dobson, 1995). Thus, the TIR radiance measured over the oceans will vary primarily with the sea surface temperature (SST) and will allow us to determine the SST accurately (Martin, 2004; Barton, 1995).

It is important to bear in mind that a satellite infrared radiometer indirectly measures the temperature of a very thin layer of about ten micrometres thick (referred to as the skin) of the ocean. The skin temperature is generally at least a few tenths of a degree cooler than the 'bulk' temperature below it because of heat loss by sensible and latent heat fluxes across this thin surface layer. For this reason, and unless an appropriate correction is applied, space- or airborne SST measurements using TIR will not be in agreement with those made *in situ* by ships or buoys.

Beginning with the launch of the Advanced Very High Resolution Radiometer AVHRR/2 on the NOAA-7 platform in 1981, there now exist nearly three decades' worth of infrared satellite SST observations. In addition to AVHRR, there are many other satellites that provide thermal infrared images (Table 3.2).

3.2.4   Radar and microwave radiometers

Radar images represent landscape and ocean surface features that differ significantly from those observed by aerial photography or multispectral scanners. A Side-Looking Airborne Radar (SLAR) irradiates a swath along the aircraft flight direction by scanning the terrain with radar pulses at right angles to the flight path. Thus, the radar image is created by pulse energy reflected from the terrain and represents primarily surface topography. As depicted in Figure 3.3, key aspects of the geometry of a radar pulse are the 'viewing' and 'depression' angles. These describe the angles between the ray path and the vertical and the horizontal, respectively. The angle of 'incidence' is the angle between the radar beam and the ground surface.

Since radar images look quite different from visible photographs, they require specialized interpretation skills. As we will see when this topic is revisited in Chapter 7, radar pulses penetrate only a few wavelengths into the soil, depending on soil moisture, salinity, surface roughness, etc. (Table 3.3). Their ability to penetrate vegetative canopies depends on the radar wavelength, the angle of incidence and the canopy density, among other things (Figure 3.4).

The range resolution of SLAR depends on the length of the radar pulse, which can be made quite short with new electronic techniques. However, the azimuth resolution is limited by the antenna size and altitude, thus preventing SLAR systems being used on satellites.

Synthetic Aperture Radar (SAR) was specifically developed to provide high-resolution images from satellite altitudes. SAR employs the Doppler shift technique to narrow down the azimuth resolution even with a small antenna. Thus, range and azimuth resolutions of the order of 3–10 metres are obtainable with SAR mounted on satellite platforms such as RADARSAT and ERS-2 (Table 3.2). The mechanics of SAR are covered in Chapter 6 (section 6.2.4).

We will see in Chapter 10 how, in oceanography, radar is used not only for imaging sea surface features such as waves and slicks, but also as an altimeter to map sea surface height, and as a scatterometer to determine sea surface winds

Table 3.2 Characteristics of some current and scheduled remote sensing systems. SAV denotes Submerged Aquatic Vegetation. Modified from Donato & Klemas (2001).

Satellite/sensor	Spectral range	Bands	Spatial resolution	Revisit time	Swath-width	Applications
AVHRR NOAA 15/16	580–12,500 nm	6	1.1 km	< 12 h	2,400 km	SST, turbidity, circulation
SeaWiFS	402–885 nm	8	1.1 km	daily	2,800 km	Ocean colour, red products
MODIS Terra/Aqua	620–14,385 nm	16 VNIR	variable	daily	2,330 km	SST, turbidity, circulation
		4 SWIR	250 m			Ocean colour
		16 TIR	1 km	< 12 h		
MISR Terra (9 camera angles)	425–886 nm	4	275 m	9 d	360 km	Ocean colour, circulation
ASTER Terra	520–11,650 nm	3 VNIR	15 m	16 d	60 km	Coral reef, SAV, vegetation
		6 SWIR	30 m			Land use, change detection
		5 TIR	90 m			Circulation, geomorphology
LANDSAT-7	450–20,80 nm	6 VNIR	30 m	16 d	180 km	Coral reef, SAV, vegetation
	10,420 nm	1 TIR	60 m			Land use, change detection
		1 Pan	15 m	16 d		Geomorphology, circulation
SPOT 1-2-4-5	500-890 nm	3 MS	20 m	26 d	60 km	Coral reef, SAV, vegetation
		1 Pan	10 m	daily		Land use, change detection
						Geomorphology, circulation
IKONOS	450–750 nm	4 MS	4 m	1–3 d	13 km	Coral reef, SAV, vegetation
		1 Pan	1 m			Littoral processes, digital elevation models
QuickBird-2	450–900 nm	4 MS	2.4 m	1 d	16.4 km	Coral reef, SAV, vegetation
		1 Pan	0.61 m			Littoral processes, digital elevation models
WorldView-2	400–1,040 nm	8 MS	1.8 m	≈ 2 d	16.4 km	Coral reef, SAV, vegetation
		1 Pan	0.46 m			Littoral processes, digital elevation models
Orbview 3	450–900 nm	4 MS	4 m	< 3 d	8 km	Coral reef, SAV, vegetation
		1 Pan	1 m			Littoral processes, digital elevation models
Orbview 4	450–2,500 nm	200 HS	8 m	< 3 d	5 km	Coral reef, SAV, vegetation
	450–900 nm	4 MS	4 m		8 km	Littoral processes
		1 Pan	1 m			Digital elevation models
ALIEO-1	400–2,400 nm	9 MS	30 m	19 d	37 km	Coral reef, SAV, vegetation
		1 Pan	10 m			Land use, change detection, geomorphology, circulation
Hyperion EO-1	400–2,400 nm	220	30 m	16 d	8 km	Coral reef, SAV, vegetation, littoral processes
NEMO/COIS	400–2,500 nm	210	30 m			Coral reef, SAV, vegetation, littoral processes
MERIS ENVISAT-1	290–1,040 nm	15	300 m	< 3 d	1,150 km	Ocean colour, circulation
ASAR ENVISAT-1	C-band 4 pol	2	30 m	< 3 d	50–100 km	Circulation, waves
AMI ERS-2(SAR)	C-band V pol	1	25 m	28 d	100 km	Circulation, waves
RADARSAT-1(SAR)	C-band H pol	1	6–100 m	1–4 d	20–500 km	Circulation, waves
RADARSAT-2(SAR)	C-band HV pol	1	3–100 m	Quick	20–500 km	Circulation, waves

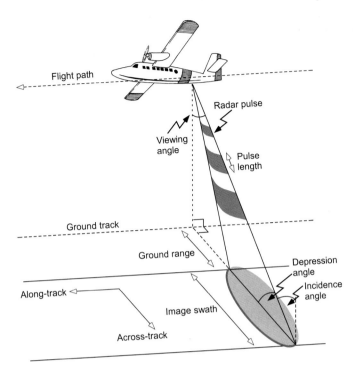

Figure 3.3 Geometric
characteristics of airborne
side-looking radar (SLAR).

Table 3.3 Canopy and soil penetration by radar.

Parameter	Deeper penetration	Less penetration
Radar wavelength	Long	Short
Viewing angle (depression angle)	Steep (large)	Shallow (small)
Polarization	Vertical	Horizontal
Soil moisture	Dry	Wet
Soil salinity	Low	High
Soil grain size	Large	Small

(Martin, 2004; Ikeda & Dobson, 1995). Radar can penetrate fog and clouds, making it particularly valuable for emergency applications and in areas where cloud cover persists. This technology is therefore also favoured for satellite missions to other planets whose surfaces are obscured by a dense cloudy and/or dusty atmosphere. Passive microwave radiometers are becoming important for measuring sea surface salinity, soil moisture (Chapter 7), polar ice thickness (Chapter 12) and a wide range of hydrology-related parameters (Parkinson, 2003; Burrage *et al.*, 2003).

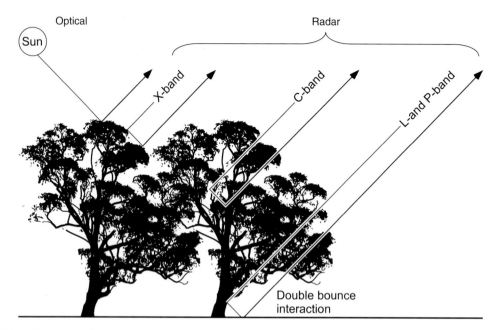

Figure 3.4 Reflectance and primary interaction of X-, C-, L-, and P-band radars with forest canopies.

3.2.5   Laser profilers

Laser systems used in measuring distances are called LiDARs (Light Detection and Ranging). Though comparatively costly, airborne LiDAR has become very useful for topographical and shallow bathymetric mapping. Laser profilers are unique in that they confine the coherent light energy within a very narrow beam, providing pulses of very high peak intensity. This enables LiDARs to penetrate moderately turbid coastal waters for bathymetric measurements, and gaps in forest canopies to provide topographical data for digital elevation models (Brock & Sallenger, 2000; Brock & Purkis, 2009). As will be explained in Chapter 9, the water depth is derived by comparing the travel times of the LiDAR pulses reflected from the sea bottom and the water surface.

3.3   Remote sensing platforms

The choice of remote sensing platform and sensors depends on the mission requirements. These can be broken down into spatial, spectral, radiometric and temporal:

- Spatial requirements include the ground resolution (minimum mapping unit) and coverage (swath-width).
- Spectral requirements include the location, width and number of the spectral bands.

Table 3.4 Typical Remote sensing requirements.

	Open ocean	Coasts/estuaries	Land
Spatial resolution	1–10 km	20–200 m	1–30 m
Extent (coverage)	2,000 km	200 km	20–200 km
Frequency of coverage	1–6 days	0.5–6 hours	0.5–5 years
Dynamic range	Narrow	Wide	Wide
Radiometric resolution	10–12 bits	10–12 bits	8–10 bits
Spectral resolution	Multispectral	Hyperspectral	Multispectral (hyperspectral)

- For radiometry we must choose the suitable dynamic range and the number of quantization (grey) levels. There are usually between 256 (8-bit) and 4,096 (12-bit) quantization levels.
- The temporal resolution is determined by the required frequency of coverage (e.g. hourly, daily, seasonal), cloud cover, sun angle and tidal conditions.

The general requirements for open ocean, coastal and upland remote sensing are summarized in Table 3.4. As shown, the spatial, spectral, radiometric and temporal resolution requirements are quite different for each of these applications and depend on the specific problem to be solved.

3.3.1 Airborne platforms

Despite the heavy use of satellite data, the very high spatial resolutions and frequent, flexible overflight times offered by airborne sensors are still required for certain projects, such as wetland monitoring and emergency response (Phinn *et al.*, 1996). Unmanned remotely controlled aircraft or drones offer an especially useful, safe and less expensive way for studying remote regions, including such projects as counting seal populations in the Arctic or determining why the edges of Greenland are melting so quickly. These unmanned aircraft can carry a wide selection of instruments, ranging from radar and LiDAR to infrared sensors and chemical analysis tools. To perform aerial surveys or observe long-term changes at local sites, helicopters and even blimps or balloons have been used successfully (Chen & Vierling, 2006).

Remote sensing aircraft are usually flown at high, medium or low altitudes, depending on the resolution and coverage requirements. High altitude flights covering large regions are normally performed by government agencies, whereas medium altitude flights are often provided by private companies. Low altitude flights may involve small aircraft and are sometimes used to supplement field data collection.

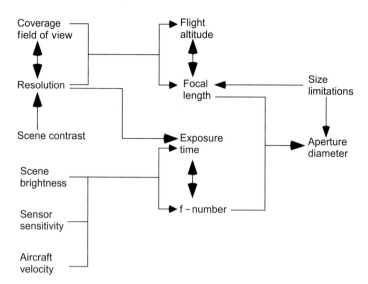

Figure 3.5 Aerial photography trade-offs.

The trade-offs one must make in selecting flight altitudes and imaging systems are outlined in Figure 3.5. For example, spatial resolution can be traded off for coverage (swath-width) by varying the flight altitude or the camera lens focal length. As we mentioned in section 3.2.2, most aerial land cover mapping would now be performed using digital cameras. However, in regional or local studies of land cover change, the time-series may include aerial film photos obtained well before digital cameras became available. Therefore, it is advisable for users to become familiar both with aerial film products and with more recent digital camera imagery.

3.3.2   Medium-resolution satellites

There are about 29 civil land imaging satellites in orbit and 34 more are planned. Among these are 18 high-resolution (0.5–1.8 metres) and 44 mid-resolution (2–36 metres) systems. A list of the more relevant satellites is shown in Table 3.2. Most of these satellites are in polar sun-synchronous orbits. As the Earth rotates beneath the polar orbiting satellite, the satellite's sensors eventually are able to observe and map every part of the globe. Its orbit can also be adjusted to be sun-synchronous, repeating its passes over a site during the same time of day (or same solar illumination angle). Geostationary orbit satellites are stationed 36,000 km above a fixed point on the equator, having the same angular velocity as the Earth; they thus continuously observe the same one-third of the Earth's surface. Geostationary satellites provide less spatial resolution (4 km) but have the short repeat cycles that are needed for tracking storms and weather fronts (every 15–30 minutes) (Jensen, 2007; Lillesand & Kiefer, 1994).

Some of the more common satellites for observing coastal land cover change and ocean productivity are described in Table 3.5. Thematic Mapper data from

Table 3.5 Some coastal zone related sensors on satellites.

Satellite	Sensor	Spectral band (nm)		Resolution (m)	Cycle (days)	Swath-width (km)
Landsat 1, 2, 3	MSS	4	500–600	80	18	185
		5	600–700			
		6	700–800			
		7	800–1100			
Landsat 4, 5	TM	1	450–520	30 band 1–6	16	185
		2	520–600	120 band 7		
		3	630–690			
		4	760–900			
		5	1,550–1,750			
		6	2,080–2,350			
		7	10,400–12,500			
Landsat 7 ETM+	TM	1	450–514	30 band 1–7	16	185
		2	525–605	60 band 6		
		3	630–690	15 band 8		
		4	750–900			
		5	1,550–1,750			
		6	10,400–12,500			
		7	2,080–2,350			
		8	520–900 (Pan)			
SPOT	HRV	1	500–590	20 band 1–3	26 (daily if camera tilted)	60
		2	610–680	10 band 4		
		3	790–890			
		4	510–730			
IKONOS		1	450–520	4 band 1–4	< 3 days	11
		2	520–600	1 band 5		
		3	630–690			
		4	760–900			
		5	450–900 (Pan)			
NOAA	AVHRR	1	580–680	1,100	2/day	2,400
		2	725–1,100			
		3	3,550–3,930			
		4	10,500–11,300			
		5	11,500–12,500			
Orbview 2	SeaWiFS	1	404–422	1,100	Daily	2,800
		2	433–453			
		3	480–500			
		4	500–520			
		5	545–565			
		6	660–680			
		7	745–785			
		8	845–885			

Landsat is often used for mapping vegetation and land cover, and this is becoming more important for change detection as the time span of available imagery becomes longer. Key features of Landsat TM data are:

- Spatial resolution: 30 metres
- Spectral resolution: seven bands, three in the visible and three in the reflective IR ranges; band 6 covers the thermal IR range at 120-metre spatial resolution
- Temporal resolution: 16 days
- Image swath-width: 185 km

Système Probatoire d'Obervation de la Terre (SPOT), the first of a series of SPOT satellites (SPOT-1) was launched by France in 1986. The orbit pattern for SPOT repeats every 26 days; however, SPOT's off-nadir pointing capability (i.e. the satellite's capacity to 'point' the sensor towards ground areas not directly below it) allows areas to be visited even more frequently. SPOT's pointable optics also provide for full-scene stereoscopic imaging (Lachowski *et al.*, 1995). ASTER, IKONOS, and Quickbird are also pointable.

The SPOT satellite can operate in either the multispectral mode or panchromatic mode. In the multispectral mode, it has:

- Spatial resolution: 20 metres
- Spectral resolution: three bands – green, red and reflective IR portions of the spectrum
- Temporal resolution: 1 to 26 days, depending on the latitude of the study area (the sensor has pointing capability)
- Swath-width: 60 km

In the Panchromatic mode it has:

- Spatial resolution: 5 m, with a super-resolution mode at 2.5 m
- Spectral resolution: one panchromatic band ranging from visible to near-IR
- Temporal resolution: 1 to 26 days, depending on the latitude of the study area (the sensor has pointing capability)
- Swath-width: 60 km

Because of the high spatial resolution of SPOT's panchromatic mode, it is often used in combination with multispectral imagery to increase the detail of ground features such as roads, buildings, and power line corridors (Read *et al.*, 2003).

3.3.3   High-resolution satellites

For several decades, environmental researchers and resource managers used the mid-resolution imagery provided by satellites such as Landsat and SPOT. However, in the late 1990s, private satellite corporations started collecting high-resolution

Table 3.6 High-resolution satellite parameters and spectral bands (Space Imaging, 2003; DigitalGlobe, 2003; Orbimage, 2003; Parkinson, 2003).

		IKONOS	QuickBird	OrbView-3	WorldView-1	GeoEye-1	WorldView-2
Sponsor		Space Imaging	DigitalGlobe	Orbimage	DigitalGlobe	GeoEye	DigitalGlobe
Launched		Sep 1999	Oct 2001	Jun 2003	Sep 2007	Sep 2008	Oct 2009
Spatial resolution (m)	Panchromatic	1.0	0.61	1.0	0.5	0.41	0.5
	Multispectral	4.0	2.44	4.0	n/a	1.65	2
Spectral range (nm)	Panchromatic	525–928	450–900	450–900	400–900	450–800	450–800
	Coastal blue	n/a	n/a	n/a	n/a	n/a	400–450
	Blue	450–520	450–520	450–520	n/a	450–510	450–510
	Green	510–600	520–600	520–600	n/a	510–580	510–580
	Yellow	n/a	n/a	n/a	n/a	n/a	585–625
	Red	630–690	630–690	625–695	n/a	655–690	630–690
	Red edge	n/a	n/a	n/a	n/a	n/a	705–745
	Near-infrared	760–850	760–890	760–900	n/a	780–920	770–1,040
Swath width (km)		11.3	16.5	8	17.6	15.2	16.4
Off nadir pointing		±26°	±30°	±45°	±45°	±30°	±45°
Revisit time (days)		2.3–3.4	1.0–3.5	1.5–3	1.7–3.8	2.1–8.3	1.1–2.7
Orbital altitude (km)		681	450	470	496	681	770

remote sensing data. The satellites from GeoEye (formerly Space Imaging & OrbImage), IKONOS and OrbView-3, Digital Globe (QuickBird and WorldView 1 and 2) are already in orbit, capturing imagery at 0.4–1.0 m ground resolution. These systems share several common specifications with respect to the spectral and spatial resolutions, as well as orbital details. Table 3.6 lists specific information about these satellite systems, including data about ground resolution, spectral coverage and swath-width.

As shown in Table 3.6, IKONOS and OrbView-3 offer 1 m resolution in a panchromatic band and 4 m resolution in four multispectral bands which are spectrally similar to the Landsat TM bands (Space Imaging, 2003; DigitalGlobe, 2003; Orbimage, 2003). One-metre colour imagery can be created using a pan-sharpening process that combines the high spatial resolution of the panchromatic images with the spectral information of the multispectral bands (Read *et al.*, 2003; Souza & Roberts, 2005).

QuickBird is another high-resolution satellite used in land observations, providing panchromatic imagery at 0.6 m resolution (Al-Tahir *et al.*, 2006). DigitalGlobe's recently launched WorldView-1 satellite possesses a panchromatic imaging system with half-metre resolution imagery. The company launched WorldView-2 in late 2009, with an even higher resolution and eight spectral bands but, as for all sensors, the imagery must be resampled to 0.5 metres for non-US Government customers.

3.3.4   Global observation satellites

Scale, resolution and the user's needs are the most important factors affecting the selection of remotely sensed data. The user's needs determine the nature of classification and the scale of the study area, thus effecting the selection of suitable spatial resolution of remotely sensed data.

Previous research has explored the impacts of scale and resolution on remote-sensing image classification (Quattrochi & Goodchild, 1997; Mumby & Edwards, 2002). In general, a fine-scale classification system is needed for classification at a local level, so high spatial resolution data such as from IKONOS and QuickBird are helpful. At a regional scale, medium spatial resolution data such as provided by Landsat TM and Terra ASTER are the most frequently used data. At a continental or global scale, coarse spatial resolution data, such as AVHRR and MODIS, are preferable (Lu & Weng, 2007).

NOAA satellites with AVHRR

The NOAA Advanced Very High Resolution Radiometer (AVHRR) was designed to collect meteorological data around the globe (Kidder & Von der Haar, 1995). The AVHRR collects data at 1.1 km resolution at the satellite sub-point (Table 3.5). Three types of imagery are produced by its radiometers: global area coverage (GAC); local area coverage (LAC); and direct read-out high-resolution picture transmission (HRPT) data (Cracknell & Hayes, 2007). The LAC and GAC data are recorded onboard and transmitted to the ground when the satellite passes over designated receiving stations. Both LAC and HRPT provide imagery at about 1.1 km spatial resolution at nadir. GAC pixels actually represent an area of 1.1 km × 4 km.

The potential of the AVHRR for vegetation monitoring was realized after the satellites became operational (Tucker *et al.*, 1991). Vegetation indices derived from NOAA-AVHRR sensors have been employed for both qualitative and quantitative studies of forest, desert and other ecosystems, including the contraction and expansion of the Sahara desert (Sellers & Schimel, 1993), a topic covered in Chapter 5. Overviews of these studies and others on the calculation of biophysical parameters for climate models are given by Prince & Justice (1991), Kondratyev & Cracknell (1998), Kogan (2002) and Kramer (2002).

Ocean colour satellites

The purpose of ocean colour sensors, such as NASA's Sea-viewing Wide Field-of-view Sensor (SeaWiFS) is to provide quantitative data on global ocean bio-optical properties to the earth science and fisheries communities. Subtle changes in ocean colour signify various types and quantities of marine phytoplankton (microscopic marine plants), the knowledge of which has many scientific and practical applications.

The ability to map the colour of the world's oceans has been used to estimate global ocean productivity (Longhurst *et al.*, 1995; Behrenfeld & Falkowski, 1997), aid in delineating oceanic biotic provinces (Longhurst, 1998) and study regional shelf break frontal processes (Ryan *et al.*, 1999; Schofield *et al.*, 2004). As shown in Table 3.2, SeaWiFS has eight spectral bands which are optimized for ocean chlorophyll detection and the necessary atmospheric corrections. The spatial resolution is 1.1 km and the swath-width 2,800 km. Due to the wide swath-width, the revisit time is once per day.

Physical oceanography satellites

Most studies of climate change require physical data as well as biological information. As described in Chapter 10, radar and thermal infrared sensors are available on aircraft and satellites for measuring and mapping the physical properties of land and ocean features and processes. For example (Martin, 2004; Elachi & van Ziel, 2006):

- Ocean surface and internal wave fields, as well as oil slicks, can be mapped with radar imagers such as the Synthetic Aperture Radar (SAR) mounted on satellites.
- Radar altimeters provide accurate sea surface height as well as wave amplitude information.
- Radar scatterometer data can be analyzed to extract sea surface winds.
- The two passive devices – microwave radiometers and thermal infrared scanners – can sense sea surface salinity and temperature, respectively.

3.4   The NASA Earth observing system

In the early 1990s, NASA developed a programme, called the Earth Science Enterprise, to acquire the environmental data needed to address specific questions posed by concerns over global environmental change. This initiated a long-term effort to study the total Earth system and the effects of natural and anthropogenic changes on the global environment.

One programme component is an integrated system of satellites, the Earth Observing System (EOS), which is designed to provide a continuous stream of data with instruments tailored to answer specific questions for a better understanding of the nature, rates and consequences of global environmental change (Campbell, 2007).

The EOS plan has included over 30 instruments designed to monitor physical and biological components of the Earth. One example of such a satellite mission is Aqua, a satellite launched in 2002. This satellite carried six distinct Earth-observing instruments to measure numerous aspects of the Earth's atmosphere, land, oceans, biosphere and cryosphere, with a focus on water in the Earth system. The six instruments include the Atmospheric Infrared Sounder (AIRS), the Advanced

Microwave Sounding Unit (AMSU-A), the Humidity Sounder for Brazil (HSB), the Advanced Microwave Scanning Radiometer for EOS (AMSR-E), the Moderate-Resolution Imaging Spectroradiometer (MODIS), and the Cloud and Earth Radiant Energy System (CERES). Each instrument has unique characteristics and capabilities, and all six serve together to form a powerful package for Earth observations (Parkinson, 2003).

The first satellite in the EOS series, Terra, was launched by NASA in 1999 to analyze the dynamic processes of Earth's land, sea and atmosphere. Several of Terra's key sensors, such as the MODIS, are described in Table 3.2.

3.5 Global Earth observation systems

3.5.1 Global Climate Observing System

Satellite data are an important component of the Global Climate Observing System (GCOS). To monitor climate change, the system must satisfy specific criteria, such as frequent repeat cycles and global coverage.

Successful examples of satellite Earth observation data in climate studies include analyses of the Earth's radiation budget via the Earth Radiation Budget Experiment (ERBE), Scanner for Radiation Budget (ScaRaB), Cloud and Earth Radiant Energy System (CERES), International Satellite Cloud Climatology Project (ISCCP), vegetation research via the Global Inventory Modelling and Mapping Studies (GIMMS), and the NOAA-NASA Pathfinder Program (Latifovic *et al.*, 2005).

3.5.2 Global Earth Observation System of Systems

To monitor long trends such as global warming and sea level rise, or short-term catastrophic events such as coastal storms and flooding, 60 countries and 40 international organizations are collaborating to develop the Global Earth Observation System of Systems (GEOSS). GEOSS is being developed to ensure that comprehensive, coordinated and sustained Earth observations are available to decision-makers in a timely and acceptable fashion (UNEP, 2005).

The US portion of the global system – the US Integrated Earth Observation System – is centred around nine societal benefits:

- to improve weather forecasting;
- to reduce loss of life and property from disasters;
- to protect and monitor ocean resources;
- to understand, assess, predict, mitigate and adapt to climate variability and change;
- to support sustainable agriculture and combat land degradation;
- to understand the effects of environmental factors on human health and well-being;

- to develop the capacity to make ecological forecasts;
- to protect and monitor water resources;
- to monitor and manage energy resources.

Nearly all of these benefits consist of a major ocean component. Therefore, developing Integrated Ocean Observing Systems (IOOS) has been a key priority in the successful completion of GEOSS (Lautenbacher, 2005).

3.5.3   Integrated Ocean Observing System

The Integrated Ocean Observing System (IOOS) is envisioned as a network of observational, data management, and analysis systems that rapidly and systematically acquires and disseminates marine data and data products describing the past, present and future states of the oceans. Existing and planned observing system elements will be integrated into a sustained ocean-observing system that addresses both research and operational needs.

IOOS is developing as two closely coordinated components – global and coastal – that encompass the broad range of scales required to assess, detect, and predict the impact of global climate change, weather and human activities on the marine environment. The global component consists of an international partnership to improve forecasts and assessments of weather, climate, ocean state and boundary conditions for regional observing systems.

Currently, the USA has nine regional coastal observing systems in place to help meet some of these goals. Data collection platforms include moored and drifting buoys; meteorological towers and stations; bottom-moored instruments; stand-alone instruments; ship survey cruises; satellite imagery; and remotely and autonomously operated vehicles. Observations collected range from water conditions such as salinity, temperature and wave height, to atmospheric conditions like barometric pressure and wind speed.

By analyzing all of these data together on a regional level, one can begin to see how slight changes in one ecological component can cause changes in others. For example, one might discover that changes in water temperatures in one area trigger a change in a particular fishery in another area. This type of research cannot be completed unless the observation system used to study the Earth is well-integrated and organized (US Department of Commerce, 2004).

The ocean and coastal observing systems will use a wide range of advanced platforms and sensors, including ships, buoys, gliders and remote sensors on aircraft and satellites. A summary of satellite remote sensing techniques used in ocean and coastal observations is provided in Table 10.1.

Biological oceanographers use colour sensors to determine ocean colour, from which one can drive chlorophyll concentrations and biological productivity (Martin, 2004). Physical oceanographers use thermal infrared (TIR) radiometers to map sea surface temperatures and ocean currents. Synthetic Aperture Radar (SAR), radar scatterometers and radar altimeters are used to obtain information on ocean wave properties, surface winds and sea surface height. SAR is also effective for monitoring ocean surface features, including oil spills. As shown in Table 10.1, passive microwave radiometers can also provide a wide range of information, including water salinity, temperature, water vapour, soil moisture, etc. (Parkinson, 2003; Burrage *et al.*, 2003; Martin, 2004).

3.6 Existing image archives

Many Earth observation systems are operated for profit, and hence the data must be purchased. This is particularly true for high spatial resolution visible and radar data delivered by commercial (as opposed to government) satellite systems. Though the commercial and research sectors may account for significant sales, the military are typically the dominant consumers of these products.

Conversely, there exist numerous satellite programmes that offer data without charge, or at a minimum cost, sufficient only to cover processing and distribution. Into this fold fall numerous remote sensing programmes run by government agencies, such as NASA, NOAA and the US Geological Service (USGS) in the United States, and ESA in the European Union. Ultimately funded by taxpayers, these agencies are not permitted to operate for profit.

Part of the challenge of creating a user-accessible image archive is that it demands standardized metadata to facilitate efficient use of the data. In this way, metadata standards are critical to the management of online records. Web publishing by federal agencies such as NOAA and the USGS creates a search challenge, in that while much more information becomes accessible, online imagery can be lost in a morass of unmanaged and undiscoverable information. The National Spatial Data Infrastructure (NSDI) metadata standard has come to the fore as the national body in the USA to facilitate development of protocols for metadata and to promote adherence to these standards.

Tendered by the USGS, The National Map Seamless Server (http://seamless.usgs.gov) is a web-based portal from which the public can explore and retrieve data, including high-resolution orthoimagery and Landsat mosaics, as well as the National Elevation Dataset (NED) and the National Land cover Dataset (NLCD). The USGS further serves as a repository for NASA 'space acquired photography', including analogue and digital photographs acquired aboard missions such as Gemini, Apollo, Skylab, the International Space Station, Space Shuttle and Shuttle Large Format Camera Missions. The USGS also offers public access to non-imaging remote sensing instruments such as LiDAR through the Internet. The Center for LiDAR Information Coordination and Knowledge (CLICK) website (http://lidar.cr.usgs.gov) serves as a portal to information for accessing and using LiDAR data. The site includes a data viewer to visualize the LiDAR point cloud data held within the database.

For ocean colour data, a large repository of imagery and derived products are archived in the NASA OceanColor Web portal (http://oceancolor.gsfc.nasa.gov).

Here, data from long-defunct sensors such as the CZCS are held, as well as more recent missions such as SeaWiFS, MODIS Aqua and Terra and the European MERIS sensor.

Of most use to a wide swath of environmental change studies is the Landsat mission archive, which dates back to the launch of Landsat 1 in 1972. This three-decade library of Earth observation imagery can be accessed from various portals such as the USGS Global Visualization Viewer (http://glovis.usgs.gov) and the Global Land Cover Facility (GLCF, www.landcover.org).

The latter web portal is sponsored by the NASA Earth Science Enterprise and currently holds more than eight terabytes of raw and derived remote sensing products. It also provides access to SRTM elevation data as well as composite MODIS scenes. Furthermore, the GLCF provides restricted access to high-resolution imagery from sensors such as QuickBird, provided by DigitalGlobe in response to the December 2004 tsunami, as well as image-derived products such as flood maps, forest change information and land cover classifications.

Various groups have also made available Landsat data that has been assembled into a mosaic for a particular purpose. For instance, produced through USA/UK collaboration and marking a major contribution to the International Polar Year, the first ever true-colour high-resolution satellite view of the Antarctic continent was released in 2007. The composite, termed the Landsat Mosaic of Antarctica (LIMA), is comprised of more than 1,000 Landsat ETM+ images acquired between 1999 and 2001, and it can be downloaded through the project's web portal (http://lima.nasa.gov).

Perhaps the largest, and certainly the most accessible, image archive is the one which can be accessed through *Google Earth* (http://earth.google.com). This innovative software application saw its debut in 2005, and it is fair to say that *Google Earth* has revolutionized public access to geospatial data, facilitating users to interact with imagery of varying resolution with global coverage. The application does not yet offer the capability to handle real-time dynamic data, but it can be updated rapidly such, for example, that images of the Louisiana coastline effected by Hurricane Katrina were available within 48 hours, allowing residents to inspect the status of their properties even though they were not permitted to return to the area of devastation.

Part of the success of *Google Earth* is that it facilitates user access to geospatial data without download. Not only does this circumnavigate many of the complications of licensing geospatial data to the public but, in the majority of cases, a physical download of information is not necessary to convey knowledge. Furthermore, more traditional data portals such as the USGS Global Visualization Viewer require a rather sophisticated user to access data. *Google Earth*, by contrast, offers familiarity of access to a much broader demographic by being simple and intuitive to operate.

Beyond simple imagery, *Google Earth* tenders a vast repository of other co-mingled data sources pertaining to the biological, chemical, geological and physical status of the planet. As these products become more scientifically advanced, Google will increasingly face the challenge of how to install public understanding of how to correctly interpret the data. What was only a few years ago restricted to science professionals is now very much in the public domain. For example, introduced in version 5.0 of *Google Earth* (February 2009), the 'Google Ocean' feature allows users to zoom below the surface of the ocean and view the 3-D bathymetry beneath the waves. Supporting over 20 content layers, it contains information from leading scientists and oceanographers. In April 2009, Google added underwater terrain data for the Great Lakes. Similar applications have subsequently been released for the Moon, Mars, and the sky in general.

In the same vein as *Google Earth* is the NOAA 'Science On a Sphere' (SOS) initiative, which is a room-sized global display system that uses computers and video projectors to display planetary data onto a two-metre diameter sphere, analogous to a giant animated globe (http://sos.noaa.gov). SOS is described as an instrument to enhance informal educational programmes at science centres, universities, and museums across the United States. The Sphere is available to any institution and is currently in operation at several tens of facilities in the US, as well as a handful in Asia and Europe.

One particularly powerful use of SOS is for the visualization of IPCC climate simulations on the sphere. This is an example of making data that are already available operable in a different and more intuitive format. The SOS can be seen as a precursor to the web-orientated direction that NOAA research is moving towards. For example, the new NOAA research ship, the Okeanos Explorer, will sail without a scientific party and will instead relay data in real time through the Internet to experts around the world. Such a trend towards online use will serve to accelerate the degree to which scientific data, including remote sensing products, are archived and distributed via the World Wide Web.

3.7 Key concepts

1 Remote sensors are designed and used as mappers (imagers), radiometers, spectrometers and profilers (rangers). Mappers produce two-dimensional images and can be used for map-making. Radiometers measure the radiant energy in a few specific bands, while spectrometers provide the energy distribution across many spectral bands. Profilers, such as radar and LiDAR, measure the distance to features, allowing us to determine the topography or bathymetry of an area.

2 Satellite imagers can be grouped into low spatial resolution (>100 m), medium-resolution (10–100 m) and high-resolution (<10 m) systems. For instance, open ocean features can typically be observed with low spatial resolution satellites, while land cover and urban studies require medium and high resolutions respectively.

3 Ocean biological productivity can be determined with multispectral satellite sensors designed to map ocean colour. Physical ocean features can be observed with satellite radar systems, including Synthetic Aperture Radar for

imaging ocean wave fields and surface slicks; altimeters for measuring sea level, wave height, and surface currents; and scatterometers for mapping sea surface winds. Thermal infrared scanners and microwave radiometers are used to map sea surface temperatures and salinity respectively.

4 Global Earth Observing Systems are designed to monitor changes on land, sea and in the atmosphere on a global scale, including such specific features as vegetation cover; the ocean's temperature; Earth's radiation budget; cloud climatology; and other climate-related variables.

5 Many online archives exist that disseminate geospatial data to the public at no cost.

4 Digital image analysis

The objective of image analysis is to create an accurate map of an area viewed by satellite sensors. The data may include time-series of satellite images used to map slow changes, such as land cover and polar ice, as well as more rapidly changing features such as coastal currents and ocean chlorophyll concentrations. After the data are collected, digitized and transmitted to the ground station, they must be processed and converted into a format that is usable by the researchers who will interpret and analyze the data.

The original format of this data is often not such that the interpreter can learn much about the target or features being observed. Usually, the data must be processed, enhanced, and manipulated to provide a useful set of information. Generally, the analysis of digital imagery includes image pre-processing, image enhancement, image classification and accuracy assessment operations.

4.1 Image data format

Image classification assigns digital image data (pixels) to categories of the classification scheme. Satellite image data are sent from the satellite to the ground station in a raw digital format, which is a stream of numerical data. The smallest unit of digital data is a binary digit or 'bit'. A bit is represented by a binary number, so that it has only two possible values – 0 or 1. This does not offer much flexibility in representing data that is more complex than a binary number, therefore, data are usually stored as collections of bits arranged in groups of eight, resulting in a unit of data called a 'byte'.

Since a byte is comprised of 8 bits, it provides a data element with up to 256 potential values (2 raised to the power 8). Radiometers that measure the intensity of EM radiation will generally convert the detected energy levels into a value that ranges from 0 to 256 and represent each of these measured energy levels with a single byte. The bytes will be strung together in a predetermined manner,

Figure 4.1 Brightness values of picture elements (pixels) in four perfectly geometrically registered spectral bands. Since this is an 8-bit system, the brightness values at each pixel location are represented by a number ranging from 0 to 255. Modified from Jensen (2007). Printed and electronically reproduced by permission of Pearson Education, Inc., Upper Saddle River, New Jersey.

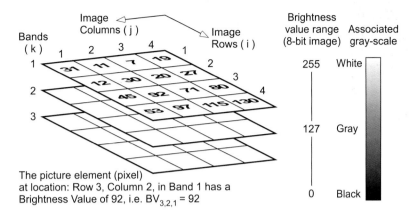

The picture element (pixel) at location: Row 3, Column 2, in Band 1 has a Brightness Value of 92, i.e. $BV_{3,2,1} = 92$

converted into a signal and transmitted to the receiving station on the ground. Here the signal is converted back into a digital stream of bytes, which can be read in and interpreted by processing software (Towson, 2009).

When a stream of bytes is received from a satellite sensor, the value of each byte is applied to a single pixel. The numerical value of the pixel, known as its Digital Number (DN), is translated into a shade of gray that ranges somewhere between white and black. These pixels, when arranged together in the correct order, form an image of the target in which the varying shades of gray represent the varying energy levels detected on the target (Figure 4.1).

4.2 Image pre-processing

Image pre-processing is performed on raw data in order to increase both the accuracy and the interpretability of the image prior to image classification (Jensen, 1996). Important pre-processing steps include:

- Radiometric correction of variations in the image resulting from environmental conditions (e.g. haze) or sensor anomalies.
- Geometric correction to compensate for the Earth's rotation and for variations in the position and attitude of the satellite.
- Terrain correction of relief distortions with the help of digital elevation data.
- Image enhancement techniques, which are used sometimes prior to image classification to improve the visual interpretability of an image.

Satellite images can be ordered at various pre-processing levels. Typical data levels for satellites such as SPOT or Landsat are shown in Table 4.1 (SPOT 2009).

Table 4.1 Data level definitions (SPOT 2009).

Level 1A	Pre-processing leaves the data in raw form, except for radiometric corrections which normalize sensor response to compensate for radiometric variations due to detector sensitivity. Level 1A data are intended primarily for experienced users who require image data that have undergone only minimal pre-processing, and who wish to do their own geometric image processing. Because minimum attention to the data has been paid by the vendor, this level is also the least expensive.
Level 1B	This is the basic pre-processing level, including both radiometric and geometric corrections. Radiometric corrections are the same as for level 1A pre-processing. Geometric corrections compensate for systematic effects, including panoramic distortion, Earth's rotation and curvature and variations in the satellite's orbital altitude. Internal distortions of the image are corrected for measuring distances, angles and surface areas. This level is specially designed for photo-interpreting and thematic studies.
Level 2A	Scenes are rectified to match a standard map projection (e.g. UTM WGS 84) without using ground control points. This level is the entry-level map product. Geometric corrections use a resampling model that compensates for systematic distortion effects and performs transformations needed to project the image in a standard map projection. It is suitable for users who want to combine different kinds of geographical information from different sources and apply their own colour processing to extract specific information. Level 2A images register directly with other layers of geographical information (vector data, raster maps or other satellite images) in the same map projection.
Level 2B	Images are georeferenced, framed in a given map projection and tied to ground control points obtained from a map or topographical surveys. The geometric corrections are similar to those of level 2A. Level 2B products are image maps, designed for use as digital maps, and they provide geographical information and global coverage. They can be used whenever relief distortions are not a major concern, i.e. when imaging relatively flat terrain.
Level 3	Imagery is georeferenced like level 2B. Level 3 products, also called 'orthoimages', are pre-processed using a digital elevation model (DEM) to correct residual parallax errors due to terrain relief. Geometric corrections consist of 'orthorectifying' imagery using a resampling model similar to level 2B. Level 3 products are image maps ideal for mapping terrain relief and are designed to offer maximum accuracy for producing and updating maps. This allows images to be registered with other kinds of data.
Level 4	Includes model outputs or results from analysis of lower level data, with variables not measured by an instrument but derived from these measurements.

4.3 Image enhancement and interpretation

One can visually interpret an image in a similar manner to interpreting aerial photographs. The interpretation tasks can include classification, enumeration, measurement, and delineation. The image analyst should understand statistical principles underlying image classification techniques, causes of variation in land cover or ocean features, causes of spectral variation in the imagery, and also needs to be able to link the variation in land cover or ocean features with the spectral variation in the imagery.

The main advantage of this approach is that the interpreter can use his or her knowledge of the area to decipher complexities in the landscape that would be impossible for a computer to interpret. The human brain is able to combine various image characteristics, including tone, texture, shape, size, shadow height and spatial relationships. Using stereoscopes, the interpreter can also observe vertical as well as horizontal spatial relationships of the features in the image.

To help the visual interpreter perform these tasks, a wide range of image interpretation equipment is available, including light tables, electronic digitizers and planimeters, stereoscopes and Zoom Transfer Scopes (Lachowski *et al.*, 1995; Avery & Berlin, 1992; Lyon & McCarthy, 1995).

The visual interpretability of an image can be improved by image enhancement techniques which increase the *apparent* distinction between the features in the scene. The process of visually interpreting digitally enhanced imagery attempts to optimize the complementary abilities of the human mind and the computer. The mind is excellent at interpreting spatial attributes on an image and is capable of selectively identifying obscure or subtle features. However, the eye is poor at discriminating the slight radiometric or spectral differences that may characterize such features. Computer enhancement aims to visually amplify these slight differences to make them readily observable (Sabins, 2007).

The range of possible image enhancement and display options available to the image analyst is virtually limitless. The most commonly applied digital enhancement techniques can be categorized (Lillesand & Kiefer, 1994) as:

- contrast manipulation, which includes gray-level thresholding, colour density slicing, and contrast stretching;
- spatial feature manipulation, which consists of spatial filtering, edge enhancement and Fourier analysis;
- multi-image manipulation, which includes multispectral band ratioing and differencing, principal components, canonical components, vegetation components and decorrelation stretching.

For example, colour density slicing converts gray levels to specific colours in an image. The human eye can only distinguish between about 16 levels of gray in an image, but it is able to distinguish between thousands of colour hues. Thus, a common image enhancement technique is to assign specific colours to specific digital number (DN) values, thereby increasing the contrast of particular DN values against the surrounding pixels in the image. An entire image can be converted from a grayscale to a colour image, or portions of an image that represent the particular DN values of interest can be coloured. Choosing the appropriate enhancement for any particular application is usually a matter of personal preference (Towson, 2009).

Figure 4.2 illustrates image enhancement by contrast stretching. The left histogram of the figure represents an imaged feature having a very narrow range of gray levels. After linear contrast stretching, the right histogram shows how the feature's contrast was expanded from a dozen to several hundred gray levels, utilizing the full capability of the display device and making it possible to identify more details.

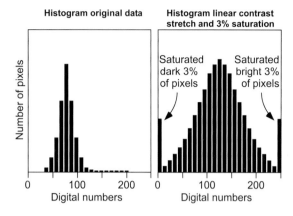

Histogram original data

Histogram linear contrast stretch and 3% saturation

Figure 4.2 Contrast stretching expands an imaged feature's gray levels (left histogram) over a much wider range (right histogram), providing more contrast and making it easier to identify details.

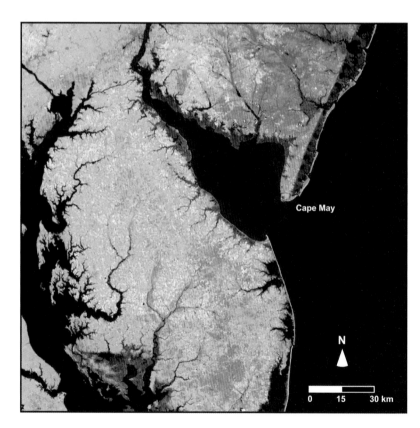

Figure 4.3 Landsat TM image of the Delaware Bay region obtained on April 24, 2002. Forests and dense vegetation are shown in dark green, cultivated agricultural fields in light green, bare fields in white and urban areas in gray. Coastal wetlands are shown in dark brown. Credit: NASA and USGS.

4.4 Image classification

Satellite images, such as the Landsat TM image in Figure 4.3, are normally analyzed using computer-aided image classification, which is a process by which a set of items is grouped into classes based on common characteristics. Classification of satellite

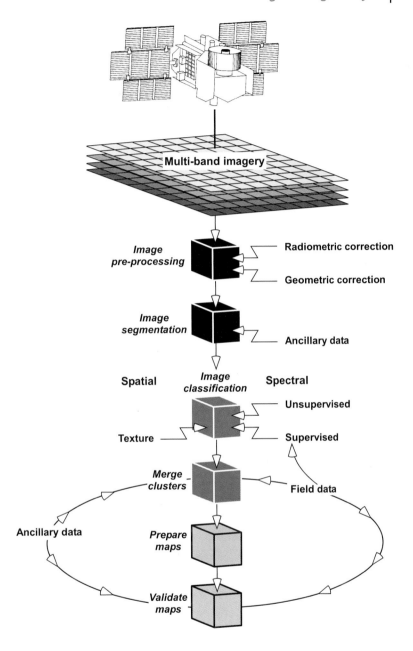

Figure 4.4 Typical image analysis approach.

image data is based on placing pixels with similar values into groups and identifying the common characteristics of the items represented by these pixels. Thus, a correctly classified image will represent areas on the ground that share particular characteristics as specified in the classification scheme (Lillesand & Kiefer, 1994).

A typical digital image analysis approach for classifying coastal wetlands or land cover is shown in Figure 4.4. Before analysis, the multispectral imagery must be radiometrically and geometrically corrected. The radiometric correction reduces the influence of haze and other atmospheric scattering particles and any sensor anomalies. The geometrical correction compensates for the Earth's rotation and for variations in the position and attitude of the satellite. Image segmentation simplifies the analysis by first dividing the image into homogeneous patches or ecologically distinct areas.

Training sites are then identified for supervised classification and interpreted via field visits or other reference data, such as aerial photographs. Supervised classification requires the analyst to select training samples from the data which represent the themes to be classified (Jensen, 1996). The training sites are geographical areas previously identified using ground-truth to represent a specific thematic class. The spectral reflectances of these training sites are then used to develop spectral 'signatures' which will be used to assign each pixel in the image to a thematic class.

Next, an unsupervised classification is performed to identify variations in the image not contained in the training sites. In unsupervised classification, the computer automatically identifies the spectral clusters representing all features on the ground. Training site spectral clusters and unsupervised spectral classes are then compared and analyzed using cluster analysis to develop an optimum set of spectral signatures. Final image classification is then performed to match the classified themes with the project requirements (Jensen, 1996). Note that throughout the process, ancillary data are used whenever available (e.g. aerial photos, maps, field data, etc.).

Training sites for supervised classification must meet well-defined criteria. They should be:

- in accessible areas;
- homogeneous with respect to vegetation/land cover;
- large enough so they could be located in satellite images;
- small enough to minimize within site variations.

There should also be multiple sites for each category of the classification scheme.

Field or ship data need to be collected for developing a spectral 'signature library' for supervised classification, calibrating remotely sensed data or training neural networks. Field checks may also have to be conducted in order to guide the interpreters during the image classification stage. Finally, field data are gathered at the end of a project to validate the remotely sensed products and statistically assess their accuracy.

A typical result of using such an image analysis approach is shown in the Delaware Bay land cover map in Figure 4.5. Note that the selection of the thematic classes of land cover is based on the specific needs of the mapping project. For example, three wetland classes are included in this case, since the emphasis of this mapping project was on wetlands.

The traditional image classification methods based on individual pixels are effective for images with relatively coarse spatial resolution. However, the recent availability of high spatial resolution imagery (0.5–4 m) from new satellites and airborne platforms has created the need for image processing routines that deal better with the local detail and heterogeneous spectral values found in this

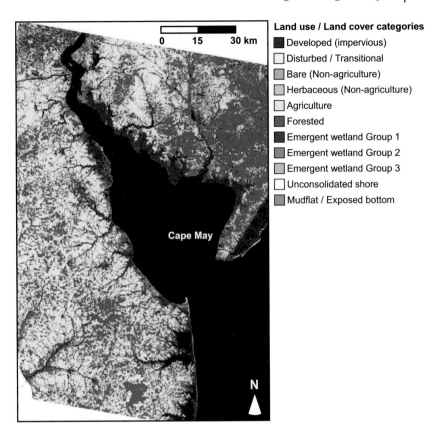

Land use / Land cover categories
■ Developed (impervious)
□ Disturbed / Transitional
■ Bare (Non-agriculture)
■ Herbaceous (Non-agriculture)
□ Agriculture
■ Forested
■ Emergent wetland Group 1
■ Emergent wetland Group 2
■ Emergent wetland Group 3
□ Unconsolidated shore
■ Mudflat / Exposed bottom

Figure 4.5 Delaware Bay land cover classification from 1993, based on Landsat TM imagery. Modified from Weatherbee (2000).

high-resolution imagery. The land cover types to be classified usually correspond to a coarser scale than individual pixels, which at this finer resolution produce a much greater within-class spectral variability, thus making separation of spectrally mixed land cover types more difficult.

One such new approach is Object-Based Image Analysis (OBIA), which segments remotely sensed imagery into meaningful landscape objects and classifies these across spatial, spectral and temporal scales (Blaschke & Hay, 2001; Burnett & Blaschke, 2003; Wang *et al.*, 2004). Homogeneous patches are created by image segmentation first, and then those patches or objects are classified into land cover classes. The image segmentation applies a homogeneity threshold to allow some heterogeneity within objects. A certain level of local homogeneity is expected, since neighbouring pixels frequently belong to the same class. Variability between pixels now defines the internal variability, or texture, of an object. Thus the OBIA approach includes segmentation into landscape objects, their spectral and textural properties, and hierarchical and contextual relationships at various scales. The OBIA methods of classifying high-resolution imagery have produced significant accuracy improvements over traditional pixel-based image analysis techniques (Kelly & Tuxen, 2009).

In every image classification attempt, ancillary data should be incorporated at all stages of the analysis. This can be done by comparing GIS layers of other types of data with the imagery to be analyzed. For example: the topography in digital elevation models can help discriminate vegetation types by altitude zonation caused by climate/temperature variations; areas having different soil types can sometimes be identified by their impact on the vegetation; road density can help differentiate urban lawns from pastures, etc.

4.5   Image band selection

Various applications on land require different spectral bands or band combinations, as shown in Table 4.2 for the Landsat Thematic Mapper bands. To conserve processing time and effort, before performing any of the image analysis functions, one should always study the statistical content of the spectral bands in order to optimize and minimize the number of bands needed for the analysis (Jensen, 1996; Lachowski *et al.*, 1995).

Table 4.2 Landsat Thematic Mapper bands (Lachowski *et al.*, 1995).

Band	Wavelength (nm)	Application
1	450–520 Blue band	Coastal water mapping, bathymetric mapping of shallow water, soil/vegetation differentiation, deciduous/conifer difference, cultural feature identification
2	520–600 Green band	Green reflectance by healthy vegetation, vigour assessment and cultural feature identification; important for discriminating vegetation types
3	630–690 Red band	Chlorophyll absorption for plant species differentiation
4	760–900 Near infrared	Biomass surveys, water delineation, vegetation types, vigour, soil moisture
5	1,550–1,750 Mid-infrared	Vegetation moisture measurement, snow/cloud difference, soil moisture measurement
6	10,400–12,500 Thermal infrared	Vegetation heat stress analysis, soil moisture, urban heat island and water surface temperature mapping applications
7	2,080–2,350 Mid-infrared	Hydrothermal mapping, mineral and rock type, vegetation moisture content

4.6   Error assessment

Error assessment of thematic map products depends on the collection of reference data, sometimes called ground-truth. Reference data represents known information of high accuracy (theoretically 100 per cent accuracy) about a specific area or process on the ground (the accuracy assessment site). This data can be obtained from ground visits or interpretation of photographs, video, etc. In a digital map, accuracy assessment sites are generally represented by groups of pixels or polygons. Accuracy assessment involves the comparison of the categorized data for these sites to the reference data for the same sites (Jensen, 2007; Lachowski *et al.*, 1995).

It should be kept in mind that error assessment is not limited to the appraisal of thematic maps but extends to biophysical variables such as, for example, the validation of remote sensing-derived estimates of chlorophyll loading in the ocean, or the concentration and spatial distribution of trace gases in the atmosphere. While the assessment of accuracy for a thematic map may only demand a brief visit to a field site to note ground cover, an assessment on the accuracy of biophysical parameters is likely to necessitate a more involved field collection backed up by laboratory measurements.

Routine error assessment of biophysical data is a key component in establishing whether a change detected through image analysis is simply an artefact or is a meaningful trend. For that purpose, error assessment for change detection may constitute two parts: first, an appraisal of the ability to differentiate change from no-change; and second, the ability to detect different kinds of change (Ridd & Liu, 1998). With the short-term expression of environmental change often being subtle, these assessments are critical.

The accuracy of completed map products can be expressed in terms of an error matrix for land cover mapping applications and as percentage error for water quality studies. The error matrix is the standard way of presenting results of an accuracy assessment for land cover classification (Story & Congalton, 1986). It is a square array in which accuracy assessment sites are tallied by both their classified category in the image and their actual category according to the reference data (Table 4.3). In this table, the columns in the matrix represent the reference (actual) data, while the rows represent the classified (interpreted) data. The major diagonal contains those sites where the classified data agree with the reference data.

The nature of errors in the classified map can also be derived from the error matrix. In the matrix, errors (the off-diagonal elements) are shown to be either errors of inclusion (commission errors) or errors of exclusion (omission errors). Omission errors are shown in the off-diagonal matrix cells that form the vertical column for a particular class. Commission errors are represented in the off-diagonal horizontal row cells. High errors of omission/commission between two or more classes indicate spectral confusion between these classes. Overall accuracy, a common measure of accuracy, is computed by dividing the total correct pixels (the diagonal elements) by the total number of pixels in the reference image or map (Campbell, 2007). Furthermore, a Kappa coefficient (Congalton, 1991) can be calculated to show how much better the map results are than a totally random labelling of the pixels in the image (Jensen, 1996; Lillesand & Kiefer, 1994).

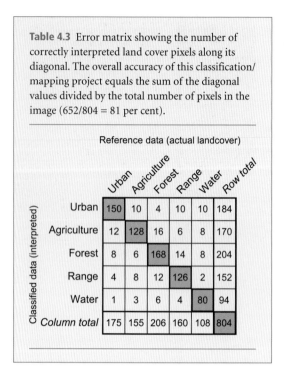

Table 4.3 Error matrix showing the number of correctly interpreted land cover pixels along its diagonal. The overall accuracy of this classification/mapping project equals the sum of the diagonal values divided by the total number of pixels in the image (652/804 = 81 per cent).

Reference data (actual landcover)

Classified data (interpreted)

	Urban	Agriculture	Forest	Range	Water	Row total
Urban	150	10	4	10	10	184
Agriculture	12	128	16	6	8	170
Forest	8	6	168	14	8	204
Range	4	8	12	126	2	152
Water	1	3	6	4	80	94
Column total	175	155	206	160	108	804

There are some practical ways of improving image classification accuracy. In addition to the spectral identification methods discussed thus far, one can use multiple remote sensing features such as spatial (texture, shape, context), multi-temporal (seasonal vegetation differences) and multi-sensor (combined multispectral and radar) data. For instance, texture can help separate coral reef habitats (Purkis *et al.*, 2006) and context can help confirm circular features (i.e. round features in arid areas may contain irrigated agricultural crops, while circular features near airports are likely to be fuel tanks). Seasonal differences are important for identifying vegetation types, especially in agriculture. Combining visible/near-IR and radar images can help differentiate between upland forests and forested wetlands (swamps). There also are some brand new image analysis approaches being developed, as discussed in detail in Lu & Weng (2007).

4.7 Time-series analysis and change detection

The dynamic nature of the Earth's surface causes it to change continuously at different time and space scales. The rate of change can be abrupt, as exemplified by fire or flooding; or gradual, such as desertification, deforestation or agricultural expansion. Since satellite sensors in their stable orbits view the same area on Earth repeatedly at fixed intervals, they are ideal for detecting and mapping global and regional changes, both natural and man-made. Time-series of satellite imagery can be used to analyze such changes and help to predict long-term trends. Typical satellite images suitable for change detection are shown in Figures 4.6 and 4.7.

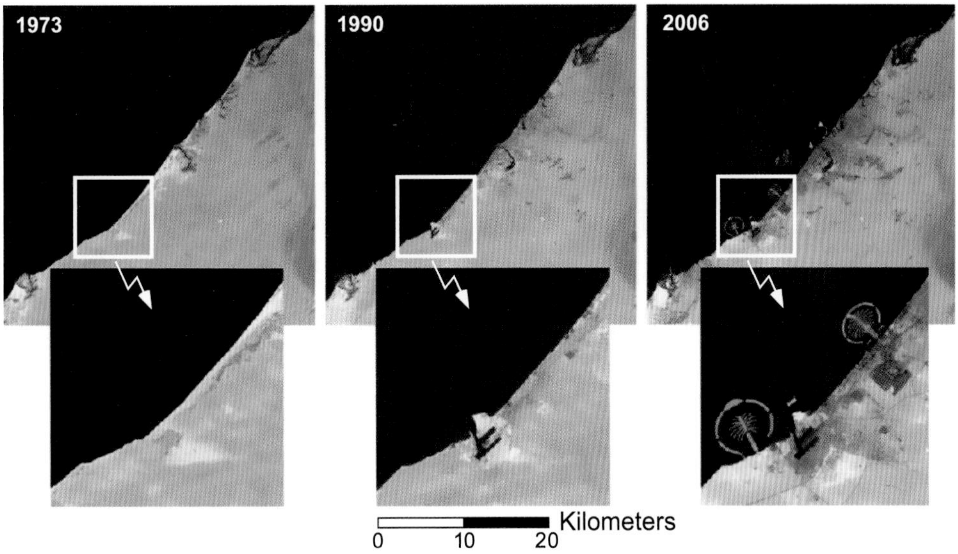

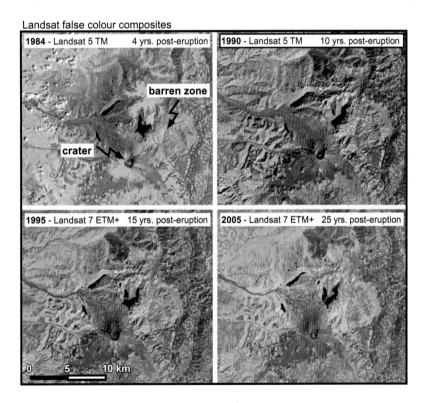

Figure 4.6 Thirty years of coastal construction imaged using Landsat on the Arabian Gulf coastline of the United Arab Emirates. North is top. Credit: NASA.

Figure 4.7 Twenty years of re-vegetation of Mt. St. Helens following the eruption on May 18, 1980, as imaged by Landsat. North is top. Credit: Landsat 7 project and EROS Data Center.

In Figure 4.6, a series of satellite images is used to demonstrate coastal construction during the last thirty years on the Gulf coast of the United Arab Emirates. The coast is seen in its natural state in 1973; by 1990, the massive seaport of Jebel Ali is clearly visible; and by 2006, this has been accompanied by two artificial 'Palm' islands, each capable of housing more than 1.5 million people.

In Figure 4.7, Landsat MSS images are used to show the abundant vegetation around Mount St. Helens in 1973, the severe loss of forest after the 1983 volcanic eruption and the rapid re-growth of vegetation by 1988. As shown in the figure, the majority of the blast was directed to the northeast of the crater, and this is where the barren zone (gray in images) is most pronounced in the ten years following the explosion. Copious quantities of ash, however, aided the rejuvenation process and accelerated vegetation growth (shades of green in the images).

Automating digital change detection is a difficult task, requiring that the satellite images in a time-series be carefully selected and pre-processed. Selecting appropriate dates for the image acquisition is critical. Choosing images close to anniversary dates is frequently attempted, since they minimize discrepancies in reflectance caused by seasonal vegetation changes and sun angle differences. However, even on annual anniversary satellite overpasses, there may be differences in atmospheric conditions, local precipitation, temperature and tidal stage.

The remotely sensed data will be used with field measurements to derive changes of a wide range of biophysical variables, such as vegetative biomass, ocean temperatures and global ice cover. This derived information will be used in a variety of models, including those for predicting global temperature trends and sea level rise, which is needed by environmental managers at local and national levels. Therefore, the assessment of the quality and accuracy of these data will be critical for determining which of these changes are real (not just image artefacts) and represent actual long-term trends.

The pre-processing of multi-date sensor imagery, when absolute comparisons between different dates are to be carried out, is much more demanding than the single-date case. It requires a sequence of operations, including calibration to radiance or at-satellite reflectance; atmospheric correction; image registration; geometric correction; mosaicking; sub-setting; and masking out clouds and irrelevant features.

In the pre-processing of multi-date images, the most critical steps are the registration of the multi-date images and their radiometric rectification. To minimize errors, registration accuracies of a fraction of a pixel must be attained. The second critical requirement for change detection is attaining a common radiometric response for the quantitative analysis for one or more of the image-pairs acquired on different dates. This means that variations in solar illumination, atmospheric scattering and absorption and detector performance must be normalized, i.e. the radiometric properties of each image must be adjusted to those of a reference image. Only with reliable radiometric calibration can one have confidence that the temporal changes are real and are not artefacts introduced by differences in the sensor calibration, atmospheric conditions or sun angle (Coppin *et al.*, 2004).

As shown in Figure 4.8, the actual digital change detection using satellite imagery can be performed effectively by employing one of several techniques, including post-classification comparison and temporal image differencing (Jensen, 1996; Dobson *et al.*, 1995; Lunetta & Elvidge, 1998). Post-classification comparison change detection requires rectification and classification of the remotely sensed images from both dates. These two maps are then compared on a

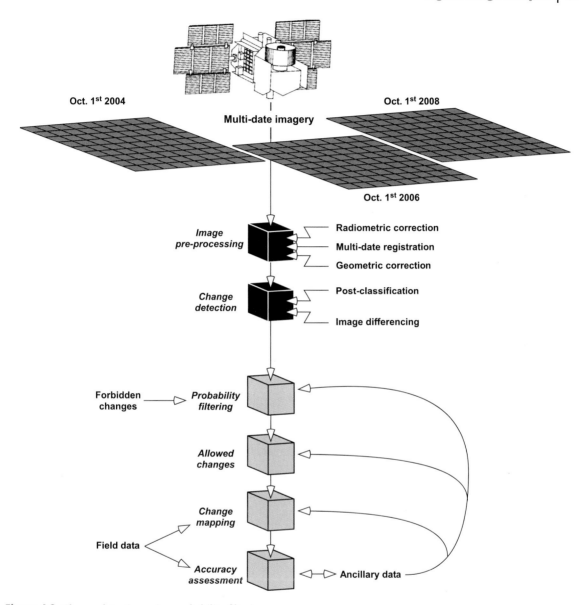

Figure 4.8 Change detection using probability filtering.

pixel-by-pixel basis. One disadvantage is that every error in the individual date classification maps will also be present in the final change detection map.

Temporal image differencing minimizes this problem by performing the traditional classification of only one of the two time-separated images. One band from both dates of imagery is then analyzed to find differences. Pixel difference values exceeding a selected threshold are considered to be changed. A change/no change binary mask is then overlaid onto the second date image, and only the pixels classified as having changed are classified in the second date imagery.

This method usually reduces change detection errors and provides detailed 'from-to' change class information (Jensen, 1996) As shown in Figure 4.8, change analysis results can be further improved by including probability filtering, allowing only certain changes and forbidding others (e.g. urban to forest). A detailed, step-by-step procedure for performing change detection was developed by the NOAA Costal Change Analysis Program (C-CAP) and is shown in Table 4.4 (Dobson *et al.*, 1995).

The main challenges facing global change monitoring from space are due to the requirements to:

1 detect modifications in addition to conversions (e.g. quantify forest cover degradation due to logging or fires);
2 monitor rapid and abrupt changes in addition to incremental changes (e.g. flooding, drought or fire versus slow expansion of agriculture);
3 separate inter-annual variability from long-term trends, given the short 20–30 year availability of satellite data;
4 integrate data obtained at different spatial resolutions;
5 match the temporal sampling rates of observations of processes to the intrinsic scales of these processes (e.g. rapidly evolving processes such as fires or floods require different sampling rates from those needed for slow changes such as deforestation or urban sprawl) (Coppin *et al.*, 2004).

To identify long-term trends and separate them from short-term variations, one needs to analyze long time-series of remotely sensed imagery. The analysis of time-series of multi-spectral imagery is a difficult task, since the time axis and wavelength axis cannot be mixed. One way to approach this problem is to reduce the spectral information to a single index, reducing the multispectral imagery into one single field of the index for each time step. In this way, the problem is simplified to the analysis of time-series of a single variable, one for each pixel of the images.

Major advantages of vegetation indices over single-band radiometric responses include their ability to reduce the data volume for processing and analysis and their ability to provide information not available in any single band. The most common index used is the Normalized Difference Vegetation Index (NDVI: see Chapter 5), which is based on the contrast between the red and infrared reflectance. Since this contrast is the most detectable spectral characteristic of green plants, this index can be related to plant biomass or stress because it combines both the abundance of green plant tissue and the surface condition of the plant leaf.

It has been shown by researchers that time-series of remote sensing data can be used effectively to identify long-term trends and subtle changes of NDVI by means of principal component analysis (Young & Wang, 2000). The NDVI is sensitive to background soil conditions (such as soil moisture) and can provide ambiguous information about the vegetation cover that consists of sparse canopies. Therefore,

Table 4.4 Steps required to conduct C-CAP change detection to extract upland and wetland information using satellite remote sensing systems.

1	**State the regional change detection problem**
a	Define the region
b	Specify frequency of change detection (1 to 5 yr)
c	Identify classes of the C-CAP Coastal Land cover Classification System
2	**Consider significant factors when performing change detection**
a	Remote sensing system considerations:
	1) Temporal resolution
	2) Spatial resolution
	3) Spectral resolution
	4) Radiometric resolution
	5) The preferred C-CAP remote sensing system
b	Environmental considerations:
	1) Atmospheric conditions
	2) Soil moisture conditions
	3) Vegetation phenological cycle characteristics
	4) Tidal stage
3	**Conduct image processing of remote sensor data to extract upland and wetland information**
a	Acquire appropriate change detection data:
	1) *In situ* and collateral data
	2) Remotely sensed data
	a) Base year
	b) Subsequent year(s)
b	Pre-process the multiple-date remotely sensed data:
	1) Geometric rectification
	2) Radiometric correction
c	Select appropriate change detection algorithm from the three C-CAP alternatives
	Apply appropriate image classification logic if necessary:
d	1) Supervised
	2) Unsupervised
	3) Hybrid
	Perform change detection using GIS algorithms:
e	1) Highlight selected classes using change detection matrix
	2) Generate change map products
	3) Compute change statistics
4	**Conduct quality assurance and control**
a	Assess spatial data quality
b	Assess statistical accuracy of:
	1) Individual date classification
	2) Change detection products
5	**Distribute results**
a	Digital products
b	Analogue (hardcopy) products

Source: Dobson *et al.*, 1995.

a wide range of vegetation indexes have been developed which allow one to make corrections for background soil variations (Qi *et al.*, 1994).

4.8 Field sampling using GPS

To validate the remote sensing results and determine their accuracy, a statistically valid sampling scheme should be selected (Figure 4.9). Simple random sampling is a powerful sampling strategy because it yields data which can be analyzed using inferential statistics. However, such sampling is often not uniform, but places

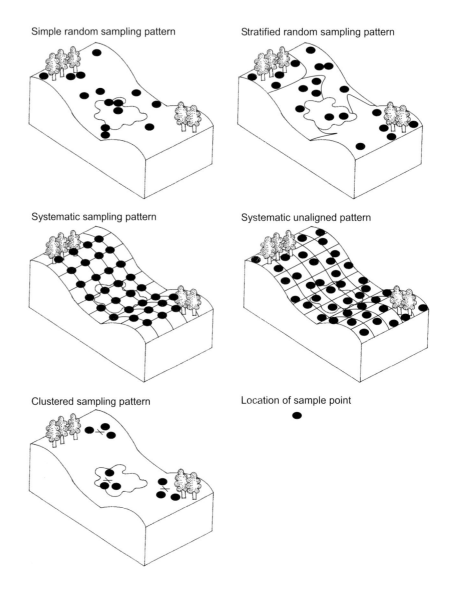

Figure 4.9 Various field sampling schemes.

observations in patterns that tend to cluster together. The stratified sampling pattern assigns observations to subregions of the image, thus insuring that the sampling effort is distributed in a rational manner. Systematic sampling positions observations at equal intervals, but it does not meet the requirements of inferential statistics for randomly selected observations. For instance, purely systematic sampling could introduce major bias errors if the features in the area to be sampled have a regular pattern (Campbell, 2007).

A stratified systematic unaligned sampling pattern combines features of both systematic and stratified samples, while at the same time preserving an element of randomness. Such a sampling strategy is especially versatile for sampling geographical distributions such as land cover (McCoy, 2005). The entire study area is divided into uniform cells using a square grid, the cells of which introduce a systematic component and form the basis for stratification. One observation is then placed in each cell, and the element of randomness is introduced by placing the observations randomly within each cell. However, sometimes one is compelled to use special sampling schemes, such as clustered sampling patterns, when access is limited or the vegetation distribution is unusual.

Field data collected for interpreting imagery and validating final mapping results must be obtained at accurately located sites. Global Positioning System (GPS) uses a network of satellites to provide both position and time information to the user on the ground. The GPS can provide the location of such sites with about 10-metre accuracy or better, which is well within one pixel size of the medium resolution satellites like Landsat TM. If the receiver signals are combined with signals concurrently collected by a nearby GPS base station for differential GPS post-processing, location accuracies down to 5–10 cm can be obtained.

A GPS receiver is generally a handheld unit designed to collect position and attribute data pertaining to geographical features. A three-dimensional position obtained from satellites can be computed by the handheld GPS receiver and the information can be exported to a GIS or plotted on maps. Although GPS is not really a remote sensing device, it complements Earth observation data by providing accurate ground location and storage of attributes that can be linked with remotely sensed imagery (Campbell, 2007).

The land cover classification process frequently requires that the spectral reflectance of the pertinent land cover be obtained in the field with spectrometers or multi-band radiometers. One can measure the reflectivity at all possible sensor and source positions using a goniometer (Jensen, 2007). This is a very time-consuming approach, so most scientists compare the reflectance of the test site with that of a large standard white panel which has a known reflectance in the range of 95 to 99 per cent (McCoy, 2005). For example, in Chapter 8 (Figure 8.13), the use of a submersible field spectrometer is demonstrated to collect reflectance signatures of various coral reef habitats.

Figure 2.4 (Chapter 2) shows a number of typical spectra for various land cover types that extend through the visible and near-infrared wavelengths, as could be acquired using a field radiometer. These reflectance spectra are taken as the ratio between upwelling radiance and downwelling irradiance. Ideally, calculation of

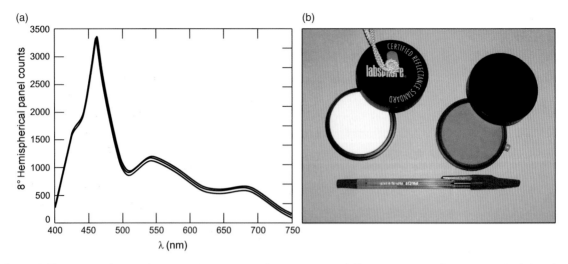

(a)

(b)

Figure 4.10 (a) Typical spectral response by wavelength for a 99 per cent diffuse Spectralon reflectance standard (b). The white panel is a 99 per cent reflectance standard; the grey is a 20 per cent standard which can be used under conditions of exceptionally bright light. These reflectance standards are manufactured by Labsphere Co., among others, and can be purchased in a multitude of different sizes.

the reflectance is based on simultaneous measurement of radiant flux towards and away from the target. This does, however, demand that the radiometer be equipped with two separate probes (one to measure radiance, the other irradiance), which is far more expensive than a sensor with a single probe. In a single-sensor set-up, spectral reflectance is instead measured by first pointing the tip of the radiometer's fibre-optic probe vertically down towards a target of known reflectance, such as a white (or gray) Spectralon reflectance standard (Figure 4.10). Being highly Lambertian, the light-flux reflecting from the Spectralon panel is reasonably equivalent to that of the downwelling irradiance from the complete hemisphere of the sky above. The subsequent radiance measurements of land cover are then made as relative to the reflectance standard.

4.9 Use of Geographic Information Systems

A Geographical Information System (GIS) is a computer system for capturing, storing, manipulating, analysing and displaying geospatial data, which are data that describe both the locations and characteristics of spatial features such as vegetation stands, land parcels, and roads on the Earth's surface. The ability of GIS to store and process geospatial data offers a practical approach for integrating spatial data from various sources, including remote sensing (Chang, 2008). Therefore GIS provides an ideal platform for data integration, synthesis and modelling to support research and decision-making – essential in most environmental applications.

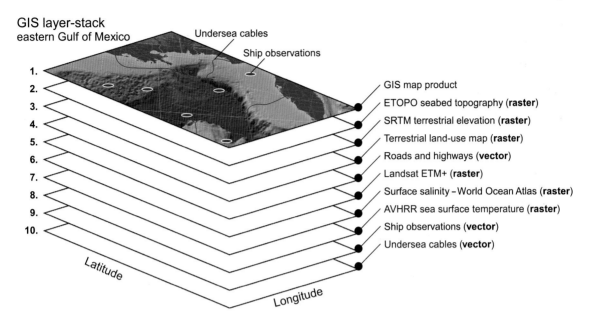

GIS layer-stack
eastern Gulf of Mexico

Undersea cables

Ship observations

1.
2.
3.
4.
5.
6.
7.
8.
9.
10.

GIS map product
ETOPO seabed topography (**raster**)
SRTM terrestrial elevation (**raster**)
Terrestrial land-use map (**raster**)
Roads and highways (**vector**)
Landsat ETM+ (**raster**)
Surface salinity – World Ocean Atlas (**raster**)
AVHRR sea surface temperature (**raster**)
Ship observations (**vector**)
Undersea cables (**vector**)

Latitude

Longitude

Figure 4.11 A Geographical Information System (GIS) layer-stack.

As shown in Figure 4.11, information is often stored in a GIS as separate layers or themes (roads, land use, elevation, etc.) that may be combined to form new layers. The information may be stored in vector form as lines, points, polygons or raster form as arrays of pixels (digital picture elements).

GIS also permits linking tabular or attribute data to spatial data. As shown in Figure 4.12, attribute data describe the characteristics of spatial features. For raster data, each cell has a value that corresponds to the attribute of the spatial feature at that location. For vector data, the amount and type of attribute data to be associated with a special feature can vary. For instance, a lake may have only the attributes of area, perimeter, depth and potability, while a soil polygon may have dozens of properties, interpretations and performance data. Therefore joining the spatial and attribute data is of key importance in the case of attribute data (Lyon & McCarthy, 1995; Chang, 2008).

Data management in a GIS is normally performed using a geo-relational database model, which is a collection of tables or relations. The connection between tables is made through a key, whose values can uniquely identify a record in a table (Figure 4.12). The feature ID serves as the key in the geo-relational data model which links spatial data and attribute data. For example, a parcel of land on a land cover layer in a GIS may have many attribute tables attached to it, including vegetative cover, soil type, slope, land ownership, etc.

A relational database is efficient and flexible for data search, retrieval, and editing. Each table in the database can be prepared, maintained and edited separately from other tables. Thus, the tables can remain separate until an

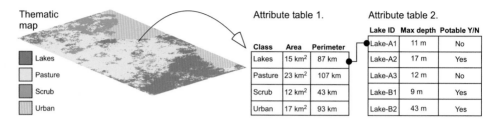

Figure 4.12 Storage of GIS attribute information in a relational database.

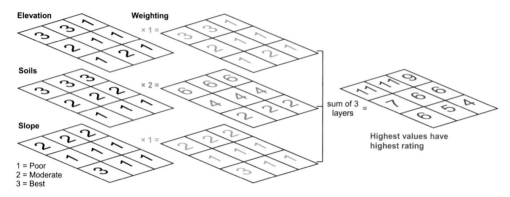

Figure 4.13 GIS overlay operation using different weights for data layers. Modified with permission from Davis (1996).

analysis requires that attribute data from different tables be linked or joined together (Chang, 2008).

Once the spatial data are properly formatted and stored in layers of the GIS database, a wide range of manipulations, analyses and modelling can be performed by combining the data layers in various ways. For instance, to analyze flooding hazards, the elevation, land use, road, Landsat TM layers and additional atmospheric/climatic, hydrologic, and population density layers would be extracted from the database and used in the modelling. The data layers can be overlaid, assigned different weights and manipulated in various ways to obtain the desired result or prediction (Davis, 1996). For instance, in Figure 4.13, the soils layer is weighted twice as much as the elevation and slope layers, to determine a best parcel of land for construction.

Some of the typical operations for GIS include map overlay, buffering, spatial interpolation, distance modelling, neighbourhood analysis, projection conversion, registration/image matching, resampling, Boolean logic operations, arithmetic operations, statistics generation, analysing different environmental scenarios, and so on. GIS have been applied to study a wide range of specific problems, including forest management, fire prediction, coastal resource management, aquatic macrophyte studies, etc. (Remillard & Welch, 1993; Green et al., 1995; Cowen et al., 1995; Lyon & McCarthy, 1995; Davis, 1996).

Since the early 1990s, computer-based GIS systems have transformed all the activities and disciplines that formerly used hard-copy maps as the basis for decision-making. This is a logical progression, considering the numerous shortcomings of maps for providing scientific insight. Computers have become ever more powerful in terms of memory and storage, and this upward trajectory in performance has been accomplished in tandem with the development of sophisticated data structures capable of representing the huge, almost infinitely detailed environment in which we live.

The provision of imagery and geospatial data through online archives discussed in Chapter 3 exemplify the state of the art of public-accessible web-based GIS. Further examples of the use of the technology in the marine realm will be given in Chapter 8 in the context of mapping and monitoring coral reefs. In addition to running in networked environments such as the World Wide Web, desktop GIS software has become a prerequisite technology in the classroom as much as in any science lab.

More recent desktop applications are advancing with the increased sophistication of geographically-enabled programming languages and standardization of geospatial data protocols. This imparts flexibility and the ability to customize the implementation of GIS and the production of novel spatial tools (Goodchild, 2003; Chang, 2008).

To summarize, GIS helps meet the challenges of global resource management by creating opportunities for:

- **Site-specific analyses**. Site-specific analysis is required for activities ranging from natural resource management to land use planning and wildlife habitat preservation. A GIS makes land use information about a specific site readily available. In a GIS, the land is spatially referenced as it ties tabular resource information (vegetation inventory, ecological units and wildlife habitat information) to a specific location on the ground and it offers resource managers and scientists the tools to use and analyze the data.
- **Visualizing spatial distributions**. Maps are integral to the handling and visualization of spatial data, which in turn reveal spatial distributions and relations. Visualization is the first step towards exploratory data analysis.
- **Spatial analysis**. GIS allows the summary of spatial distributions and the solving of spatial problems. Both are invaluable in the process of site-specific decision making.
- **Spatial databases**. Database systems provide the engines for GIS. The computer is used as a facilitator of data storage and sharing, and also allows the data to be modified and analyzed while stored. A successful geo-database GIS interface is as capable of dealing with a query submitted by a single-time user as it is with the unpredictable requirements of numerous simultaneous users. The evolution of database technology has accelerated the dissemination of GIS through the Internet. The most well-known such application is *Google Earth*, a technology discussed previously in Chapter 3 in the context of image archiving.
- **Repeatable and flexible analyses**. GIS allows resource managers to consider many options and alternatives, to perform sensitivity analyses and modelling and to document and defend analyses using site-specific information.

- **Ecosystem management.** GIS can be the repository of a wealth of information on ecosystems varying in size, scale, and complexity. It also has the capability to combine and integrate extensive but separate databases to formulate new perspectives on land resources (Lachowski *et al.*, 1995).
- **Modelling uncertainty.** Spatial information is rife with uncertainty for a number of reasons, and the correct conceptualization of this uncertainty is fundamental to the correct use of spatial data. GIS offers a platform for the visualization and quantification of spatial error, vagueness, and ambiguity.
- **Integration of GIS and remote sensing technologies.** Remotely sensed images can be used as both a source of spatial data within a GIS and to exploit the functionality of GIS in processing these data. GIS forms the backbone of environmental simulation models, and hence is highly relevant to the science of global change.

4.10   Key concepts

1 Digital satellite images must be pre-processed to correct radiometric and geometric distortions before they can be analyzed. Satellite images can be ordered at various levels of pre-processing, ranging from raw data to radiometrically and geometrically corrected and georeferenced images.

2 Digital images can be enhanced to aid visual interpretation by using contrast manipulation, spatial feature manipulation or multi-image manipulation.

3 Multispectral and hyperspectral images can be classified based on their spectral content using computers to map and study such features as land cover, vegetative biomass, wetland losses, urban expansion, water pollutants and ocean productivity.

4 Time-series analysis and change detection requires careful selection of the multi-date images to minimize seasonal, atmospheric, precipitation, temperature and sensor calibration variations between dates to be compared. Anniversary dates are frequently chosen for the image-pairs.

5 Change detection also requires extensive pre-processing of the multi-data images, including atmospheric correction, geometric correction, image registration and masking out clouds and irrelevant features, etc. After pre-processing, changes can be detected using several techniques, including post-classification comparison and temporal image differencing. Geographical Information Systems (GIS) and the Global Positioning System (GPS) have enabled environmental modellers to manipulate many layers of spatially georeferenced data, including ancillary data, to improve classification of satellite images and to model a wide range of environmental processes.

5 Monitoring changes in global vegetation cover

5.1 EM spectrum of vegetation

Some of the earliest applications of civilian remote sensing have been in agriculture and forestry. The reason for this rich lineage is that chlorophyll, and the plants that contain it, have a distinct, readily detectable spectral signature in the visible and near-infrared portions of the electromagnetic spectrum (Figure 5.1). Space- and airborne remote sensing systems have therefore been used for many decades to estimate crop health, predict agricultural yield and vegetation vitality, map deforestation and quantify the damage inflicted by disease and insect infestation on crops and forests. Since vegetation is also sensitive to changes in climate, including temperature and rainfall, it can be used as an indicator of global climate change.

As shown in Figure 5.1, plants, by virtue of their chlorophyll content, have a very distinct spectral signature which differs markedly from other land cover types such as soil, water and bare earth. The vegetation spectrum has two primary chlorophyll absorption bands in the blue and red regions which are due to pigments such as chlorophyll a, chlorophyll b and β-carotene, which reside in the upper layers of the leaf (palisade parenchyma). By contrast, there is strong reflectance at near-infrared wavelengths, caused by scattering in the spongy mesophyll, the deeper layers of the leaf. In the middle infrared region, three water absorption bands express themselves as a function of how hydrated the leaf tissue is. Thus, if the plant is stressed or drying up, there will be less absorption by the chlorophyll and water bands, as well as less reflection in the near-infrared.

The ability of the near-infrared to penetrate deep into a leaf canopy makes it also useful for estimating plant biomass. Figure 5.1 suggests that it is easy to distinguish vegetated areas from bare ones; however, since the spectra of various plants are somewhat similar, it is not a trivial task to identify plants to the species level on the basis of their spectral signature alone (Lachowski *et al.*, 1995; Jensen, 2007).

The remote sensing of vegetation also benefits from the fact that the wavelengths of visible light to which chlorophyll is sensitive occupy an atmospheric 'window',

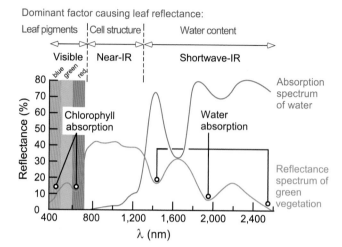

Figure 5.1 Spectral response characteristics of green vegetation. Chlorophyll contained in a leaf has strong absorption at 450 nm and 670 nm and high reflectance in the near-infrared (700–1,200 nm). In the shortwave-IR, vegetation displays three absorption features that can be related directly to the absorption spectrum of water (blue line) contained within the leaf.

i.e. the light passes relatively unhindered through the atmosphere to the Earth's surface. Even if positioned outside the Earth's atmosphere, any remote sensing system that is tuned to image these wavelengths will enjoy a relatively high signal, and this is the reason that so many sensing systems utilize the visible spectrum. More than 45 per cent of the radiant energy downwelling on the Earth's surface during the day is in the visible wavelength region, and it is also no coincidence that this is the portion of the spectrum that plants have evolved to initiate photosynthetic reactions.

5.2 Vegetation indices

All photosynthetic organisms have one or more pigments capable of absorbing visible radiation. Two portions of the electromagnetic spectrum are strongly absorbed by the chlorophyll pigment. These correspond to blue (450 nm) and red light (670 nm), and the magnitude of the absorption is directly related to the amount of radiant energy utilized by the plant. The upshot is a small peak at 500–600 nm (visible green), which explains why our eyes perceive most vegetation to be green. As shown by Figure 5.1, the near-infrared region (700–1,200 nm) acts inversely, displaying a pronounced peak in reflectance. It is a fairly unique characteristic of green vegetation that the reflectance in the near-IR is several times as large as that in the visible band. Since the transition from low reflectance in the red to high in the near-infrared is very abrupt, it is often termed the 'red edge'.

The red edge position (REP) shifts according to changes of chlorophyll content, leaf area index (LAI), biomass and hydric status, age, plant health levels and seasonal patterns. For healthy plants with high chlorophyll content and high LAI, the red edge position shifts toward the longer wavelengths; when a plant suffers from disease or chlorosis and low LAI, it shifts toward the shorter wavelengths (Figure 5.2).

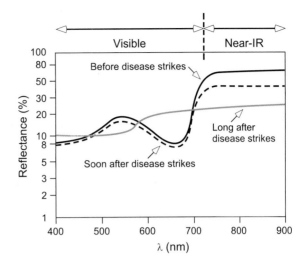

Figure 5.2 Spectral reflectance changes in some plants impacted by diseases, drought, pollution or insect infestations.

The high absorption of the red and blue spectral regions, with high reflectance in the near-infrared, is unique to vegetation, setting it apart from other land cover types – even those that are green in colour but lack chlorophyll. This phenomenon has obvious military applications when true vegetation must be separated from artificial camouflage. Since the inception of satellite remote sensing, it is this differential absorption of visible-red to near-infrared light that has been employed as a way to differentiate between surfaces with various amounts of vegetation cover. Changes in the red absorption and infrared reflection have been used for decades to detect early stress in plants directly. As shown in Figure 5.2, a decrease in infrared reflection may start in potato plants impacted by blight, or forests defoliated by gypsy moth infestation, well before it is apparent in the visible bands, thus giving farmers and foresters sufficient advance warning to allow them to try to save the plants.

Many techniques have been developed to study quantitatively and qualitatively the status of vegetation from satellite images. The vegetation index concept was developed as a method of reducing the number of variables present in multispectral measurements down to one unique parameter. Vegetation indices are combinations of spectral channels, blended in such a way that they strengthen the spectral contribution of green vegetation. This is achieved by minimizing the disturbing influences of soil background, irradiance, solar position, yellow vegetation and atmospheric attenuation variations.

Most vegetation indices utilize the red and near-infrared bands. For example, based on the reflectance difference that green vegetation displays between the visible region (10 per cent) and the near-infrared region (50 per cent) of the electromagnetic spectrum (e.g. in channels 1 and 2 of the AVHRR images of the NOAA satellites), the Normalized Difference Vegetation Index (NDVI) can be expressed as:

$$NDVI = \frac{NIR - R}{NIR + R}$$

where R is the red band and NIR is the near-infrared band reflectance.

Figure 5.3 The NDVI algorithm capitalizes on the fact that green vegetation reflects less visible light and more near-IR, whereas sparse or less green vegetation reflects a greater portion of the visible and less near-IR. By combining these reflectance characteristics in a ratio, NDVI provides a useful index of photosynthetic activity.

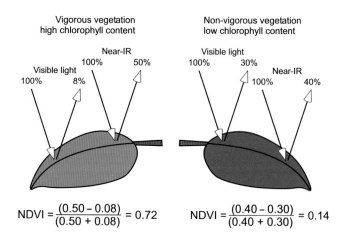

$$NDVI = \frac{(0.50 - 0.08)}{(0.50 + 0.08)} = 0.72 \qquad NDVI = \frac{(0.40 - 0.30)}{(0.40 + 0.30)} = 0.14$$

The range of values obtained by the NDVI is between −1 and +1. Only the positive values correspond with the vegetated zones, and the higher the index, the greater the chlorophyll content of the target. Negative values, generated by a higher reflectance in the visible region than in the infrared region, are due to clouds, snow, bare soil and rock. The value of the NDVI can change depending on the land use, the season, the hydric situation and the climate of the area (Figure 5.3) (Goward *et al.*, 1991; Leprieur *et al.*, 1994; US Geological Survey, 1997).

An important point is that NDVI is sensor-dependent, since the value of the index depends not only on the net sensor system gain in the channels used, but also on the specific bandwidth and location. Further differences are seen, depending on whether radiance data (which are affected by the solar irradiance illumination curve) or reflectance data are used. NDVI thus provides a 'relative' indication of the vegetation rather than an absolute measure. Furthermore, one must make sure that the relative size of the near-infrared band response compared to that in the visible band is large enough to confirm that the land cover is vegetation and only vegetation.

The NDVI obtained from the AVHRR images of the NOAA satellites has a pixel size of 1.1×1.1 km in the nadir, which lets one work on a global scale (1 : 2,000,000). The coverage of the NOAA satellites is of one image every 12 hours for each satellite. Only the daytime images are useful for obtaining NDVIs.

As shown in Figure 5.4, using Landsat, fields with particularly high plant vigour have high NDVI values and are depicted in warm colours. Plots that have been recently ploughed can be distinguished in the true-colour image (upper pane of the figure) by their brown colour. Lacking any vegetation, they correspondingly have negative NDVI values and cool colours in the lower pane.

Any satellite or aircraft sensor that has spectral bands positioned in the visible red and infrared can be used to calculate an NDVI. Those with a fine spatial resolution usually have a poor temporal resolution and hence cannot be used for daily monitoring. Satellites in this class include Landsat, the French SPOT sensor, GeoEye's IKONOS, and DigitalGlobes's QuickBird.

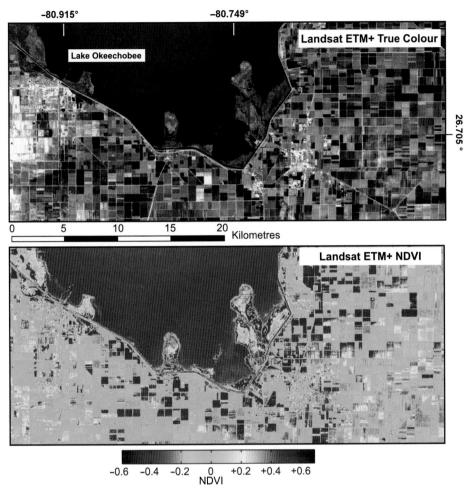

Figure 5.4 Upper pane is a true-colour Landsat ETM+ image of agricultural land surrounding Lake Okeechobee in central Florida. The lower pane is the NDVI of the same image. North is top. Credit: Landsat 7 project and EROS Data Center.

WorldView-2, a newly launched satellite from DigitalGlobe, and one that will be discussed several times through the course of this book, packs five spectral bands into the visible spectrum (in order of increasing wavelength termed 'coastal', 'blue', 'green', 'yellow' and 'red'), and three into the infrared ('red edge', 'near-IR 1' and 'near-IR 2'). WorldView-2 delivers 1.8 m² resolution pixels and a hitherto unparalleled capability to monitor vegetation by virtue of the judicious placement of a full eight multispectral bands spread through the visible and near-infrared, as compared to the typical four for similar satellites such as QuickBird and IKONOS. All eight bands key into areas of the spectrum relevant to the chlorophyll signature. It is believed that this enhanced spectral capability, coupled with a high spatial resolution, will allow the separation of coniferous from hardwood trees and the identification of invasive plant species, as well as a host of other applications related to discerning plant health, vigour, age and species.

Positioned at nearly twice the altitude of its predecessor QuickBird, the WorldView-2 satellite has shorter revisit times, and this higher orbit also allows the sensor to collect repeat data at look-angles that are closer to nadir than QuickBird. Imagery acquired at nadir, or close to it, is typically of better quality than that collected at more oblique look-angles, since the instrument is imaging through a thinner layer of atmosphere. Also, when imaging the ocean (and as demonstrated in Figure 8.6), it is preferable to acquire data close to nadir in order to minimize bright reflection of the sun off the water's surface. This is referred to as 'sunglint' and will be treated in detail in Chapter 8. Because of its greater altitude and less extreme look-angle, the WorldView-2 instrument is less likely than QuickBird to collect glinted images.

In 2009, Britain launched the imaging satellite UK-DMC2, which took its place in a larger group of orbital instruments termed the Disaster Monitoring Constellation (DMC). This cluster is discussed further in Chapter 14 and is used to obtain rapid information about areas struck by natural calamities. Aboard the same rocket was a twin spacecraft, Deimos-1, operated by Spain. Both UK-DMC2 and Deimos-1 have a $22\,m^2$ spatial resolution and bands placed in the green, red and near-infrared. During its projected five-year lifetime, the mission is expected to be capable of monitoring precision farming and deforestation via vegetation indices.

5.3   Biophysical properties and processes of vegetation

Through its relative simplicity of calculation from a diverse suite of satellite sensors, the NDVI remains a well-used index of vegetation. However, the measure remains an artificial variable with often only a tenuous link to the real-world biophysical processes at play within a plant. For this reason, natural resource managers and policymakers strive to retrieve more fundamental properties of vegetation via remote sensing data. One such parameter is the structure of vegetation, such as tree height, density, canopy architecture and biomass.

LiDAR, a sensing method that is discussed frequently in this book, is a particularly powerful technology that can be used to retrieve both the three-dimensional structure of vegetation as well as sub-canopy topography. Full-waveform (FW) airborne LiDAR is required for such studies, meaning that the system is capable of recording the entire emitted and backscattered signal of each laser pulse. By contrast, more conventional (i.e. non-FW) instruments only capture clusters of 3-D points.

As the FW LiDAR laser energy penetrates into the vegetation canopy, the returned waveforms are directly linked with the vertical and horizontal structure of the target (Figure 5.5). For instance, direct relationships exist between the full LiDAR waveforms and parameters such as tree height, stem diameter

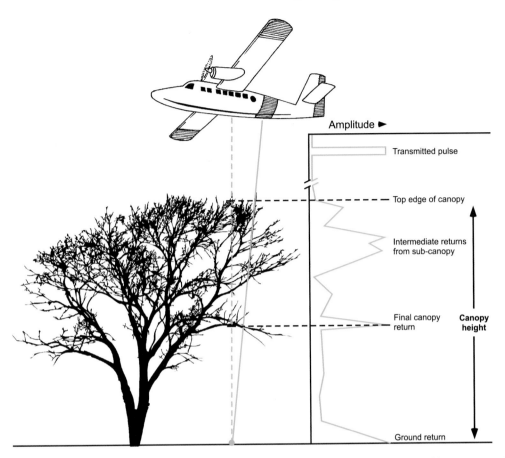

Figure 5.5 Principles of measurement of canopy structure using airborne LiDAR. Incident pulses of laser energy reflect off various portions of the canopy, resulting in a return waveform where the amplitude of the pulse at a given height is a function of the canopy structure and the last large-amplitude spike is the ground return.

and above-ground biomass (Blair *et al.*, 1999; Dubayah & Drake, 2000; Harding *et al.*, 2001; Lefsky *et al.*, 2002). The processed LiDAR echoes can be used to produce high-resolution regional maps of canopy properties (Neuenschwander *et al.*, 2009). LiDAR technology is treated in detail in Chapters 3, 6, 8 and 9 in the context of its application in urban and submarine environments.

With LiDAR now routinely used to monitor three-dimensional canopy structure, and with imaging spectrometers having been routinely used since the early 1980s for land cover identification and biochemical information, and given that the two are both aircraft mounted, they are often used in synergy (Thomas *et al.*, 2008). For example, the National Ecological Observatory Network (NEON), a National Science Foundation programme that focuses on continental-scale ecology, employs an aircraft that carries both a shortwave infrared imaging spectrometer and a scanning small-footprint FW LiDAR (Kampe *et al.*, 2010). This blending of spectroscopy and FW LiDAR delivers sufficient information, for example, to differentiate invasive plants from native ones on the basis of their spectral and structural properties (Asner & Vitousek, 2005; Asner *et al.*, 2008).

The challenge when fusing data in this way lies with the accurate registering of both datasets to a common geographical coordinate system during processing. This requires the relative alignment of the two sensors to be accurately known and to remain stable during flight. In the case of the NEON airborne platform, the payload position of the LiDAR and spectrometer are measured via GPS for the duration of each flight (Kampe *et al.*, 2010).

One particularly mature canopy measure derived via remote sensing is the Leaf Area Index (LAI). In its most simple form, this is the ratio of total upper leaf surface of vegetation divided by the surface area of the land on which the vegetation grows. However, definitions of the index vary, and the general term now encompasses several permutations such as the Total Leaf Area Index (ToLAI), the Projected Leaf Area Index (PLAI), the Silhouette Leaf Area Index (SLAI), the Effective Leaf Area Index (ELAI) and the True Leaf Area Index (TLAI).

LAI is a dimensionless value, typically ranging from 0 for bare ground to 6 for a dense forest (Gower & Norman, 1991; Chen & Black, 1992). Strictly speaking, LAI is not a true canopy structural property as it does not capture geometric detail such as leaf angle distribution, canopy height, or shape. However, the index is popular as it is one of the few that can be directly derived from LiDAR (Lefsky *et al.*, 1999; Harding *et al.*, 2001), as well as indirectly using passive visible-NIR spectral imagery (Zheng & Moskal, 2009), with the latter being based on the close coupling between radiation penetration and canopy structure. It is likely that a relationship also exists for LAI in the microwave spectrum, though the use of radar remote sensing to derive the index has not received a great deal of attention (Brakke *et al.*, 1981; Prevot *et al.*, 1993; Durden *et al.*, 1995).

The LAI is used by ecological and climate modellers because it is related to a variety of canopy processes such as water interception, evapo-transpiration, photosynthesis, respiration and the seasonal fall of leaf litter. These can be considered as intermediate variables useful for calculating terrestrial energy, carbon, water cycling processes and the biogeochemistry of vegetation at the regional scale (Wulder & Franklin, 2003).

Remote sensing is also a capable tool for deriving certain chemical properties of vegetation, such as the concentrations of photosynthetic and non-photosynthetic pigments. These pigment-protein complexes are relevant in controlling the ability of a plant to harvest light and in turn, control primary production. Hence, remotely sensed concentrations of pigment within vegetation can be used to audit underlying physiological processes and quantities such as photosynthesis, transpiration and net-primary production.

The measurement is best accomplished using hyperspectral audits of plant canopy reflectance in the visible-NIR spectrum (Tucker & Sellers, 1986; Landsberg *et al.*, 1996). Though methodologies differ, the algorithms are broadly based on the premise that photosynthetically active radiation (PAR) (400–700 nm) is absorbed strongly within green leaves by chlorophyll-a, chlorophyll-b and carotenoids. The same scattering mechanisms necessary for photosynthesis result in high values of leaf reflectance in the 700–1,300 nm spectral region, where little absorption occurs. While the NDVI ratio (previous section) can be calculated using the broad spectral bands of an instrument such as Landsat, a more evolved interrogation of the complete spectrum of reflected light field can yield relationships that link a leaf's pigment complement to its reflectance (Jago *et al.*, 1999; Broge & Mortensen, 2002; Ustin *et al.*, 2004). The same idea can be extended to measure pigment concentrations in corals (Hochberg *et al.*, 2006), albeit confined to the shorter visible EM wavelengths as the longer are rapidly absorbed by water.

To quantify pigments in vegetation, hyperspectral data with high radiometric and spatial resolution are required. To this end, airborne instruments such as the AVIRIS (Airborne Visible/Infrared Imaging Spectrometer) and HyMap have been used, as well as more recent experimental spaceborne imaging spectrometers such as the Compact High Resolution Imaging Spectrometer (CHRIS) and Hyperion.

To extract pigment information, the range of other factors which also influence vegetation reflectance spectra must first be taken into account. On the scale of a single leaf, these include the internal plant structure and absorption of light by non-photosynthetic pigments and biochemicals and moisture. On the scale of an entire canopy, factors such as the LAI, the orientation of leaves and their coverage, as well as the presence of shadow and reflectance from the ground, all serve to obscure the relationship between spectral reflectance and concentrations of individual pigments (Blackburn, 2006). Despite these many factors that must be attended to, several groups of spectral variables have been identified that empirically relate spectral reflectance to pigment concentrations (Blackburn, 1999; Peñuelas *et al.*, 1995).

An alternative, more advanced, group of approaches for quantifying pigments from hyperspectral data are based on the numerical inversion of physically-based leaf and canopy radiative transfer (RT) models (le Maire *et al.*, 2004; Coops & Stone, 2005). The RT strategy is attractive in that it is a more generic approach than the empirical calibrations, though it still suffers from conceptual problems and is not yet routinely proven to be more accurate (Combal *et al.*, 2002).

5.4 Classification systems

Before performing image analysis for thematic land cover or vegetation mapping, one must choose or develop a classification system which meets the needs of the problem to be addressed (Jensen, 2007). A workable classification system must meet at least the following criteria, i.e. it must be:

- exhaustive, i.e. all possible land covers or conditions must be assigned to a class;
- mutually exclusive, i.e. any land cover must be assignable to one and only one class;

Table 5.1 USGS Land Use and Land cover Classification System for use with Remote Sensor Data.

Level I		Level II
1 Urban or built-up land	11	Residential
	12	Commercial and services
	13	Industrial
	14	Transportation, communications and utilities
	15	Industrial and commercial complexes
	16	Mixed urban or built-up land
	17	Other urban or built-up Land
2 Agricultural land	21	Cropland and pasture
	22	Orchards, groves, vineyards, nurseries and ornamental horticultural areas
	23	Confined feeding operations
	24	Other agricultural land
3 Rangeland	31	Herbaceous rangeland
	32	Shrub-brushland rangeland
	33	Mixed rangeland
4 Forest land	41	Deciduous forest land
	42	Evergreen forest land
	43	Mixed forest land
5 Water	51	Streams and canals
	52	Lakes
	53	Reservoirs
	54	Bays and estuaries
6 Wetland	61	Forested wetland
	62	Non-forested wetland
7 Barren land	71	Dry salt flats
	72	Beaches
	73	Sandy areas other than beaches
	74	Bare exposed rock
	75	Strip mines, quarries, and gravel pits
	76	Transitional areas
	77	Mixed barren land
8 Tundra	81	Shrub and brush tundra
	82	Herbaceous tundra
	83	Bare ground tundra
	84	Wet tundra
	85	Mixed tundra
9 Perennial snow or ice	91	Perennial snowfields
	92	Glaciers

Source: Anderson *et al.* (1976).

- hierarchical, i.e. begin at a general level and divide each class into subclasses at the next lower level;
- flexible, i.e. allow the production of many different specialty maps (e.g. data layers of a GIS) and be able to respond to changing needs.

As shown in Table 5.1, in the USA the most commonly used land cover classification system is the USGS land use and land cover classification system for use with remote sensor data (Anderson *et al.*, 1976). Most projects use the top classes of the Anderson scheme and define lower classes based on the needs of the specific project (Jensen, 2007). There are many other classification schemes in use by programmes such as the USFWS National Wetlands Inventory, the USGS GAP Analysis Program and the NOAA Coastwatch Change Analysis Program (Cowardin, 1978; Cowardin *et al.*, 1979; Klemas *et al.*, 1993; Wilen & Bates, 1995; Jensen, 2007). Most of these classification systems have subclasses that can still be aggregated into the upper classes of the USGS Anderson Classification System.

The most recent effort to standardize vegetation inventory procedures in the U.S. has been conducted by the U.S. Geological Survey and the National Park Service, resulting in the Standardized National Vegetation Classification System (NVCS). The NVCS categorizes and describes ecological communities, including information on species composition, community structure, and physical conditions such as climate, hydrology, slope, aspect and soils (Nature Conservancy, 1994).

5.5 Global vegetation and land cover mapping programmes

Global vegetation and land cover can change in various ways and at differing timescales. To understand the significance of the changes and whether they constitute a long-term global trend, remote sensors are required to answer a series of questions (Mayaux *et al.*, 2008):

- Are the changes continuous – such as intensification of agriculture, expansion of urban settlements, forest exploitation – or do they totally change the predominant land cover?
- Are these changes short-lived, such as due to flooding and forest fires, or are they long-term changes, such as urban sprawl and deforestation?
- What is the intensity and spatial distribution of these changes, i.e. tropical deforestation is massive in Southeast Asia, yet very limited in Central Africa? Spatial patterns of deforestation can be massive, diffuse, linear, insular, etc.

Over the past several decades, remote sensors on satellites have provided the information needed to answer many of these vital questions.

Most Earth-scale mapping programmes for vegetation owe their existence to a highly successful joint NASA-NOAA project that concluded in 1976. The three-year Large Area Crop Inventory Experiment (LACIE), using Landsat MSS imagery, first demonstrated that global monitoring by satellite of food and fibre production was possible. LACIE's mission was to prove the feasibility of Landsat for yield assessment of one crop (wheat), but this activity later extended to the monitoring of multiple crops at the global scale. The LACIE experiment highlighted the potential impact that a credible crop yield assessment could have on world food marketing, administration policy, transportation and other related factors.

In 1977 it was decided to test the accuracy of the Soviet wheat crop yield data by using Landsat-3 to assess the total production from early season to harvesting in Russia. Remarkably, the Russian wheat production in 1977 was predicted with an accuracy of better than 90 per cent (Cracknell & Hayes, 2007).

Today, Landsat, along with SPOT, continues to be the most common medium-resolution satellite for mapping vegetation and general land cover on a regional scale. As shown in Table 3.2, these satellites have multispectral scanners which provide spatial resolutions of 10–30 metres and cover swaths from 60–180 km wide. However, their repeat cycle, even without cloud cover, is only every 16 to 26 days. For global land cover mapping, the NOAA AVHRR sensors seem to be more efficient, having 2,400 km swath-widths and 1.1 km spatial resolutions.

5.5.1   NASA Pathfinder global monitoring project

Global vegetation change and drought mapping have been the objectives of various projects sponsored by UNEP, NASA, NOAA and other agencies. Two of the most relevant ones have been the Landsat Pathfinder Project for mapping global forest resources and the NOAA/NASA project producing global maps of the Normalized Difference Vegetation Index (NDVI) on a monthly basis (Chomentowski et al., 1994; Kogan, 1997; 2001; 2002).

The NASA Pathfinder project is demonstrating the use of Landsat data for global change research and monitoring the world's forest resources. One of five Pathfinder projects initiated by the US National Aeronautics and Space Administration (NASA), Landsat Pathfinder is a first step toward establishing a global monitoring system using moderate-resolution satellite imagery. The imagery is stored, managed and analyzed with a GIS. Each Pathfinder project centres around currently available data from different satellites. Landsat provides the largest Pathfinder database in terms of data volume, consisting of approximately 3,000 multispectral scanner (MSS) and thematic mapper (TM) images.

The Landsat Pathfinder Science Working Group has defined several projects to address land cover change, such as:

1 The Humid Tropical Forest Project (HTFP), which is generating a forest/non-forest dataset showing areas of deforestation in three moist tropical forest regions (the Amazon, Central Africa and Southeast Asia).
2 The North American Landscape Characterization (NALC) project which is a collaborative effort between the Environmental Protection Agency (EPA) and the U.S. Geological Survey (USGS) to provide complete coverage of the conterminous U.S. and Mexico for the purposes of mapping land cover and land cover change. (Maiden & Greco, 1994)

The NOAA-NASA Pathfinder Normalized Difference Vegetation Index (NDVI) dataset is a widely used dataset with an 8 km × 8 km spatial resolution and a 10-day temporal resolution. The NDVI being mapped by NOAA satellites on a global scale is a measure of the amount and vigour of vegetation at the surface. The index is derived from data collected by NOAA satellites and is processed by the Global Inventory Monitoring and Modelling Studies (GIMMS) project at NASA (D'Souza *et al.*, 1996).

Vegetation indices derived from the NOAA AVHRR sensor have been employed for both qualitative and quantitative studies (e.g. on the contraction and expansion of the Sahara desert (Tucker *et al.*, 1991) and on the calculation of biophysical parameters for climate models (Sellers & Schimel, 1993).

5.5.2 *International geosphere-biosphere program*

Vast areas of Africa, Asia, and South America still remain poorly, and often incorrectly, mapped. A UNESCO-sponsored project has completed a series of land use maps at scales of 1 : 5,000,000 to 1 : 20,000,000. Although invaluable as a general record of land use (and for agriculture, hydrology and geology), these maps have insufficient detail to assist resource managers in many of their decisions. Furthermore, frequent changes in land use are difficult to plot at such coarse scales. Satellites are able to contribute significantly to the improvement of this situation (Friedl *et al.*, 2002).

To remedy this situation, two global one-kilometre land cover datasets have been produced from 1992–1993 AVHRR data: the International Geosphere-Biosphere Program Data and Information System (IGBP-DIS) DIS-Cover; and that of the University of Maryland (Hansen & Reed, 2000). An update to these datasets for the year 2000 (GLC2000) has been produced by the EU Joint Research Centre (JRC), in collaboration with over 30 research teams from around the world (Bartholmé & Belward 2005).

The general objective of the GLC2000 initiative was to provide a land cover database over the whole globe for the year 2000, that year being considered a reference year for environmental assessment in relation to various activities. The GLC2000 data was derived from 14 months of pre-processed daily global data acquired by the VEGETATION instrument on board the SPOT 4 satellite, called the VEGA 2000 dataset (VEGETATION data for Global Assessment in 2000).

In order to provide the scientific community with more precise information on the global distribution of habitat types, the European Space Agency, in collaboration with an international network of partners, has started the production of a global land cover map for the year 2005 at 300 m spatial resolution using MERIS fine resolution mode imagery acquired between December 2004 and June 2006. This new land cover product updates and complements the GLC2000 data as well as other comparable global data products (Mayaux *et al.*, 2008).

The AVHRR data also complement the output from newer satellite Earth observing systems, such as the Moderate Resolution Imaging Spectrometer (MODIS), Medium Resolution Imaging Spectrometer (MERIS) and Vegetation (VGT) sensors, which provide data at similar or finer (250–500 m) spatial resolutions. AVHRR will also overlap with future US National Polar-Orbiting Operational Environmental Satellite System (NPOESS). This overlap between AVHRR and other sensors can be used to ensure consistency of satellite climate data records and to bridge the gap between historical AVHRR observations and data from future systems (Latifovic *et al.*, 2005).

5.5.3 *Application of new satellites and radar*

Improvements in mapping global vegetation and land cover are expected from new medium-resolution satellites (150–300 m) that have been launched during the past ten years. The improved spatial resolution of these sensors meets land cover needs at global, regional and sub-regional levels, while at the same time providing a global or regional view of the Earth's surface. Hansen *et al.* (2003) demonstrate the advantages of the new satellite imagery by mapping the percentage of tree cover on a global scale.

Similarly, investigators at Boston University have begun using Moderate Resolution Imaging Spectrometer (MODIS) data from the NASA Aqua and Terra satellites to enhance and update the IGBP DIS-Cover and the University of Maryland global 1 km datasets (Hansen & Reed, 2000; Cracknell & Hayes, 2007). MODIS is a more recent sensor which is used for both ocean colour and land cover observations (Gallo *et al.*, 2005). As shown in Table 3.2, this sensor has 36 spectral bands, ranging from 620 to 14,385 nm, and a variable ground resolution from 250 m to 1 km. Its swath-width is 2,330 km, with a daily revisit time (Jensen, 2007; Mayaux *et al.*, 2008).

Although radar in general has not been widely used for vegetation mapping, radar systems such as side-looking airborne radar (SLAR) and the spaceborne SAR have proven useful for forestry and other land cover application (Lillesand & Kiefer, 1994). Some important characteristics of radar imagery are:

- Radar can penetrate the atmosphere under variable conditions such as clouds, haze, light rain and smoke, depending on the wavelength and other variables.
- Microwave reflections or emissions are interpreted differently from visible light or thermal energy emissions; therefore, radar helps image interpretation by producing a distinctly different 'view' of the environment.
- The sensor's viewing angle can be altered, which means radar images can be produced in stereo pairs.

Radar is most useful in analyzing surface texture, shape and orientation of objects on the Earth, and it can penetrate dry soil to depths of several wavelengths (Chapter 7). Radar has been used to map rocks, geologic features, vegetation, agricultural crops, ocean currents, oil spills, snow distribution and sea ice.

The radar return from flooded forests is usually enhanced by comparison with returns from non-flooded forests. The enhancement is due to the double bounce interaction, where the signal penetrating the canopy is reflected off the water surface and subsequently reflected back toward the sensor by a second reflection off a tree trunk (Hess *et al.*, 1990 – see also Figure 3.4 in Chapter 3). Radar can detect flooded coastal marshes because they usually provide a weaker radar return than non-flooded ones. The marsh grasses may calm the water surface, accentuating specular reflection but without the grasses providing the double bounce (Ramsey, 1995).

5.6 Remote sensing of vegetation as a monitor for global change

Vegetation cover is a good indicator of global change. Since the early 1980s, global satellite mapping of the biosphere has generated long time-series measurements of vegetation that can be used as proxies for understanding the dynamics of variability for Earth's vegetative cover. In this short period, major changes have been witnessed, brought about by drought and desertification, deforestation, ecosystem stress and urbanization. The recent IPCC report shows a distinct drying trend during the 20th century over most of the globe and, in particular, over some of the world's major agricultural areas, such as the Mediterranean basin, southern Africa, the Australian cereal belt, northeast China and much of southeast Asia. Agricultural growth in these areas has been largely achieved through expansion of cultivated areas and not productivity gains (IPCC, WMO/ UNEP, 2007).

Changing climate is predicted to alter the global distribution of precipitation, increasing flooding in some areas, yet making rainfall unpredictable in areas that have traditionally been used for agriculture. With limited access to irrigation, the agricultural production of many developing countries is highly vulnerable to variations in rainfall. Also, the destruction of urban and agricultural infrastructure due to heavy flooding will have long-lasting negative effects (Noble, 2008). The most effective way to monitor these dramatic changes on a global scale is to use remote sensors on satellites orbiting the Earth.

CASE STUDY: Desertification in the African Sahel

The Sahel, a semi-arid belt that straddles Africa from the Atlantic to the Red Sea, forms an ecological transition between the Sahara desert to the north and the humid tropical savanna to the south. The area is particularly susceptible to chronic rainfall shortage, such as the Sahelian drought that began in 1968 and was responsible for the deaths of between 100,000 and 250,000 people. The 1990s, by comparison, saw anomalously high rainfall in the Sahel.

This erratic distribution of precipitation through time, combined with changing land use, has led to severe local desertification as the Sahara expands southward into areas previously dominated by forest and bushland. It is a misconception that this shift is present in all areas or that it can be solely attributed to natural climate change. Man-induced processes, and in particular slash-and-burn farming and overgrazing, have lead to acute nutrient depletion of the soils in this region. This, in turn, reduces the prevalence of the natural vegetation that once served to hold the desert at bay. In some areas of the Sahel, the annual southerly march of the Sahara exceeds 40 km, engulfing vast tracks of arable land in its path.

The true mechanics of the process only become apparent when the entire region can be viewed simultaneously. Satellite remote sensing is well poised to monitor this phenomenon as the advancing desert is marked by an increase in albedo. The associated decline in biological diversity has a clear signature in the NDVI vegetation index (e.g. Justice, 1986; Prince & Justice, 1991; Kogan, 2002). Working on this premise, Anyamba & Tucker (2005) used NOAA's AVHRR satellite data to assemble for the region an NDVI history that stretched from 1981 to 2003 (Figure 5.6). From their work, it became clear that anomaly patterns in the NDVI data correspond well to documented rainfall patterns across the Sahel in the 23-year time period considered. Values of NDVI correspondingly increased during years of substantial rainfall.

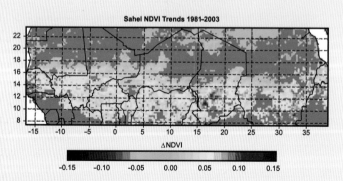

Figure 5.6 Changes in NDVI across the Sahel from 1981–2003 derived from AVHRR imagery. Yellow to red colours indicate areas of significant change at 90 per cent confidence level, while gray indicates no significant trend. Modified from Anyamba & Tucker (2005) with permission from Elsevier.

The NDVI time-series in Figure 5.6 depicts several sites in the central Sahel that have experienced a pronounced increase in NDVI from 1981 to 2003. The change can be attributed to a recovery through the 1990s from chronic drought conditions that peaked in 1984. The study is powerful in that it counters the long-held paradigm that the Sahel is witnessing a simple trend of increasing aridity and advancing desertification. The AVHRR work instead shows that the vegetation vigour in the area is dynamic on a decadal timescale.

However, care must be taken to not place blind faith in the satellite-observed pattern, as even subtle changes in the spacecraft have the potential to impact a study's reliability. For instance, over the last two decades, the overpass times of AVHRR have shifted. This shift leads to a periodicity through time of the solar zenith angle under which the imagery was collected. When such long time-series are considered, meticulous quality assessment of all parameters related to the data is crucial.

In the case of the Sahel, it seems that the rise in NDVI through the nineties is, indeed, related to a greening of the region. Such an understanding of the true dynamics of the region is important if the impacts of drought and desertification are to be fully understood and effectively tackled. The Sahel and the fate of the soil that blows out of the region are again considered in Chapter 11 in the context of remotely sensing airborne dust plumes.

Several interesting case studies of land cover changes in Africa are presented by Mayaux *et al.* (2008). The first case describes studies of deforestation in the Congo Basin of Central Africa. The deforestation estimates were derived from coarse- and medium-resolution satellite imagery. Since this approach could not detect small forest changes, a new cost-effective approach was developed to derive area estimates of land cover change by combining a systematic regional sampling scheme, based on high spatial resolution imagery, with object-based unsupervised classification techniques. As a result, the annual gross deforestation rate for Central Africa's tropical forest was estimated at 0.21 per cent per year.

In the other two case studies, Mayaux *et al.* (2008), discuss agricultural cropland expansion in Africa and land cover changes in Senegal. The studies used a wide range of imagery, including medium- and high-resolution satellite images and both visual and automated image analysis techniques. One study showed that in the year 2000, sub-Saharan Africa was covered by about 17 per cent of agriculture, 20 per cent of forest, 60 per cent of non-forest vegetation such as wood- and shrub-lands and savannas, over 2 per cent of barren land, excluding permanent deserts, and less than 1 per cent of water.

CASE STUDY: Deforestation of Amazonia

The Amazon region contains more than five million square kilometres of closed tropical forest which serves as habitat for an immense number of plant and animal species. Though the majority of the forest is contained within Brazil, the region extends into the territory of nine other nations. More than ten per cent of Earth's terrestrial primary productivity can be attributed to Amazonia and it stores approximately the same percentage of ecosystem-based carbon dioxide. However, an influx of migrants, agriculture, cattle ranching, mining and other economic activities have resulted in the deforestation of over half a million square kilometres of forest. The estimated average deforestation rate from 1978 to 1988 was $15,000\,km^2$ per year.

The extent of the Amazon and similar tropical forests makes it difficult to access and survey with traditional aerial photography. Therefore, remotely-sensed satellite imagery has been used extensively (Chomentowski et al., 1994; Souza & Roberts, 2005). As for the previous case study focusing on the Sahel, NDVI products obtained from the NOAA's AVHRR satellite have enabled the surveillance of regional trends of Amazon deforestation at 8 km resolution. This coarse evaluation has been supplemented with local-scale observations made using SPOT and Landsat at higher spatial resolution, but with a temporal resolution of several weeks (e.g. Skole & Tucker, 1993). Such a multi-scale approach is necessary as the annual clearing of forest can be on the scale of hundreds of square kilometres in Brazil, down to fragmented but highly numerous plots of just a few km^2 in French Guiana.

More important than the total area that has been deforested is the rate of clearing. It is this that shows whether the problem is becoming more acute or if efforts to mitigate the deforestation are effective. Determination of rate requires multi-temporal analysis of satellite data. Although much of the clearing is conducted at a scale that is sub-pixel for the MODIS sensor, the instrument has proved surprisingly reliable for detecting changes in Amazon land cover.

Situated in the tropics (which receive two-thirds of the world's rainfall), the region is frequently obscured by clouds and an instrument such as Landsat, with a repeat interval of 16 days, may go several months without capturing a useful image of the forest. MODIS, by comparison, which is carried by both the Terra and Aqua satellites, can obtain imagery up to four times per day. Cloud contamination is further reduced since these daily data are in turn assimilated into short-term composites created using several images. A time-series of cloud-free NDVI data are thus much easier to obtain from MODIS, with its minimum pixel size of 250 metres, than from using a spatially more capable instrument such as Landsat, SPOT or IKONOS.

The utility of MODIS was highlighted by Morton *et al.* (2005) who, through comparison with Landsat data, demonstrated that the sensor could detect a significant portion of Amazon deforestation (Figure 5.7). Once a clearing was the size of three MODIS pixels, a change-detection algorithm was consistently able to identify areas as deforested.

The results were confirmed by field studies in the state of Mato Grosso, in which areas that MODIS flagged as deforested were verified by scientists on the ground. The scientists discovered that the number of large forest clearings (areas greater than 10 MODIS pixels) doubled in Mato Grosso during the period from August 2003 to August 2004, as compared to the period August 2001 to August 2002, from 750 large clearings to 1,500.

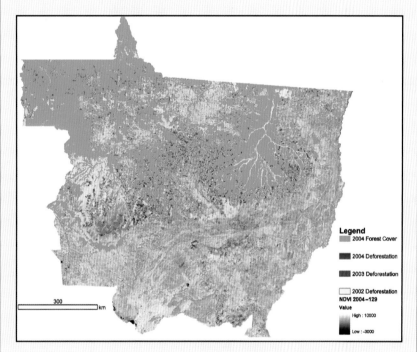

Figure 5.7 Amazon deforestation as imaged by the MODIS 16-day composite NDVI for the state of Mato Grosso, Brazil. In this example, a red reflectance threshold has been used to identify large deforestation events; the 2004 NDVI layer is displayed to provide an indication of cover types in Mato Grosso. After Morton *et al.* (2005). Copyright American Meteorological Society. Reprinted with permission.

Building on the results of this study in 2005, Brazil's federal government, through the National Institute for Space Research (INPE), developed a near real-time monitoring application for deforestation. This system, known as DETER (Deforestation Detection in Real Time), publishes monthly alert maps that highlight both areas that have been completely deforested and those in the process of being cleared.

The programme reports that 26,130 km^2 of forest were lost in the year up to August 1, 2004. This figure was produced on the basis of 103 satellite images covering 93 per cent of the so-called 'Deforestation Arc' – the area in which most of the trees are being cut down. The study shows that, in spite of its limited spatial resolution, analysis of MODIS imagery could form the basis for a wide array of regional studies in a highly automated fashion, with both scientific and decision-making utility.

5.7 Remote sensing of wetlands change

Wetlands are a highly productive and critical habitat for a wide variety of plants, fish, shellfish and other wildlife. Wetlands also provide flood protection; shelter from storm and wave damage; water quality improvement through filtering of agricultural and industrial waste; and recharge of aquifers. After years of degradation due to dredge and fill operations, impoundments, urban development, subsidence/erosion, toxic pollutants, eutrophication and sea level rise, wetlands are finally receiving public attention and protection (Morris *et al.*, 2002; Odum, 1993).

Recently, studies have been initiated to determine the impact of climate change on coastal wetland habitat losses, especially due to sea level rise, increasing temperatures and changes in precipitation. Climate change is considered a cause for habitat destruction, shift in species composition and habitat degradation in existing wetlands (Baldwin & Mendelssohn, 1998). Coastal wetlands have already proven to be susceptible to climate change, with a net loss of 33,230 acres from 1998 to 2004 in the United States alone (Dahl, 2006). This loss was primarily due to conversion of coastal salt marsh to open salt water. Rising sea levels can not only cause the drowning of salt marsh habitats, but can also reduce germination periods (Noe & Zedler, 2001). Furthermore, warmer mean low temperatures have induced forested mangroves to spread northward and replace emergent salt marshes in some of the Gulf states.

The impact of global change in the form of accelerating sea level rise and more frequent storms is of particular concern for coastal wetlands. Salt marshes can often recover from slow disturbances such as sea level rise but, when such changes are coupled with erosion and destruction due to more frequent coastal storms, the damage can be irreversible (Klemas, 2009).

The heightened awareness of the value of wetlands has resulted in the need for better understanding of their function and importance and for finding ways to manage them more effectively as environmental conditions change. To accomplish this, at least two types of data are required (Daiber, 1986; Klemas, 2005):

1 Information on the present distribution and abundance of wetlands;
2 Information on the trends of wetland losses and gains.

In order to obtain the required information and get a better view of wetland ecological patterns and processes, researchers and managers need synoptic, spatially-referenced information acquired over entire watersheds. Remote sensing, GIS and GPS techniques make it possible to collect, organize and analyze such spatial information.

The US Environmental Protection Agency defines wetlands as areas where saturation with water is the dominant factor determining the nature of soil development and the types of plant and animal communities living in the soil and on its surface. Common diagnostic features of wetlands are hydric soils and hydrophytic vegetation. Coastal wetlands can be divided into four major types: salt marshes, coastal forests, scrub-shrub wetlands and tidal flats. Each of these types has different hydrologic requirements and is dominated by one or several different types of vegetative cover.

Factors such as varying plant densities, different species having similar spectral reflectances, and the same species producing different spectral reflectances under different conditions, make wetland remote sensing quite challenging. Upland freshwater wetlands are patchy and more diverse, including swamps, bogs, potholes, sloughs, etc., and have a mixed vegetative cover producing even more complex, composite spectral signatures. They are therefore even more difficult to map remotely than coastal wetlands (which occur in larger clusters that are frequently dominated by a single vegetative cover), and this is the reason why the high spatial resolutions and frequent flexible overflights offered by aircraft sensors are still required for certain projects, such as wetland monitoring (Phinn *et al.*, 2000).

Nevertheless, the recent availability of high spatial and spectral resolution satellite imagery, combined with new image processing and classification algorithms, make wetlands mapping from satellites increasingly accurate, automated and cost-effective (Kelly & Tuxen, 2009; Yang, 2009).

Coastal estuaries exhibit complex spatial patterns and ecological processes in response to multiple physical and biological gradients. Therefore, estuarine applications of remote sensing require significantly finer spatial, temporal and spectral resolution than is necessary for open ocean or land observations. This is particularly true for studies of rapidly varying tide-dominated coastal/estuarine processes, which require time-series of high-resolution images to be used for modelling and predicting near-term behaviour and long-term trends. As a result, one must employ a combination of satellites and aircraft with multispectral and hyperspectral sensors which can provide the required temporal, spatial, and spectral resolution (Table 3.4). Furthermore, since the repeat cycles of many satellites are not frequent enough for observing rapidly varying estuarine processes and features such as estuarine fronts, plumes and oil slicks, these may have to be tracked by observers in small aircraft and airborne sensors (Klemas, 2009).

Coastal watershed models require land cover or land use as an input. Knowing how the land cover/use is changing means that these models, together with a few other inputs like precipitation, can predict the amount and type of run-off into rivers, bays, and estuaries (Jensen, 2007; Donato & Klemas, 2001).Techniques for

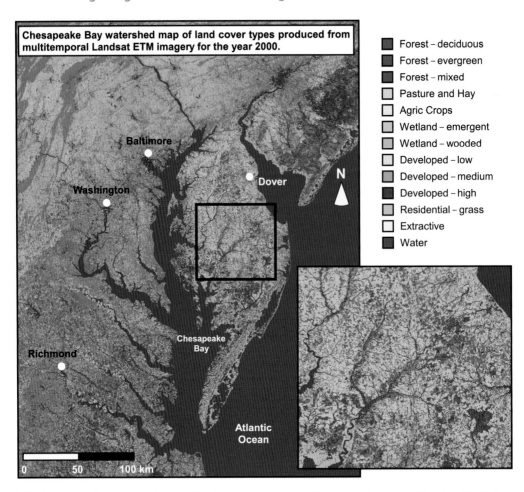

Figure 5.8 Chesapeake Bay watershed map of land cover types produced from multi-temporal Landsat ETM+ imagery for the year 2000. Modified with permission from Goetz *et al.* (2004).

detecting changes in land cover using time-series of satellite images are presented in Section 4.7 of this book.

Two of the more common medium-resolution satellites for mapping coastal watersheds on a regional scale have been the US Landsat and French SPOT (Le Systeme pour l'Observation de la Terre). As shown in Table 3.5, the satellites have multispectral scanners which provide spatial resolutions of 10–30 metres and cover swaths from 60 to 185 km wide. The Landsat Thematic Mapper (TM), with its 30 m resolution, has provided the most reliable data for monitoring land cover changes in large coastal watersheds such as Chesapeake Bay (Lunetta & Balogh, 1999).

Chesapeake Bay is the largest estuary in the United States, encompassed by a watershed extending 168,000 km² over portions of six states and Washington, DC. Figure 5.8 shows a land cover map of the Chesapeake Bay watershed obtained from Landsat TM imagery. As shown in Figure 5.8, among the 13 mapped land cover classes there are two wetland types. Smaller, patchy upstream wetlands could not be detected by Landsat ETM, since they require higher spatial and spectral resolution.

The classification scheme usually employed for the top level classes (i.e. Table 5.1) is the Anderson (USGS) system (Anderson *et al.*, 1976; Jensen, 2007). Researchers develop their own classification for the more detailed levels, such as the C-CAP Classification System (Klemas *et al.*, 1993; Dobson *et al.*, 1995). A very detailed wetlands classification system is the one developed by Cowardin *et al.* (1979). However, this classification system proved to be too complex for satellite remote sensing.

Some of the coastal ecosystem health indicators that can be observed by remote sensors include percentage of impervious areas; natural vegetation cover; buffer degradation; wetland loss and fragmentation; wetland biomass change; invasive species, etc. (Klemas, 2000; Lathrop *et al.*, 2000; Lunetta & Balogh, 1999). A typical digital image analysis approach for classifying coastal wetlands or land cover is shown in Figure 4.4. Note that throughout the process, ancillary data are used whenever available (e.g. aerial photos, maps, field data, etc.).

When studying critical wetland sites or small watersheds, one can use aircraft or high-resolution satellite systems. Airborne georeferenced digital cameras, providing colour and colour infrared digital imagery, are particularly suitable for mapping or validating satellite data; such digital imagery can be integrated with GPS information and used as layers in a GIS for a wide range of modelling applications (Lyon & McCarthy, 1995). Small aircraft flown at low altitudes (e.g. 500 metres) can be used to supplement field data. High-resolution imagery (0.6 m to 4 m) can also be obtained from satellites, such as IKONOS and QuickBird (Table 3.2). However, cost becomes excessive if the site is larger than a few hundred square kilometres.

Wetland species identification is difficult; nonetheless, some progress is being made using hyperspectral imagers (Schmidt *et al.*, 2004; Porter, 2006; Kelly & Tuxen, 2009).

Since coastal wetlands are legally defined in terms of mean high tide levels (MHW), which are difficult to map, it is fortunate that there are specific marsh plants, shown in Table 5.2, which are related to tidal inundation levels and can be readily mapped by remote sensors. The marsh vegetation averages out the tidal fluctuations and provides a good indication of marsh boundaries. Based on this principle, there have been many projects designed to map wetlands in response to various wetland protection laws. The US federal government and most coastal states have prepared wetlands maps using both aircraft and satellite imagery. Most of the major wetlands mapping programmes conducted by the United States Geological Survey (USGS), the National Oceanic and Atmospheric (NOAA) and the Environmental Protection Agency (EPA) and other agencies are described in Kiraly *et al.* (1990).

Traditionally the US Fish and Wildlife Service (FWS) has played a key role. Its first nationwide wetlands inventory was conducted in 1954 and focused on waterfowl wetlands. In 1974, the FWS established the National Wetlands Inventory Project (NWI) to generate scientific information on the characteristics of US

Table 5.2 Typical wetland species and inundation levels for the Delaware Bay region. Note that the Low Marsh is dominated by salt water cord grass (*Sp. alterniflora*), while the High Marsh is dominated by salt marsh hay (*Sp. patens* and *D. spicata*).

Low Marsh (below MHT)	Intermediate Marsh (across MHT)	High Marsh (above MHT)
Spartina alterniflora	*Spartina cynosuroides*	*Spartina patens*
Salt marsh cordgrass	Big cordgrass	Salt-meadow cordgrass
Highly productive		Salt hay
Many animals and birds:	*Scirpus olneyi*	Drained marsh
muskrats, ducks and geese, song	Sedge	
sparrows, wrens, clapper rail, etc	Less productive	
		Distichilis spicata
		Salt grass
	Scirpus robustus	Salt hay
	Saltmarsh bulrush	Drained marsh
	Less productive	Less productive
		Phragmites australis
		Common reed
		Of little use
		Crowds out other plants
		Grows near disturbed areas
		Iva frustescens
		Marsh elder
		Baccharis halimifolia
		Sea myrtle
		Hibiscus moscheutos
		Marsh hollybrook

wetlands, including detailed maps and status/trend reports. The maps are available as 7.5 min quads at a scale of 1 : 24,000. Most have been digitized, converting them from paper maps to a GIS-compatible digital line graph (DLG) format (Wilen & Bates, 1995).

When looking at coastal land cover changes or beach erosion over long time periods, it is important to review historical aerial photographs held by local, state and federal agencies. The US Geological Survey and the USDA Soil Conservation Service have useable aerial photos of the coast dating back to the 1930s, as well as various maps, including planimetric, topographical, quadrangle, thematic, ortho-photo, satellite and digital maps (Rasher & Weaver, 1990).

Biomass is one of the indicators of wetland vigour and health. It can be derived from multispectral imagers on aircraft or satellites. The spectral bands used are primarily the red band (which, as we saw earlier in this chapter, is absorbed by the chlorophyll in the upper leaf layers) and a near-infrared band (which is reflected from the inner leaf structure, yet still penetrates several leaf layers and thus provides information on the canopy thickness and density) (Daiber, 1986). These spectral bands are combined into an NDVI to provide an estimate of above-ground plant biomass in grams dry weight per square metre (Jensen *et al.*, 2002). To validate biomass mapping results in the field, it is necessary to cut quadrats of wetland grass, dry it and weigh it. This approach has been particularly effective for mapping the biomass of tidal marshes (Hardisky *et al.*, 1984; Gross *et al.*, 1987).

Vegetation indices and biomass have long been used in remote sensing for monitoring temporal changes associated with wetland vegetation (Lyon *et al.*, 1998). For instance, a particularly effective method for remotely sensing wetland changes uses biomass as an indicator. To detect biomass changes, the Modified Soil Adjusted Vegetation Index (MSAVI) can be used with red and near-infrared reflectances derived from Landsat TM images (Qi *et al.*, 1994; Eastwood *et al.*, 1997). This biomass algorithm is applied to a time-series of Landsat TM images and used with selected thresholds to detect wetland changes. To minimize natural variations between images in the time-series (e.g. atmospheric, annual, seasonal, etc.), it is assumed that the relative distribution of biomass in each sub-basin will remain essentially constant over time. Wetland pixels whose MSAVI deviation from the sub-basin mean changes from its previous deviation by more than a selected threshold value are considered as having changed. Threshold selection determines whether many small changes or only the more significant ones are detected. To minimize data costs, only changed sites 'flagged' by Landsat/TM are studied in more detail with high-resolution systems such as IKONOS or airborne scanners (Klemas, 2007; Porter, 2006).

5.8 Fire detection

Monitoring the status of global vegetation while it is living is important because of the vast shift in carbon storage, so it is also critical to map the process of its demise. On a daily basis, fire clears vast swaths of vegetated land. Wildfire changes the surface cover type and releases gases and particulate matter into the atmosphere, affecting ecosystems and atmospheric chemistry on a rapid and intense scale. In terms of an agent of climate change, however, they should not be thought of as modern, let alone an anthropogenic phenomenon; wildfires have scorched the Earth's surface for 400 million years.

Although wildfires are a natural part of the Earth system, humans now strive to suppress them. However, many managers and scientists would rather they be left to burn out than be interfered with – an approach that may in some cases be the most effective in promoting biological community diversity (Richards *et al.*, 1999).

If there is increased CO_2 in the atmosphere, several models predict the vigour of plant growth to rise, increasing the amount of material available to burn. Also, a warming climate promotes aridity, again leading to increased fire occurrence. Even early humans radically altered the landscape, and the advent of agriculture had a particular influence on natural patterns of biomass burning (Marlon *et al.*,

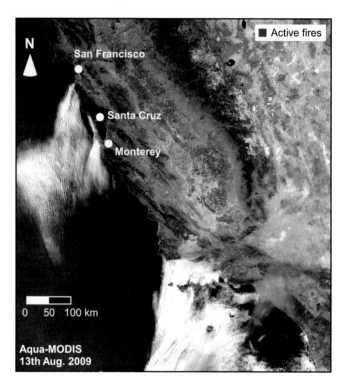

Figure 5.9 The La Brea Fire in southern California as imaged by Aqua-MODIS in August 2009. Areas in red indicate where MODIS has detected unusually warm surface temperatures associated with active fires. Credit: NASA.

2008). Hence, it is now very difficult to separate any influence of climate on the frequency of wildfires from the impact of humans.

Large-scale burning of biomass has six important impacts:

- changes of physical state of vegetation and release of greenhouse gases;
- the release of chemically reactive gases;
- the release of soot and other particulate matter;
- changes in the exchange of energy and water between the land surface and the atmosphere:
- changes in plant community development and the nutrient, temperature and moisture content of soils:
- impacts on cloud development and reflectivity.

The main contributions of remote sensing to fires may be grouped into three categories according to the three phases of fire management: risk estimation (before the fire); detection (during the fire); and assessment (after the fire). Vegetation indices (considered earlier in this chapter) are critical in the risk estimation and assessment arena.

The MODIS sensor has a data product that is dedicated to the detection of fire; it uses calibrated radiances and is an amalgamation of a cloud mask, vegetation indices, land surface temperature and land cover type. Figure 5.9 shows fires in southern California as imaged by MODIS. With the instrument carried aboard both the Terra and Aqua platforms, MODIS data can provide four observations per day, giving a good sampling of the diurnal cycle of fire activity. An additional 'thermal anomaly' product is provided by MODIS and includes day/night fire occurrence, fire location and an energy calculation for each fire. The product also includes composite eight-days-and-nights fire occurrence at full 1 km resolution, along with a lowered 10 km resolution averaged monthly composite. Post-fire, MODIS has the spatial and radiometric capabilities for burn-scar detection (van der Werf *et al.*, 2006). In summary, MODIS offers significant improvements over previous fire-related products which have traditionally been derived from either the NOAA AVHRR or GOES systems.

Volcanic eruptions also have short- and long-term consequences on both the local land surface and the regional and global atmosphere, including destruction of crops. For example, Figure 4.7 in the previous chapter depicts the loss (and recovery) of vegetation around Mount St. Helens following the May 18, 1980 eruption. For some eruptions, global tropospheric temperatures may be lowered to such a point that the short-term effects of climate warming are masked.

Higher in the atmosphere, in the middle and upper stratosphere, the effect of volcanic emissions is instead typically a force for warming. The rise in temperature results from increased aerosol absorption of terrestrial long-wave and solar near-infrared radiation. It is important to note that both the cooling of the troposphere and the warming of the middle and upper stratosphere can impact climate change.

5.9 Key concepts

1 By virtue of its pigment composition, chlorophyll has a distinct spectral signature with absorption bands in the blue (430 nm) and red (600 nm) regions of the visible spectrum. This offers an effective means of differentiating between green plants and other land cover types that do not contain photosynthetic pigments.

2 Several 'vegetation indices' have been developed for use with remote sensing data. These metrics capitalize on the unique spectral signature of chlorophyll. The most widely used is the Normalized Difference Vegetation Index (NDVI), which operates on the fact that green vegetation has low reflectance in the red and a high reflectance in the near-infrared. NDVI provides a useful index of photosynthetic activity and scales between −1 and +1. Only the positive values correspond with vegetated zones.

3 Many satellite sensors have been designed with band-sets that are capable of calculating vegetation indices. Programmes to note are the US-run AVHRR, Landsat and MODIS, and the French SPOT systems. AVHRR is notable for having provided nearly 30 years of continuous data with a potential repeat time of only 12 hours.

4 Vegetation cover is a good indicator of global change. There is a definite trend towards decreasing cover by rain forests and the expansion of deserts. The present trajectory of climate warming is predicted to yield changes in the distribution of rainfall worldwide. This will be a dominant factor driving the rearrangement of global vegetation and it will impact the current division of agricultural zones.

5 The monitoring of agriculture was one of the earliest applications of civilian satellite remote sensing and the impetus behind the development of the Landsat programme. The LACIE experiment served as testament to the ability of satellites to predict agricultural yield for entire nations.

6 Wetlands and estuaries exhibit complex spatial patterns and ecological processes in response to multiple physical and biological gradients. Therefore, estuarine applications of remote sensing require significantly finer spatial, temporal and spectral resolution than is necessary for open ocean or land observations. This is particularly true for studies of rapidly varying tide-dominated coastal/estuarine processes.

7 Remote sensing techniques have been used to show that a combination of sea level rise and coastal storms have contributed to salt marsh and estuarine aquatic bed losses in various coastal areas, including the US Gulf coast, while increasing temperatures have induced mangrove forests to move northward and replace salt marshes.

6 Remote sensing of urban environments

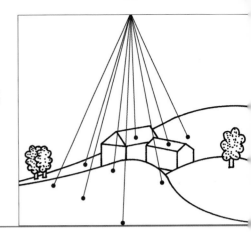

6.1 Urbanization

Urban remote sensing entails the collection and analysis of imagery of heavily populated areas. Cities, composed of dense networks of artificial structures, are unique in that they are characterized by a trend of permanent expansion and growth. They are also the point-source of the majority of anthropogenic pollution.

Although only 1.2 per cent of the Earth's surface is presently urbanized, over 50 per cent of the world's population reside in these areas. Fuelled by several decades of population explosion, the pace of migration to the cities is accelerating such that by the year 2025, 60 per cent of the world's population may live in the cities (UNFP 1999). Such profound urbanization has serious environmental and socio-economic impacts in developing and developed countries alike (Berry, 1990; Leakey & Lewin, 1995; Longley, 2002). Urban growth 'consumes' the surrounding land resources that are generally non-renewable in terms of surface- and ground-water, and food resources (Netzband *et al.*, 2005). In arid climatic zones, the threat of dwindling water resources is especially serious. Urbanization can be considered as an extreme case of land cover and land use change, and it lends itself well to remote sensing.

Human population had grown from half a billion in 1600 to a billion by 1800; by 1940 it had reached almost 3 billion and now exceeds 6.5 billion. With population having quadrupled in less than a century, there has been immense migration into cities. The 2005 revision of the UN World Urbanization Prospects report described the 20th century as witnessing 'the rapid urbanization of the world's population' as the global proportion of urban population rose dramatically from 13 per cent (220 million) in 1900 to 29 per cent (732 million) in 1950 and to 49 per cent (3.2 billion) in 2005. The same report projected that the figure is likely to rise to 60 per cent (4.9 billion) by 2030, equivalent to the entire 1987 world population living in an urban centre.

Industrialization accompanies urbanization as both are presently fuelled by hydrocarbon energy. As presented by Leakey & Lewin (1995), the energy

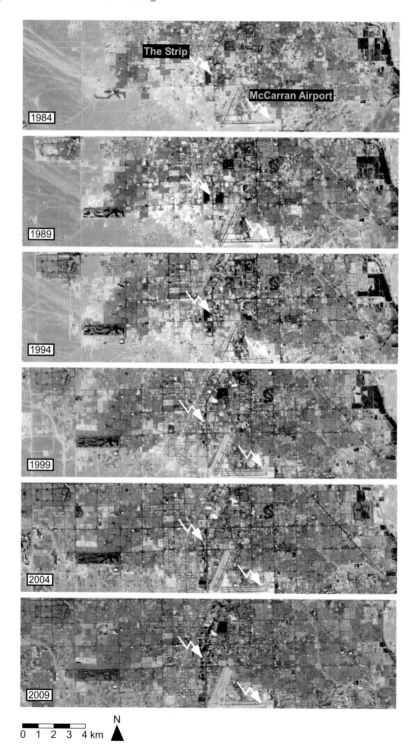

Figure 6.1 Two and a half decades of urban expansion in Las Vegas (Nevada, USA) documented by Landsat imagery. The tan-coloured desert to the west of the scenes in images acquired prior to 1994 is transformed into a network of roads and buildings. Credit: NASA's Earth Observatory.

consumption per person in North America or Western Europe is approximately equivalent to 10 tonnes of coal per year; in India or China it is approximately 0.2–0.3 tonnes of coal per year. As the developing countries attempt to industrialize and urbanize to achieve a standard of living comparable with that in Europe or North America, their energy demands will rise hugely, irrespective of population growth. In all likelihood, their increasing energy demands will be met by burning fossil fuels, and the rate of increase of CO_2 production is going to be tremendous (e.g. Auffhammer & Carson, 2008).

As cities grow, they impact climate in several ways. First, intensive energy consumption, industrial growth and the expansion of transport networks lead to greatly increased production of greenhouse gases (Fuglestvedt *et al.*, 2008). Second, change in land use can promote surface warming as the paving of formerly vegetated land reduces albedo, with the result that a smaller fraction of incoming sunlight is reflected. Both greenhouse gas emission and altered albedo promote climate warming, and the relative contribution of each is difficult to decouple (Gallo *et al.*, 1996; Gallo & Owen, 1999), though greenhouse gases are perceived to be by far the strongest effect.

Separation of the effects of increasing greenhouse gas emissions from the direct impact of urbanization has been hindered by the fact that different measures of urban growth return different estimates of the areal coverage of rural and urban areas (Gallo *et al.*, 1999; Hansen *et al.*, 2001). The commonly employed techniques for urban classification are population data (Easterling, 1997) or satellite measurements of night light (Gallo *et al.*, 1996, 1999). Furthermore, monitoring of air quality in cities has traditionally been ground-based. Such measures provide continuous data but at discrete locations, so they offer temporal resolution but their spatial coverage is poor. Satellites, on the other hand, offer tremendous spatial coverage but an unsatisfactory temporal component. In view of these difficulties, Kalnay & Cai (2003) attribute a 0.27 °C mean surface warming per century in the continental United States due to land use changes.

6.2 Urban remote sensing

Remote sensing is a powerful technology for the study of urban phenomena. Space- and airborne images offer a freeze-frame view of spatio-temporal variations in urban patterns. The power of having an aerial view of a city became apparent as early as the 1840s, when the photographic camera started to be combined with technology such as hot air balloons or arrays of kites. A striking example of this is the 1906 photograph obtained of San Francisco following the great earthquake of that year (Figure 6.2).

From its early beginnings, aerial photography has provided a means of mapping urban extent and dynamics that simply cannot be obtained from administrative survey. The science of remote sensing cities today forms a major focus of aerial and spaceborne campaigns.

Photograph of San Francisco in ruins from Lawrence Captive Airship, 2000 feet above San Francisco Bay overlooking water front. Geo. R. Lawrence Co. May 28th 1906

Figure 6.2 A 1906 oblique aerial image of earthquake-ravaged San Francisco. The image was captured by George R. Lawrence using an array of kites to hoist aloft a camera. The image was acquired at an altitude of approximately 600 metres and an electric wire controlled the shutter to produce a negative.

Urban landscapes are spatially complex, consisting of an intricate mosaic of land cover types with little systematic orientation. Imaging such a landscape demands sensors with high spatial resolution. This requirement has led to a traditional dominance of aerial photography as standard imaging input to urban studies (Ehlers, 2007). The advent of satellite systems that can provide spatial data at a resolution exceeding 1 m heralded a new era of urban remote sensing as the cost of repeated monitoring was lowered substantially (Table 6.1; Figure 6.3).

In addition to spatial heterogeneity, urban landscapes are spectrally complex, composed of many different materials, each with a unique spectral signature. Cities are therefore most effectively imaged by sensors capable of delivering high-resolution data in both the spectral and spatial realm (Figure 6.3). As we discussed in Chapter 2, spatial resolution defines the level of spatial detail depicted in an image and may be described as a measure of the smallness of objects on the ground that may be distinguished as separate entities within it. In contrast, spectral resolution defines the number of contiguous or non-contiguous spectral bands that a sensor is capable of collecting data on. The spectral resolution of a remote sensing system can be described as its ability to distinguish different parts of the range of measured wavelengths, and this is covered in Chapter 2 (see Figure 2.8).

Table 6.1 High-resolution satellite systems useful to urban sensing.

Satellite	Spectral properties	Image characteristics
IKONOS	Panchromatic 1 m edge-pixel	11 km swath-width
	525–929 nm	11 × 11 km total image size
www.geoeye.com	Multispectral 4 m edge-pixel	11 bits per pixel
	445–516 nm (blue)	
	506–595 nm (green)	
	632–698 nm (red)	
	767–853 nm (near-infrared)	
OrbView-3	Panchromatic 1 m edge-pixel	8 km swath-width
	450–900 nm	8 × 8 km total image size
www.geoeye.com	Multispectral 4 m edge-pixel	11 bits per pixel
	450–520 nm (blue)	
	520–600 nm (green)	
	630–690 nm (red)	
	760–900 nm (near-infrared)	
QuickBird	Panchromatic 0.61 m edge-pixel	16.5 km swath-width
	450–900 nm	16.5 × 16.5 km total image size
www.digitalglobe.com	Multispectral 2.44 m edge-pixel	11 bits per pixel
	450–520 nm (blue)	
	520–600 nm (green)	
	630–690 nm (red)	
	760–900 nm (near-infrared)	
WorldView-1	Panchromatic 0.50 m edge-pixel	17.6 km swath-width at nadir
	400–900 nm	60 km × 110 km Max contiguous area collected in a single pass (at nadir)
www.digitalglobe.com	No multispectral	11 bits per pixel
WorldView-2	Panchromatic 0.46 m edge-pixel	16.4 km swath-width at nadir
	450–800 nm	65.6 km × 110 km Max contiguous area collected in a single pass (at nadir)
www.digitalglobe.com	Multispectral 1.84 m edge-pixel	11 bits per pixel
	400–450 nm (coastal)	
	450–510 nm (blue)	
	510–580 nm (green)	
	585–625 nm (yellow)	
	630–690 nm (red)	
	705–745 nm (red edge)	
	770–895 nm (near-infrared-1)	
	860–1,040 nm (near-infrared-2)	
Formosat-2	Panchromatic 2 m edge-pixel	24 km swath-width
	450–900 nm	24 × 24 km total image size
www.nspo.org.tw	Multispectral 8 m edge-pixel	11 bits per pixel
	450–520 nm (blue)	
	520–600 nm (green)	
	630–690 nm (red)	
	760–900 nm (near-infrared)	
SPOT-5	Panchromatic 5 m edge-pixel	60 km swath-width
	480–710 nm	60 × 60 km total image size
www.spotimage.com	Multispectral 10 m edge-pixel	8 bits per pixel
	500–590 nm (green)	
	610–680 nm (red)	
	780–890 nm (near-infrared)	
	1580–1750 nm (swir)	

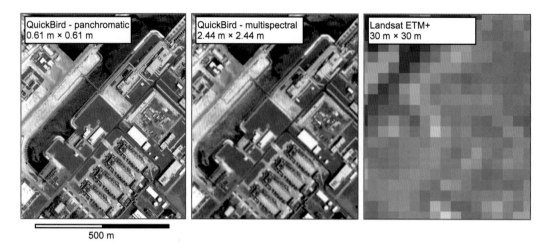

Figure 6.3 Satellite images of an industrial complex in the United Arab Emirates acquired using the QuickBird panchromatic channel (left), the Quickbird multispectral channels (middle), and Landsat ETM+ (right). The degree to which urban features can be resolved is dependent on the spatial capability of a remote sensing instrument. Satellite images: DigitalGlobe.

The method by which a remote sensor captures an image creates an inherent trade-off between spectral and spatial resolution. By virtue of the high velocity at which a satellite or aircraft sensor travels over the Earth's surface, an image must be captured very rapidly, typically in a matter of seconds. The length of time during which data are collected over the area of a single pixel is termed the 'dwell time'. The smaller the pixel, the more frequently a scan line needs to be generated and the faster the scan-rate needs to be in order to image successive pixels within a scan line. Both of these factors shorten the dwell time, and a shorter dwell time reduces the total energy detected for each image pixel, increasing the noise in the data.

In order to ensure that the image signal is not swamped by noise, the dwell time must be partitioned to collect sufficient signal, even if the reflectance of the target is low. For this reason, the majority of the dwell time can be utilized to collect spatial or spectral information; both cannot be captured simultaneously. This is why sensors of high spatial resolution have low spectral resolution and vice versa.

Airborne instruments such as CASI can offer hyperspectral (typically defined as more than 21 spectral bands) with fine spatial resolution (on the order of $4\,m^2$), since dwell time is increased by virtue of the slow speed at which a light aircraft can be flown. Even so, configuring the instrument to deliver a smaller pixel size will require a drop in the number of acquired spectral bands. For urban remote sensing, by virtue of advances in classification algorithms based on the textural component of imagery (a derivative of the spatial arrangement of features – Myint *et al.*, 2007), a lower spectral resolution and higher spatial resolution is preferable (Figure 6.3).

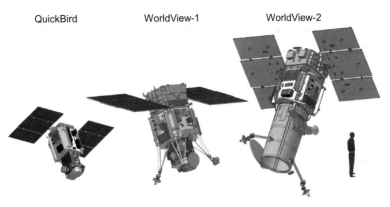

QuickBird   WorldView-1   WorldView-2

Figure 6.4 A comparison of DigitalGlobe commercial satellites. The WorldView programme is currently replacing the ageing QuickBird sensor. This new campaign offers marginally finer spatial resolution but boasts better spectral capability and improvements in daily revisit time, which offers a greater chance of successful image collection in high cloud areas.
Source: DigitalGlobe.

	QuickBird	**WorldView-1**	**WorldView-2**
Operational altitude	450 km	450 km	770 km
Weight class	2000 lbs	5700 lbs	5700 lbs
Panchromatic resolution at nadir	0.6 m	0.5 m	0.5 m
Multispectral resolution at nadir	2.4 m	2.0 m	2.0 m
Average revisit at 1 m resolution (40° lat target)	2.5 days	1.7 days	1 day
Swath-width	16.5 km	16 km	16 km
Operation schedule	Lofted Oct. 2001	Lofted Sept. 2007	Lofted Oct. 2009

Of the satellite products currently available, the two most relevant in terms of urban feature extraction are the IKONOS and QuickBird sensors. The former is now commercially available at a spatial resolution of 1 m and the latter at 0.6 m. These data have been used in a variety of urban studies (Sawaya *et al.*, 2003), and both data sources are adequate to provide three-dimensional positioning and building extraction with sub-metre accuracy (Fraser *et al.*, 2002). The utility of space-based imagery for this work is bright, with the recent launch of DigitalGlobe's WorldView satellites that will offer spatial resolution as fine as 0.5 m (Fig. 6.4).

Accompanied by technological advances in spaceborne platforms, digital airborne scanners have also become available, offering sub-metre spatial resolution coupled with hyperspectral imaging capability. Relying on gyroscopes to monitor the attitude of the aircraft, past airborne scanners were unable to deliver imagery with sufficient geometric precision for use in the urban environment. However, by employing differential GPS and inertial navigation equipment, modern systems are able to overcome these limitations and can provide data on a par with satellite sensors.

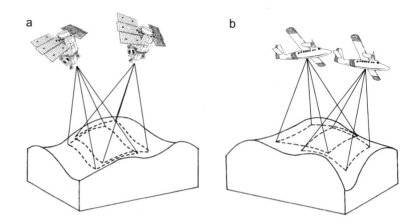

Figure 6.5 Stereo imaging provided by the acquisition of stereo-paired satellite images acquired using a pointable sensor (a) or overlapping along-track aerial photography (b). In each case, parallax is resolved within the areas of image overlap.

6.2.1 Three-dimensional urban model generation

As was possible with traditional analogue aerial photography, the ability to generate digital elevation models (DEMs) from digital imagery has had a profound impact on urban monitoring. Cities are three-dimensional, and being able to visualize them as such offers a way of understanding the underlying systems of a city and how it functions with regard to its structures. Being able to present a policymaker with a 3-D visualization and flythrough of an urban landscape is a powerful tool indeed. Three-dimensional reconstruction of cities is routinely used and has been made possible by advances in three key remote sensing technologies – stereo imaging, LiDAR, and imaging radar.

6.2.2 Stereo imaging

Provided that data are collected in stereo with sufficient overlap (typically 50 per cent or more), adjacent tiles of imagery can be used to derive topography within a scene. Referred to as 'stereo pairs', the overlapping tiles provide two different perspectives of the same area. Elevations and slope lengths and inclinations can be determined from stereo pairs by application of parallax principles. Stereo parallax is the physical condition that refers to the apparent displacement of a point or a feature of some height in a photo or image caused by a shift in the position of observation.

Observation geometry suitable for parallax can be achieved using a satellite sensor capable of looking both directly down at the Earth and, later, sideways, to obtain an oblique image of the same area (Figure 6.5a). It can also be achieved from a line of overlapping imagery traced on the ground beneath an aircraft during acquisition of data (i.e. a nadir view; Figure 6.5b). The former technique is favoured by satellite systems able to image a common part of the Earth from different orbital tracks. They are said to have a 'pointable' capability and examples

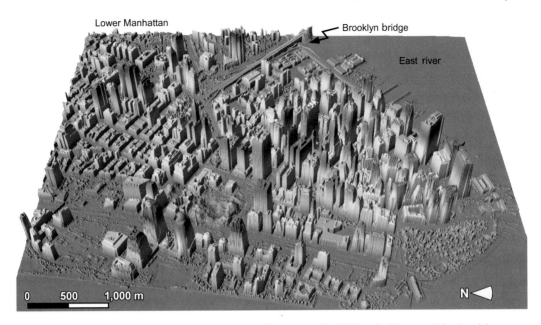

Figure 6.6 Click LiDAR image of lower Manhattan, New York, rendered Sept. 27, 2001 by the US Army Joint Precision Strike Demonstration from data collected by NOAA. Credit: NOAA/US Army JPSD.

include SPOT, ASTER, and IKONOS. The stereo capability of ASTER is further considered in Chapter 7 in the context of the ASTER Global Digital Elevation Model (GDEM; Figure 7.4).

6.2.3 LiDAR

Light detection and ranging (LiDAR) sensors mounted aboard aircraft are another source of three-dimensional information for urban monitoring (Figure 6.6). LiDAR, a technology discussed frequently in this book, is an active remote sensing technique that employs pulses of laser light directed towards the ground to measure distance between the sensor and a target (Figure 6.7). An accurate clock records the round-trip travel time between the transmitted and received pulse. Dividing this travel time by two and multiplying the result by the speed of light yields the distance between the LiDAR unit and the target. Since the laser pulses are travelling at the speed of light, the timing mechanism must be accurate to within a few nanoseconds to ensure a topographical accuracy of 10 cm or less.

The use of LiDAR for accurate determination of terrain models began in the late 1970s but was limited by its complexity, cost-effectiveness, and georeferencing capability (Lillesand *et al.*, 2004). In the last decade, the technology has overcome these hurdles following the availability of more reliable electronics and inertial measurement units that bring georeferencing accuracy to the

Figure 6.7 (a) Basic functioning of a topographical LiDAR. During the flight, the LiDAR sensor pulses a narrow, high-frequency laser beam toward the Earth through a port opening in the bottom of the aircraft's fuselage. The LiDAR sensor records the time difference between the emission of the laser beam and the return of the reflected laser signal to the aircraft. With the speed of light known, this information can be translated to elevation data. (b) LiDAR elevation model of Port Everglades, Ft. Lauderdale, Florida. LiDAR is an 'active illumination' technique that, unlike photography, does not depend on ambient illumination. It works equally well during the day or at night.

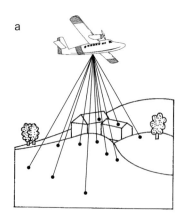

sub-metre level, as well as commercial lasers capable of pulsing at rates as high as 50 kHz.

Coupled with such a high scan rate, low aircraft speed and altitude allow a LiDAR setup to cover the ground with a high density of sample points. This density is critical in the urban environment, where the terrain is highly variable and buildings are closely packed together. Depending on the reflective property of the target, a transmitted pulse may be returned multiple times to the receiver, because the pulse is distended when it travels through the air and reaches the target. As a result, if part of the pulse hits the roof edge of a building, the rest of the pulse may continue to travel to reach a tree, or the ground, or perhaps both (Shan & Sampath, 2007). At present, some LiDAR receivers are able to record many such multiple returns which, through post-processing, can be useful in determining the physical structure of sparse structures such as tree canopies. This latter application is explored in Chapter 5, Section 5.3 (see Figure 5.5).

Modern LiDARs have the capability to capture reflectance data from the returning pulse, in addition to the three-dimensional coordinates of the laser returns (Lillesand *et al.*, 2004). Like the strength of radar returns, the intensity of LiDAR 'echoes' varies with the wavelength of the source energy and the composition of the material returning the incoming signal. Such radiometric information (termed LiDAR intensity) can be invaluable in identifying different building materials (e.g. concrete has a much higher reflectance than asphalt, etc.). Modern LiDAR instruments collect intensity data in addition to the more traditional elevation soundings. This application is discussed in Chapter 8 (Section 8.9) in the context of bathymetric LiDAR.

6.2.4   Synthetic Aperture Radar (SAR)

Synthetic Aperture Radar (SAR) is a remote sensing technology that utilizes imaging radar. As it supplies its own source of microwave illumination with wavelengths of several centimetres, SAR is unhindered by poor weather and night – factors that confound passive optical instruments. Another advantage of SAR is the opportunity to record large areas in a short time from a large distance.

These properties render SAR as an attractive alternative to costly LiDAR surveys. The radar reflections from terrain are mainly determined by the physical characteristics of the surface features (such as surface roughness, geometric structure, and orientation), the electrical characteristics (dielectric constant, moisture content, and conductivity) and the radar frequency of the sensor. As is the case for the visible bands, urban landscapes are spectrally diverse in these different properties.

SAR instruments can be mounted on both air- and spaceborne platforms. The basic mode of operation is to illuminate large areas on the ground with the radar signal and to sample the backscatter. Like LiDAR, given that the microwave signal travels at the speed of light, the range between the sensor and target can be calculated from the difference in time between transmitting a signal and receiving its reflection.

SAR takes rapid radar samples looking sideways along a flight path. Just as in an optical instrument, the resolution of a radar system is affected by the size of the aperture: a larger aperture gives a finer spatial resolution. However, scaling up to the long wavelengths of radar has problems. For example, to achieve an azimuth (along-track) resolution of 100 m with 5 cm wavelength radar from a range of 800 km requires an aperture some 400 m long. Such an antenna size is impractical.

As the radar is attached to a rapidly moving platform, be it a satellite or aircraft, it is possible to combine reflected signals from along the flight path to synthesize a very long antenna. The aperture, or area used to receive signals, is subsequently created artificially during the signal processing and takes advantage of the fact that the target area of interest is essentially stationary during the fly-by time.

The principle of the SAR depends upon the fact that objects on the ground experience illumination by the radar over an interval of time as the sensor moves along its flight path or orbit. Throughout this interval the reflected radar pulses returned from the ground are recorded by the SAR instrument. The return signals are subsequently processed as if they were all received by a single, much larger, antenna, as opposed to an actually rather small one that, in the example of Figure 6.8, has moved through 13 separate positions.

The challenge to this strategy lies in the fact that it is necessary to assign the numerous returned signals their correct positions with relation to one another to yield a viable image. This ordering is feasible because of the different Doppler shifts in frequency imparted to the returning signal; a factor dependent on an objects position on the ground with relation to the passage of the sensor across the sky. Relative to the transmitted frequency, the Doppler shift will impart an increase in frequency for beams scattering from ground objects ahead of the sensor. Similarly, objects behind the aircraft or satellite will return beams with a lower frequency than that transmitted. Since the frequency of the radar beam transmitted by the SAR is known, it is possible to compare the frequencies of the transmitted and reflected signals to determine the nature and amount of frequency shift. With the shift quantified, the correct relative positions of the recorded beams can be retrieved and an image constructed.

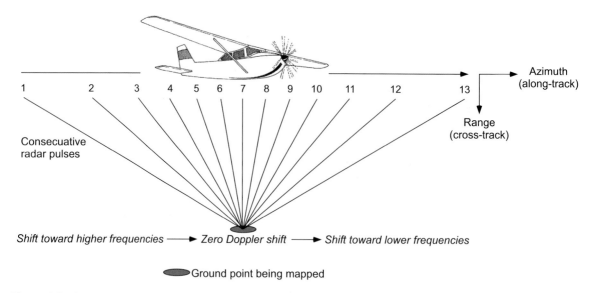

Figure 6.8 The azimuth (along-track) resolution of SAR is controlled by the motion of the sensor against a stationary target. Multiple consecutive radar pulses are used to emulate a large antenna and produce a fine azimuth resolution. The numbers along the top of the figure denote the position of the aircraft along the flight path from which the SAR illuminated the ground point being mapped and subsequently received returns. The technology relies on full knowledge of the frequency of the transmitted signal and the Doppler effect to assign reflections to their correct position in the recorded image. The range (cross-track) resolution is dictated by the duration of the microwave pulse, which should be as short as possible.

By using sensor motion to synthesize a large antenna, the SAR instrument obtains a fine azimuth resolution. This mathematical reconstruction, termed 'focusing', is computationally time consuming on even advanced workstations, but it typically results in a thousand-fold improvement in spatial resolution as compared to acquisition using a real aperture radar. By contrast, the range (cross-track) resolution is not artificially synthesized and is controlled by the duration of the transmitted microwave pulse – narrow pulses yield fine range resolution. Pulse length is inversely proportional to bandwidth, so it is advantageous to utilize as high a bandwidth as possible.

Buildings have a pronounced signal in SAR images because their simple geometric shapes give rise to characteristic reflections. Vertical walls orientated perpendicular to the incoming radar signal, for example, reflect particularly strongly, creating in the image a line of bright scattering in the azimuth direction at the building footprint (Stilla & Soergel, 2007). This occurs since the radar signal undergoes so called 'double-bounce scattering' between the ground and the building wall. At the opposite side of the building, the ground is partly occluded from the building shadow. This region appears dark in the SAR image because no signal is reflected.

The appearance of a building in high-resolution SAR data is governed to a large extent by its roof structure. Different roof geometries lead to special shapes in the

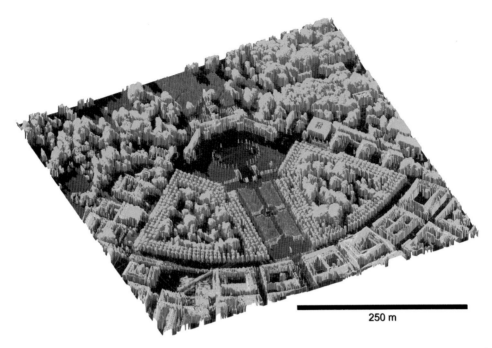

250 m

Figure 6.9 A 3-D representation of a stereoscopic InSAR acquisition of the Karlsruhe Palace, Germany. Modified with permission from Brenner & Roessing (2008). Copyright 2008 IEEE.

reflected radar pulse. Flat-roofed buildings cast L-shaped shadows; a building with sloped roof may cause a trapezoidal shadow; and a gabled roof building typically casts a hexangle-shaped shadow (Stilla *et al.*, 2005). Analysis of the character and geometry of the returned signal allows detection and reconstruction of buildings, including extraction of a DEM.

The Sun's illumination reaching the Earth is incoherent (scrambled), meaning that only amplitude information can be used for remote sensing. This leads to another important difference between optical and SAR systems. A SAR system records both the amplitude and the phase of the back-scattered microwave radiation, making it a coherent imaging process. In many cases, the coherent phase information transmitted by a SAR satellite or aircraft is reflected from the surface back to the sensor with the phase more or less intact. A subsequent satellite pass, several days to several years after the initial pass, may also retain the phase information.

Using a technique termed Interferometric SAR (InSAR), the phase of two time-separated images are compared. InSAR has two uses with regard to remote sensing urban environments:

- First, it can be used to construct three-dimensional city models, providing that the two images are acquired with slightly different geometry so as to produce a stereoscopic effect (Figure 6.9). InSAR derivation of a DEM demands that the images are offset by a few hundred metres.

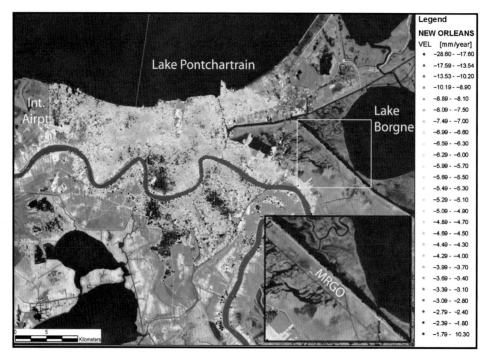

Figure 6.10 Map showing rate of subsidence for permanent scatterers in New Orleans and vicinity during 2002–05. Adapted from Dixon *et al.* (2006) *Subsidence and flooding in New Orleans*, Nature [**441**], 587–588 by permission of Macmillan Publishers Ltd, copyright 2006.

- Second, paired image-sets acquired on exactly the same orbital path may reveal subtle shifts in the position of the Earth's surface (closer to or further from the satellite). As covered in the following section, the latter function of InSAR makes it a powerful tool for detecting subsidence (Figure 6.10). This application is termed differential interferometry DInSAR.

Until recently, the application of stereoscopic InSAR to urban mapping has been somewhat limited by its coarse resolution, and data fusion with more traditional technologies such as LiDAR and aerial photography have been favoured (Tupin & Roux, 2003). Recent commercial airborne SAR sensors, however, can now offer a spatial resolution on the order of a metre (Brenner & Roessing, 2008), and currently developed spaceborne systems will likely achieve similar values (Roth, 2003). This improvement in resolution makes SAR an increasingly attractive technology for mapping the topography of urban centres.

6.3 Microwave sensing of subsidence

Subsidence, regardless of its cause, is a major hazard in urban centres. The Earth's surface can sink for purely natural reasons such as sediment compaction, thawing permafrost, or the chemical dissolution of carbonate rocks and salt deposits by rainwater infiltrating into the subsurface. Alternatively, subsidence can be induced artificially by the excessive extraction of water (Woldai *et al.*, 2009), oil or natural gas (Fielding *et al.*, 1998), or else through tunnelling for mining or transport (Wright & Stow, 1999).

Subsidence does not always occur instantaneously, with a catastrophic collapse of the ground, though this may occur, for example, when the roof of a buried sinkhole fails. Slow subsidence may persist over years to decades; this can have equally dire economic consequences, but spread out over a long time period. For example, a decrease in land elevation under urbanized areas causes structural damage to buildings that is costly to replace or repair and also greatly increases flooding potential (Fielding *et al.*, 1998; Figure 6.10). It is in this respect that remote sensing is a powerful tool as it offers the ability to flag an area as suffering gradual subsidence prior to the sag becoming noticeable on the ground. This gives time to intervene and remedy the problem, by for example, structurally reinforcing foundations with concrete, or else halting the extraction of groundwater, hydrocarbons, etc. before the problem becomes acute. A final solution is to abandon development of the subsiding land entirely.

As discussed in the previous section, differential interferometry (DInSAR) is a well-proven technology for monitoring the deformation of Earth's surface (Amelung *et al.*, 1999; Crosetto *et al.*, 2003; Tomás *et al.*, 2005). The technique demands a pair of images, separated in time, that are analyzed for inconsistencies in the phase-state of the recorded signal. The resulting 'interferogram' is a contour map of the change in distance between the ground and the radar instrument (Massonnet & Feigl, 1998). This is constructed on the premise that a difference in phase indicates that the distance between the sensor and target has altered. Provided that the orbital path and direction of observation of the satellite for the time-separated images are identical, the change in distance can be attributed to a movement of Earth's surface.

DInSAR is a powerful technique for three reasons:

1 Coupled with rigorous processing, estimates of Earth's elevation change can be precise to ≈1 cm. DInSAR can detect shifts in the order of a small fraction of a wavelength (Table 6.2).
2 The repeat-cycle of the key SAR satellites is relatively short (several days to several weeks) which, combined with the all-weather and night-time capability of the microwave spectrum, eases the assembly of a time-series of data.
3 The interferogram provides excellent spatial sampling, far exceeding that which could be achieved through ground-based methods such as optical levelling. New X-band sensors such as TerraSAR-X, for example, have metre-scale resolution over a $50\,km^2$ scene.

Detection of subsidence via DInSAR is not without its limitations. Not least of these is that sensors such as the Canadian RADARSAT (1 and 2) were optimized to observe sea ice and ocean dynamics (covered in Chapters 10 and 12, respectively)

Table 6.2 Radar satellites suitable for interferometry. Note the lack of a satellite between the 1978 demise of SEASAT and the 1991 launch of ERS-1. The glut of launches in the early 1990s heralded a productive decade for the geophysical applications of interferometry. Modified after Massonnet & Feigl (1998).

Sensor	Launch year	Max. resolution (m²)	Wavelength (mm)	Band
SEASAT	1978–1978	20	235	L
ERS-1	1991–2000	15	56.7	C
ERS-2	1995–ongoing	15	56.7	C
JERS-1	1992–1998	18	135	L
RADARSAT-1	1995–ongoing	25	57.7	C
RADARSAT-2	2007–ongoing	3	57.7	C
SIR-C	1994 (Shuttle mission)	30	–	X, C, L
ENVISAT	2002–ongoing	150	56	C
SRTM	2000 (Shuttle mission)	90	56	C
ALOS	2003–2008	2.5	235	L
TerraSAR-X	2007–ongoing	1	31	X

and not to be used as an interferometric tool. Generic limitations to the technology can be summarized as follows:

- DInSAR demands the existence of 'permanent scatterers' – points on the ground that strongly reflect radar in a persistent manner over many satellite revolutions. Examples of permanent scatterers are dense vegetation, buildings and their roofs, roads, etc. Coherent scatterers such as buildings are orientated perpendicular to the radar look direction. Parks or other vegetation are termed incoherent areas and require more in-depth analysis to yield phase differences, via a process termed Permanent Scatterer InSAR (PSInSAR) (Mora *et al.*, 2001; Dixon *et al.*, 2006). DInSAR is therefore particularly well poised to detect subsidence in urban centres, due to the guaranteed high density of well-defined coherent radar targets.
- The state of the atmosphere will differ because the two images are acquired at different times. Inconsistencies in the atmosphere (in particular water vapour in the troposphere and charge in the ionosphere) will impart noise into calculations that is several orders of magnitude greater than the subsidence signal being detected. This noise stems from the variable path-delay of the microwave signal and is particularly pronounced in humid regions. By exploiting several characteristics of radar scattering and atmospheric decorrelation, PSInSAR is able to discern surface displacement even under suboptimal atmospheric conditions.
- Though not a major problem in urban environments, there must be strict stability in the reflective characteristics of the permanent scatterers used in

the interferometric analysis. Problems in this respect occur if an area becomes inundated by water, or a building's roof is replaced, or if a vegetated plot is altered (e.g. through ploughing or irrigation).

- The considered image-pairs must be processed identically so as not to spoil the interferometric effect (Massonnet & Feigl, 1998). This also precludes the combination of images acquired with different sensors. ERS-1 and ERS-2 (Table 6.2) are exceptions to this rule as they carry identical radar instruments (e.g. Fielding *et al.*, 1998; Wright & Stow, 1999; Woldai *et al.*, 2009). Initiating in 1991 and still active at the time of writing, the ERS mission sets the gold standard for interferometry. The mission accrues 0.12 terabytes of data daily while imaging 5 million km^2 (Massonnet & Feigl, 1998).

- InSAR measurements are, by definition, relative (ambiguous), so determination of subsidence requires calibration with one or more ground control points of known elevation and motion (Dixon *et al.*, 2006). This calibration is typically defined in reference to a GPS-monitored datum on the ground, or else over an area in the image which is known to have remained stationary during the time of observation (which is likely to be several years).

- The spatial resolution of the image is limiting. The interferometric measurement is meaningless on a single pixel because it includes an unpredictable quantity of noise. Successful interpretation thus depends on the structure of the image and the agreement of several (typically $\approx$10) neighbouring pixels.

Since the application of DInSAR to detect urban subsidence demands a high spatial resolution, new instruments such as the TerraSAR-X, lofted from Kazakhstan in 2007, will perhaps represent an advancement in the state of the art. This instrument arose out of a public/private partnership between the German Aerospace Centre (DLR) and the European Aeronautic Defence and Space Company (EADS). In its 'spotlight' mode, TerraSAR-X can acquire images of an area measuring 5 km $\times$ 10 km with a resolution as fine as 1 m^2.

Using a time-series of 33 RADARSAT scenes, Dixon *et al.* (2006) generated a subsidence map for the city of New Orleans. Their work is particularly noteworthy in clearly demonstrating that the city underwent rapid subsidence in the three years before Hurricane Katrina struck in August 2005. As depicted in Figure 6.10, subsidence was highest along the southern shore of Lake Pontchartrain and east of the city in the vicinity of the airport. The former area has been intensively augmented with engineered fill, while the latter were wetlands that had been drained and urbanized. Wetlands have highly organic soils that desiccate, oxidize and compact when drained; the result is subsidence. Of the levees that breached during Katrina (e.g. Figure 9.3, Chapter 9), many were situated in areas that were subsiding at rates exceeding 20 mm yr^{-1}. Dixon *et al.* provide strong evidence that subsidence played a role in weakening the levees that failed and were overtopped as the hurricane made landfall.

Figure 6.11 Typical training samples of urban land cover types extracted from a grayscale QuickBird image. The textures of the classes are clearly different. Satellite images: DigitalGlobe.

6.4 Textural metrics

If delineating urban areas is a difficult task, then classifying different types of urban land use is even more so. The urban environment is characterized by a mixture of spectrally differently diverse materials (e.g. plastic, metal, rubber, glass, cement, wood, etc.) concentrated in a small area (Jensen & Cowen, 1999). Such materials are ubiquitous in many land cover classes, such as buildings, commercial infrastructures, transportation networks and parks. Because they are combinations of spectrally distinct land cover types, mixed pixels in urban areas are frequently misclassified as other land cover classes.

Similarly, the definition of an 'urban' spectral class will usually incorporate pixels of other, non-urban classes. Such spectral heterogeneity severely limits the applicability of standard classification techniques, where it is assumed that the study area is comprised of a number of unique and internally homogeneous classes. Even with metre-scale resolution, per-pixel spectral diversity is typically high enough to confound the traditional classification algorithms used for digital image processing (Myint *et al.*, 2002). This restriction is exacerbated by the difficulty in defining spectrally homogeneous training areas with which to drive a classifier. Training sets need to be homogeneous groups of pixels representing a single class.

A single class in an urban environment, as we have just seen, is rarely homogeneous (Figure 6.11), due to variation in the spectral response of their component surface covers in each class. For example, rooftops alone do not represent a suitable training sample for a residential class, because other land covers, such as road surfaces, footpaths, driveways, swimming pools, grass, scrubs, exposed

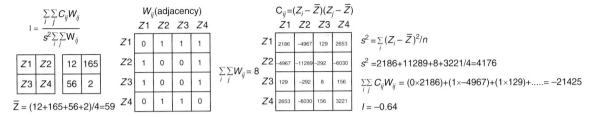

Figure 6.12 A worked example of the calculation of Moran's I autocorrelation statistic for a 2×2 pixel moving window. W_{ij} is the weight at distance d so that $W_{ij} = 1$ if point j is within distance d from point i; otherwise, $W_{ij} = 0$; C_{ij} are deviations from the mean (i.e. $C_{ij} = (Z_i - \bar{Z})(Z_j - \bar{Z})$, where $\bar{Z}$ is the mean brightness value in the local window. s^2 is the mean of the major diagonal of C_{ij}.

soil and shrubs or trees can also be part of a residential class. If a single land cover (e.g., tile rooftop) is defined as the training sample, only that cover type will be identified as residential, and all other land cover types within the residential area will not be classified as residential class. If, however, a cover type such as grass is identified as the training sample, the algorithm will identify all grasses as residential – including those in parks, in commercial complexes, and golf courses. Alternatively, if all land covers within a land use are selected, the standard deviation of the training set will be very large and the data distribution of the class will likely violate one of the key assumptions of the statistical classifiers (e.g., maximum likelihood) that the pixel values follow a normal distribution (Barnsley *et al.*, 1991; Sadler *et al.*, 1991). Hence, the likelihood of accurate image classification will be very small.

In view of these limitations, and although it is extremely labour intensive, manual interpretation of digital imagery remains a commonly employed technique for mapping city structure. The most efficient way to automate urban classification is to learn from the way that an expert is able to recognize different land cover types and emulate the process using a computer.

It is in this realm that geospatial approaches have been particularly successful by relying on the 'textural' differences between land use classes to improve urban mapping (Myint, 2003, 2006; Thomas *et al.*, 2003). Texture can be generally characterized as fine, coarse, smooth, rippled, mottled, irregular or lineated. In fact, texture is an inherent property of all surfaces (e.g. the grain of wood, the weave of a fabric), including all types of land use and land cover classes (e.g. the pattern of crops in a field, the crown features of trees in a dense forest). It contains important information about the structural arrangement of surfaces and their arrangement to the surrounding environment.

The texture of a land cover type is elegantly summarized by its degree of spatial autocorrelation (the similarity/dissimilarity of pixel values within a moving window). An autocorrelation metric that has been particularly successful in urban differentiation is Moran's I (Figure: 6.12).

Moran's I assesses how dispersed, uniformly distributed or clustered points are in space (weighted by their attributes) and whether or not this pattern has occurred

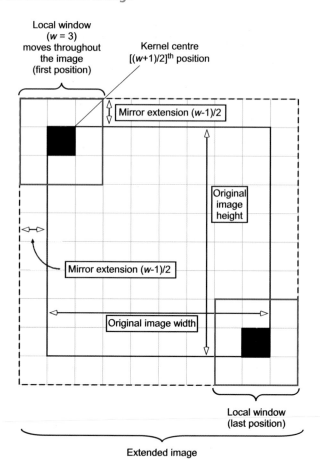

Figure 6.13 The use of a local moving window in remote sensing data is a powerful technique for calculating local statistical functions such as texture and edge detection. This figure shows the start and end position of a 3 × 3 pixel window in a hypothetical image. Mirror extension of the image boundaries has been used to grow the image by a single pixel in each dimension, so as to allow the moving window to access the pixels at the periphery of the original dataset.

by chance. The statistic is based on a comparison of the attribute values of neighbouring units, with strong positive spatial autocorrelation indicating similarity between those units.

Local moving windows (e.g. Figure 6.13) are commonly used in digital image classification approaches to define the local information content around a centre pixel. Moran's I autocorrelation statistic is calculated using a moving window (worked example: Figure 6.12), and the size of the window alters the result of the calculation significantly. In an urban setting, accuracy should increase with a larger local window size because it contains more information and provides more complete coverage of spatial variation, directionality and spatial periodicity of a particular texture than a smaller window size does. However, this is not necessarily true in a real-world situation because, in most cases, we are dealing with more than two homogeneous texture features while the local image moves across the image (Myint, 2003). Hence, it should be noted that a larger window can produce better accuracy only when dealing with a homogeneous texture feature (one window does not cover more than one texture). Window size is typically optimum in the range of 10 m × 10 m to 20 m × 20 m in urban settings (Myint, 2003).

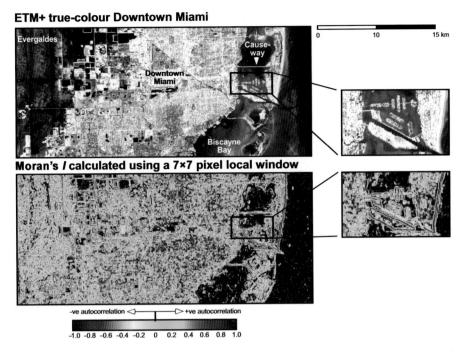

ETM+ true-colour Downtown Miami

Moran's *I* calculated using a 7×7 pixel local window

-ve autocorrelation ◁——————▷ +ve autocorrelation

-1.0 -0.8 -0.6 -0.4 -0.2 0 0.2 0.4 0.6 0.8 1.0

Figure 6.14 Moran's *I* calculated for a Landsat ETM+ scene of downtown Miami using a 7×7 pixel moving local window. The Moran's result scales between 1.0 (a perfect positive correlation) to −1.0 (a perfect negative correlation). North is top. Credit: NASA.

Figure 6.14 depicts a Landsat ETM+ image (top) and the result from the calculation of the Moran's *I* statistic (bottom). The values in the lower pane scale −1 through +1 and are colour-coded to represent levels of autocorrelation within a local moving window (i.e. Figure 6.13) that has been imposed on the Landsat scene. Areas of negative autocorrelation are depicted in hot colours and equate to pixels that have disparate values as compared to their neighbours. For example, the causeways that link downtown Miami to the offshore Keys have high values of Moran's *I* and hence are coloured red. This is because these narrow roadways have very different (higher) spectral values compared to the surrounding waters of Biscayne Bay. Conversely, the deep ocean to the east of the scene, which is characterized by wide expanses of pixels with similar low values and hence high autocorrelation, is coloured blue.

6.5 Monitoring city growth

The rate of urban growth is one of the questions that social scientists, urban planners and decision-makers deal with most frequently. The direct impacts of urban expansion on physical, ecological and social resources have made research on urban sprawl of increased interest (Linehan & Gross, 1998; Leitão & Ahern, 2002).

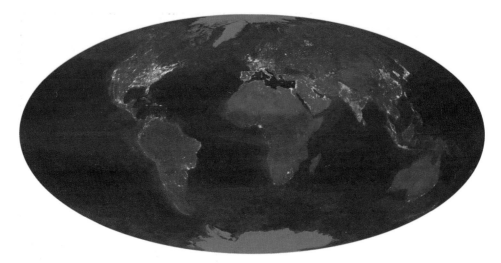

Figure 6.15 This image of Earth's city lights was created with data from the Defence Meteorological Satellite Program (DMSP) Operational Linescan System (OLS). Originally designed to view clouds by moonlight, the OLS is also used to map the locations of permanent lights on the Earth's surface. The brightest areas of the Earth are the most urbanized but not necessarily the most populated. Credit: NASA.

Traditional census sources are extremely useful in that they capture changes in the socioeconomic and demographic structure of cities, but they lack spatial details and are not frequently updated. Remote sensing, on the other hand, makes available a vast amount of data with continuous temporal and spatial coverage, and it can therefore provide a successful means for monitoring urban growth and changes (Masek *et al.,* 2000; Weng, 2002; Xian & Crane, 2005).

Contrary to the majority of satellite observations that are made during daylight hours using reflected solar radiation, visible spectrum imagery acquired at night has been found to have great utility in urban monitoring. Night-time satellite scenes provided by the Defence Meteorological Satellite Program's Operational Linescan System (DMSP OLS) provides a unique view of urban areas around the globe (Sutton *et al.,* 2007). Designed as cloud imagers, the DMSP OLS consists of a pair of satellites stationed in low altitude polar orbits. The two instruments image 14 orbits per day in the visible and thermal spectrum, such that they image the Earth twice every 24 hours with a 3,000 km swath-width. When operated at night, a photomultiplier tube intensifies the visible band signal to enable the detection of moonlit clouds and, by proxy, artificial lighting.

Data products derived from DMSP OLS allow for important advances in the assessment of urban density, economic activity (a variable correlated with the brightness of night-time lights: Balk *et al.,* 2005; Elvidge *et al.,* 2007) and land cover and land use change (Imhoff *et al.,* 1997; Sutton *et al.,* 2001). Night light imagery can be considered in unison with more typical visible day-acquired sensors such as Landsat. Integrated studies rely on remote sensing observations to

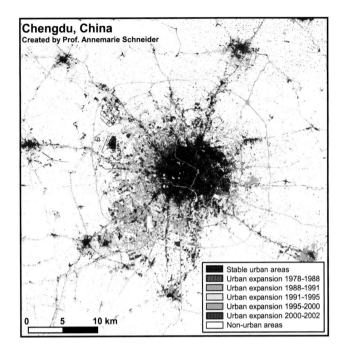

Figure 6.16 Phases of growth of Chengdu, southwest China, for the period of 1978–2002 derived from a time-series of Landsat images. Adapted from Schneider *et al.* (2005) by permission from Pion Publishing Ltd, London, copyright 2005.

provide static freeze-frames of the spatio-temporal pattern associated with urban change. A time-series of sequential snapshots can be used to generate quantitative descriptors of how settlement patterns are changing.

Figure 6.16 shows the 2,000+-year-old city of Chengdu, in the Sichuan Province of southwest China. Chengdu has grown dramatically, particularly on its western margins, since the late 1980s (Schneider *et al.*, 2005; Schneider & Woodcock, 2008). In Figure 6.16, purple depicts areas of the city that have been stable during the period 1978–2002. Growth around this core has been monitored using Landsat in the intervals depicted in the legend. Following reforms initiated in 1990, Chengdu has expanded rapidly, primarily along the lines of the many major roads that radiate from the city centre. As depicted in green, between 1995 and 2000, two concentric ring roads were built to encircle the city.

6.6   Assessing the ecology of cities

Remote sensing of urban environments has traditionally been preoccupied with the mapping of land cover types, without a great deal of regard to the ecology of city environments. As espoused by Ridd (1995) however, the satellite images being used to assess urban composition also contain a great deal of biophysical and ecological information. The advent of high spatial resolution sensors in the last decade has only served to increase the degree to which environmental parameters such as morphology, ecology, energy, moisture and vegetation can be monitored in parallel to land use.

Ridd (1995) addressed the disjoint between ecology and land cover monitoring by breaking the urban environment into three simple but fundamental components that could be repeatedly and accurately mapped from diverse remote sensing instruments: green vegetation; impervious surface material; and exposed soil. The three were selected as having significant impact in urban environments on the dynamics and distribution of energy and moisture (Jensen & Cowen, 1999; Weng, 2009). Furthermore, development of urban centres is accompanied by a change in composition between these fundamental building blocks of the city land cover mosaic. This evolution led to the development of the vegetation-impervious surface-soil (V-I-S) model of Ridd (1995), an approach that has received considerable attention from the remote sensing community (Slonecker *et al.*, 2001; Phinn *et al.*, 2002; Lu & Weng, 2006) and has shown to be more reliable in the urban setting than more traditional vegetation measures such as NDVI (Yuan & Bauer, 2007). The NDVI is one of the several vegetation indices covered in Section 5.2 of this book.

The V-I-S methodology went on to evolve into a base tool for urban energy exchange models and several commonly used energy/water flux models such as *CITYgreen* (www.americanforests.org/productsandpubs/citygreen), which were in turn used for environmental management and urban planning. *CITYgreen* is a software package that conducts complex analyses of ecosystem services and delivers dollar benefits for the services provided by trees and other green spaces in urban environments. The package is raster-based and compatible with industry-standard GIS software. The analysis is based on a land cover dataset, typically derived from aerial photography or satellite, which is provided by the user. The input imagery must be of fine spatial resolution ($<4\,m^2$) and pre-classified into the V-I-S land cover categories.

6.7 Urban climatology

Not to be confused with global warming, an urban microclimate is the distinctive climate of a small-scale area where the weather variables (temperature, rainfall, wind, humidity, etc.) differ from the conditions prevailing for the region as a whole. Though the microclimate effect may be subtle, it is of interest as it affects many people. The clearest local indicator of climate changes due to urbanization is a well-known urban/rural convective circulation known as the urban heat island (UHI) (Bornstein, 1968), which is defined as a dome of high temperatures observed over urban centres as compared to the relatively low temperatures of rural surroundings (González *et al.*, 2007). The difference may be subtle or, in the case of a large city, pronounced and complex.

Compared to surrounding rural areas, the heat island effect typically raises North American summer urban temperatures by 1 °C to 5 °C. This difference is as extreme as 10 °C in Mexico City, Mexico (Akbari *et al.*, 1992). Local urban warming heats the populace, who consume vast amounts of energy powering air conditioners, which in turn emit heat and atmospheric pollutants, compounding the problem of city heat stress. In winter, weak urban planning and inappropriate architecture in certain cities often create efficient wind tunnels that allow bitterly cold winds to pick up speed and amplify the wind chill. Energy is again consumed to bring buildings up to a comfortable temperature, with an associated production of greenhouse gases.

Surfaces that absorb electromagnetic energy from the solar spectrum heat up. It is a common misconception that the wavelength of radiant energy must be in the thermal infrared to initiate heating; absorbed energy in the visible spectrum will also cause heating. Earth's surfaces typically absorb shortwave energy (e.g. visible light) and then re-radiate it at longer wavelengths in the 8,000 nm to 12,000 nm band. Earth's atmosphere is largely opaque to energy of these wavelengths but there is a narrow band, between 10,400 nm and 12,500 nm, where the atmosphere displays reasonably high transmittance (an atmospheric window).

It is this region, sandwiched between the atmospheric absorption bands of O_3 and CO_2, that Landsat's Thematic Mapper band 6 is situated. Band 6 produces images that show the relative differences in emitted thermal energy from surfaces of varying composition and orientation. The channel is commonly used to image the temperature differences arising from urban microclimates and also airborne pollution (see next section). The spatial resolution of Landsat imagery in this thermal channel ($120 m^2$) is rather lower than that of the shorter wavelength channels ($30 m^2$), but it is sufficient to resolve patterns of thermal radiance in urban landscapes.

With a 1.1 km spatial resolution at nadir, NOAA's Advanced Very High Resolution Radiometer (AVHRR) provides four- to six-band multispectral data from a series of polar-orbiting satellites. Similar to Landsat, AVHRR has provided a fairly continuous global coverage since June 1979, with morning and afternoon acquisitions available. The sensor has been used extensively to sense the macro-scale thermal structures of urban areas (Roth *et al.*, 1989; Streutker, 2002).

Among the factors that cause the formation of an urban heat island is the replacement of natural vegetation with man-made materials (asphalt and concrete for roads, buildings, and other structures necessary to accommodate growing populations) that have a significantly different energy and water balance (Avissar, 1996). The partitioning of the sun's energy by these materials results in a change from mostly latent heat fluxes (evaporation) to sensible heat flux (absorbance) and into storage (González *et al.*, 2007). The effect is amplified by the presence of many tall buildings that provide multiple surfaces for the reflection and absorption of sunlight, increasing the efficiency with which urban areas are heated. Furthermore, urban surfaces are largely impervious and do not allow infiltration of precipitation into the soil (Weng, 2004). This reduces evaporative cooling and promotes higher air temperatures (as well as increasing the likelihood of flooding).

Less well constrained, yet perceived to be of importance, is the 'urban canyon effect' (Oke, 1982). In the spaces between buildings, the long-wave radiation emitted by the surface at night is absorbed into the walls, resulting in trapped energy and higher temperatures.

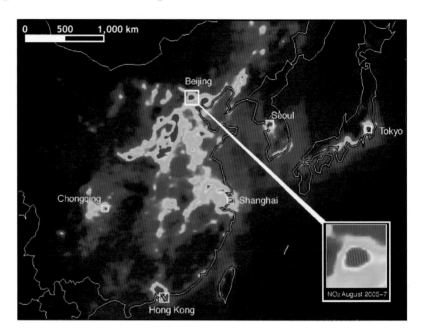

Figure 6.17 Average nitrogen dioxide (NO$_2$) levels for the period August 2005–2007 over eastern China acquired using the Aura OMI (Ozone Monitoring Instrument). High levels of pollution are evident around the major cities of the region. OMI makes daily measurements of the total column of both SO$_2$ and NO$_2$.
Credit: NASA.

City climates have a noticeable influence on plant growing seasons up to 10 km away from a city's edges. Using MODIS data, Zhang *et al.* (2004) showed growing seasons in 70 cities in eastern North America to be about 15 days longer in urban areas than in rural areas outside of a city's influence.

6.8 Air quality and air pollution

Cities are a point source for pollution. By virtue of their induced local climatology (see previous section), pollution in the air may linger for longer than would be the case in a rural setting. This compounds the degradation of air quality to a point that it threatens the well-being of residents (McCubbin & Delucchi, 1999; World Health Organization, 2003). As the polluted air dissipates, prevailing winds spread it over the surrounding countryside. The endurance of this plume is differential as certain gases, such as sulphur dioxide (SO$_2$) and nitrogen dioxide (NO$_2$), which are both major by-products of coal-fired power plants and key ingredients of acid rain, have particularly long lifetimes in the atmosphere (Figure 6.17).

Air pollution emanating from urban and industrial centres takes two common forms: first, that caused by raised concentrations of aerosols and trace gases; and second, the increased prevalence of airborne particles. Both are produced from internal combustion engines and power stations. Automobile exhaust and industrial emissions release a family of nitrogen oxide gases (NO$_x$), methane (CH$_4$) and carbon monoxide (CO), as well as volatile organic compounds (VOC), all by-products of burning gasoline and coal.

As we will see in Chapter 11, high concentrations of tropospheric ozone are a product of photochemical reactions of these precursors. During sunny, high-temperature conditions, NO_x and VOC combine chemically with oxygen to form ozone. Such pollution is hence most acute in cities with sunny, warm, dry climates and a large number of motor vehicles.

The health effects of breathing ozone pollution can be immediate, causing wheezing, coughing, asthma attacks, and arguably reducing life expectancy (Wang & Mauzerall, 2004; Ashmore, 2005). Particulate air pollution similarly affects the respiratory system. Since particle size is a main determinant of where in the respiratory tract the particle will come to rest when inhaled, small particles penetrate into the deepest reaches of the lungs and are particularly dangerous (Lifpert, 1994; Pope *et al.,* 1995).

Both particulates and ozone combine with high concentrations of sulphur dioxide from vehicular and industrial emissions to form city smog. This may, in turn, take the more sinister form of photochemical smog through the action of sunlight. This is particularly detrimental to human health, as the constituent chemicals are highly reactive and oxidizing. Photochemical smogs, despite their known dangers, continue to affect the majority of low- and mid-latitude cities for the greater part of the year.

Urban air quality is traditionally assessed via a network of ground stations. As with many environmental parameters considered in this book, this discrete measurement of a spatially continuous variable must be interpolated. The inevitable consequence is a poor assessment of air quality across areas not populated by monitoring stations. Since variations in air quality are complex in time and space, remote sensing is a relevant monitoring technology for several reasons:

- Satellite imagery can provide a complete survey of a city and its surroundings.
- Imagery has the capacity to identify point sources of pollution, together with their distribution.
- Analysis of remote sensing data in combination with GIS can indicate where efforts should be preferably made to decrease the level of pollution.
- This aids further analysis by highlighting relationships that might exist between city features (e.g. roads, power plants, industrial sectors, etc) and the spatial and temporal distribution of air pollution.
- Remote sensing provides a basis for redesigning the location of ground monitoring stations to increase their effectiveness.

The primary hindrance to the application of satellite data is that the instrument must be capable of delivering an accurate audit of air quality at a spatial and temporal resolution that is relevant to the scale of a city. Many of the sensors discussed in Chapter 11 record data for global assessments of atmospheric pollution but are too coarse for urban application. The methods of determining the quality of the air are the same, however, and based on the differential absorption of light by constituents in the atmosphere.

Such spectroscopy assigns concentrations of pollutants based on a known relationship between their prevalence and the absorption of energy at specific wavelengths. For both ozone and sulphur dioxide (and hence, smog, too), the ultraviolet ($\approx 200\,$nm) is the most useful portion of the electromagnetic spectrum, and NASA's Earth Probe Total Ozone Mapping Spectrometer (TOMS) satellite (discussed in Chapter 11) has been used with varying success over cities. With a spatial resolution of $39 \times 39\,$km at nadir and daily coverage, it is feasible to measure urban tropospheric ozone with Earth Probe TOMS (Varotsos *et al.*, 1998), though the instrument is limited by its poor performance under cloudy conditions and the fact that it is not night-capable.

Also UV-based and often used in unison with TOMS, the OMI (Ozone Monitoring Instrument) aboard NASA's Aura satellite maps several key tropospheric pollutants, including nitrogen dioxide (NO_2), sulphur dioxide (SO_2) and ozone (Figure 6.17). Unfortunately, the mean backscattering altitude is located several kilometres above the Earth's surface, so assumptions must be made about the radiative transfer through the lower troposphere, including the boundary layer, which can be polluted, especially over heavily populated areas. Consequently, OMI shows a significant positive bias at polluted locations (Tanskanen *et al.*, 2005; McKenzie *et al.*, 2008).

Though not yet at an operational stage, future technologies involving active laser remote sensing conquer many of the limitations of passive instruments such as TOMS and OMI. Of particular promise is a technique termed 'time-resolved absorption spectroscopy' that uses ultra-short laser pulses in the visible, near-infrared and ultraviolet spectral regions. Unlike the majority of passive remote sensing systems that are optimized for the monitoring of a single gas and rely on scattered solar radiation, this system can simultaneously detect multiple gases in the atmosphere, including trace-gas concentrations such as ozone (Rairoux *et al.*, 2000).

Despite the future promise that lasers may hold, there is currently no dedicated satellite instrument in orbit capable of direct pollution measurements in the lower troposphere. This is a disjunct in satellite technology, as trace gases in the stratosphere are well catered for, albeit with only coarse spatial resolution, by several missions such as the Microwave Limb Sounder (MLC) carried by the Aura platform and the Cryogenic Limb Etalon Spectrometer (CLAES) aboard NASA's Upper Atmosphere Research Satellite (UARS). Other instruments that do image the troposphere, such as the Along Track Scanning Radiometer (ATSR), provide column-integrated aerosol data, but again at too coarse a resolution to be useful for urban studies.

Operated by the Canadian Space Agency and lofted in 1999 aboard NASA's Terra spacecraft, the Measurements of Pollution in the Troposphere (MOPITT) sensor, as the name suggests, does return trace gas information for the troposphere. Although MOPITT was not designed to detect city pollution, and despite its large pixel size ($22 \times 22\,$km) and long revisit time, some progress has been made in its application to the identification of carbon monoxide plumes originating in urban centres (Clerbaux *et al.*, 2008). This passive sensor utilizes the thermal infrared spectrum via a technique termed 'gas correlation radiometry'.

As a result of this sensor gap, remote sensing research into city air quality has had to make do with instruments designed primarily to observe Earth's surface and not its atmosphere. Over the last two decades, a variety of techniques using satellite data with spatial resolutions varying from coarse, such as Meteosat or AVHRR (Kaufman *et al.*, 1990, Holben *et al.*, 1992, Costa *et al.*, 2002, Ignatov & Stowe, 2002, Retalis *et al.*, 2003), to fine, such as SPOT or Landsat (Sifakis & Deschamps, 1992, Sifakis *et al.*, 1998, Retalis *et al.*, 1999, Wald *et al.*, 1999), have come to the fore.

As is evident in a photograph of a polluted city, smog and haze scatters energy in the visible spectrum. If nitrogen dioxide is present in the pollution cloud, absorption by this gas will give the plume a yellow-brownish tinge (Waggoner & Weiss, 1985). As such, urban air pollution can be quantified in terms of atmospheric turbidity which can be derived through the radiometric comparison of two (or more) satellite images of similar geometry acquired under dissimilar atmosphere and pollution conditions. Ideally, one scene, the reference image, should be acquired on a day with little to no pollution.

While this method of estimating 'aerosol optical thickness' using time-separated imagery is reliable over water, it is problematic over land because the ground's albedo can change for reasons unrelated to the atmosphere, such as seasonal fluctuations in vegetation, or land cover alteration, etc. These activities are common in urban areas which witness a constant cycle of tree cutting and replanting, and demolition and construction.

One solution to this problem is to consider paired images acquired in quick succession in order to minimize error from land cover changes. Retalis *et al.* (1999) followed this route using analysis of the blue and green bands of paired Landsat images to asses the distribution of aerosols over Athens (Greece). An alternative approach is to use textural analysis to filter out the noise arising from temporal variations in ground reflectance between observations. Also using multiple Landsat scenes over Athens, Sifakis *et al.* (1998) describe this procedure. Once filtered, spectral comparison between the visible and infrared bands of time-separated images yielded an estimate of urban sulphur dioxide and smoke that honoured that measured independently by ground stations.

A similar solution, also over Athens but using SPOT imagery, delivered favourable results (Sifakis & Deschamps, 1992), proving the strategy to be transferable between sensors. In an alternative tact using Landsat, Wald *et al.* (1999) used a thermal signature as a means of assessing urban air pollution. The radiance of Landsat band 6 (120 m^2 pixels) was shown to be highly correlated with the amount of airborne black particulates. It is postulated that sulphur dioxide would also be correlated with Landsat band 6, but the radiometric resolution of the instrument is likely too coarse to yield reliable data.

6.9   Climate change as a threat to urbanization

The impacts and economics of climate change are particularly acute in cities. Here, people live at high densities and a perturbation to the norm, such as increased flooding, impacts a great number of people simultaneously. Considering that the majority of Earth's population now live in cities, and the majority of these are within a few kilometres of the ocean, sea level rise represents a significant threat. A rise of one metre, roughly equivalent to the rise predicted by several hundred years of the thermal expansion of seawater – or a considerably lesser period if continental ice sheets melt – is more than sufficient to swamp cities all along the U.S. eastern seaboard. A 6-metre sea level rise would almost entirely submerge Florida.

Even a subtle rise in sea level becomes threatening when a storm surge is superimposed upon it. This results when water is pushed toward shore by a storm's various forces. As discussed in Chapter 9, surge height depends on the storm's forward speed, its wind speed and central pressure. The surge propagates shoreward as a dome of water that may be several metres above normal sea level and 50–100 km in diameter.

The surge preceding category-4 hurricanes may exceed 5 m. Certainly this was the case when Katrina made landfall on the US Gulf Coast in 2005. Here, New Orleans was particularly vulnerable since it was low-lying and, as we have seen earlier in this Chapter, large-scale subsidence was acting to exacerbate relative sea level rise (Dixon *et al.*, 2006; Figure 6.10). Despite formidable flood defences, the combination of surge atop an already rising sea level and a subsidence-weakened levee system served to allow inundation of the city, with disastrous impact on life and property.

Thankfully, slow-moving hurricanes have a tendency to make landfall during low tide, since they are influenced by atmospheric pressure changes induced by tidal movements, and this usually somewhat lessens the devastating impacts of storm surge.

Many cities are particularly at risk from changing climate patterns due to their having been sited originally to capitalize on a particular property of the landscape, examples including being adjacent to rivers or estuaries for transport, or proximity to agricultural lands. Rising sea levels and flooding obviously jeopardize urban centres, but so does a shift towards aridity.

Increases in rainfall variability resulting from changes in global climate can rapidly reduce productivity and alter the composition of forests, grasslands and agricultural land. Prolonged drought conditions promote aridity and can quickly swing once fertile lands into the infamous dustbowl conditions that depressed the American economy in the 1930s. The modern day West African Sahel region is succumbing to a similar fate (see Chapter 5), as is China, whose land mass is already about 25 per cent composed of deserts. China is undergoing desertification at such a rate that millions of acres of land are engulfed annually. The capital, Beijing, is subjected to millions of tonnes of sand each year that frequently reduces visibility to a point such that soaring skyscrapers can barely be seen and air traffic is grounded.

Recent work using remote sensing also demonstrates a link between rapid city growth and altered rainfall patterns. Kaufmann *et al.* (2007) showed that as Chinese cities get bigger, there is a negative impact on precipitation patterns, such that in the winter season there is a reduction in rainfall as an effect of urbanization. This can primarily be attributed to the conversion of vegetated land to asphalt, roads and buildings. As a result, the soils have significantly less ability to absorb water, so in the winter months there is less moisture in the atmosphere and therefore a reduction in precipitation.

6.10 Key concepts

1 Immigration into cities is accelerating and they are expanding. This rapid growth is termed 'urbanization' and is accompanied by unprecedented levels of land use change. Construction radically reduces the land's albedo as formerly vegetated areas become paved. This results in a net warming of cities, termed an 'urban microclimate'. In addition, intensive energy consumption, industrial growth and the expansion of transport networks lead to greatly increased production of greenhouse gases.

2 As urban centres are spatially complex, the most important feature for a remote sensing instrument to successfully monitor them is a high spatial resolution. As a result, sensors that offer metre-sized pixels have been favoured, such as IKONOS, Quickbird, and WorldView. A high spectral resolution is also useful for work in the spectrally diverse urban environment, but no single spaceborne sensor can satisfy the necessary spatial resolution while offering a high number of spectral bands. Airborne instruments can be used to fill this gap.

3 Even with metre-scale pixels, the degree of spectral variability may be so great in an urban environment that traditional spectral-based classification algorithms are of little use. This restriction is exacerbated by the difficulty in defining spectrally homogeneous training areas with which to drive a classifier. In such cases, textural differences between land use classes can be used to improve urban mapping.

4 Urban remote sensing strives to capture the three-dimensionality of cities, and three key technologies have emerged as particularly useful: stereo imaging, LiDAR and imaging/interferometric radar.

5 Remote monitoring of urban air quality is hampered by lack of a dedicated sensor to retrieve trace gas concentrations in the troposphere. Current solutions rely on coarse-resolution instruments capable of returning a 'column averaged' (i.e. stratosphere plus troposphere) assessment of pollution load, or the analysis of time-separated images captured by higher spatial resolution satellites in the visible and infrared wavelengths.

7

Surface and ground water resources

Surface waters encompass streams, rivers, and lakes, which are all easily monitored through EM remote sensing. More challenging to sensing is subsurface water, which is defined as water stored below Earth's surface but above the water table (e.g. soil moisture), and groundwater, a term describing waters stored below the water table. Shallow-depth soil moisture may be monitored via passive microwave sensors but groundwater, being situated at considerable depth beneath the ground, is invisible to all but gravitational remote sensing data.

Though several of today's satellites have the necessary spatial resolution necessary to examine processes occurring at groundwater seeps, rivers, and small lakes, they are hampered by a lack of spectral resolution. Conversely, instruments that are spectrally powerful are inappropriate by virtue of their large pixel sizes. In saying this, numerous sensors do exist that have found application for the investigation of ground and surface waters (Table 7.1).

The global water cycle describes the continuous movement of water on, above and below the surface of the Earth. The cycle consists of reservoirs and fluxes of moisture moving from one reservoir to another. The major fluxes that complete the cycle are evaporation; condensation; precipitation; infiltration; percolation; transpiration; and run-off.

Storage of water within some reservoirs have residence times of only a matter of days, as is the case for the atmosphere and rivers, while water molecules held within lakes or groundwater may reside for several years before moving to an alternative reservoir. Once incorporated within an ocean or polar ice sheet, residence times extend beyond thousands of years.

Human activity increasingly controls run-off and surface water storage and serves to alter the global water cycle. The thermodynamic effect of climate change has a particularly pronounced impact because, among many other effects, it decreases snow cover at high latitudes, alters global precipitation patterns and increases evaporation to the point of inducing desertification in arid regions (Lawford, 2008).

Remote Sensing and Global Environmental Change, First Edition. Samuel Purkis and Victor Klemas.
© 2011 Samuel Purkis and Victor Klemas. Published 2011 by Blackwell Publishing Ltd.

Table 7.1 A list of spaceborne sensors that report data for potential use in the study of terrestrial surface and ground waters (after Becker, 2006).

Sensor	Launch year	Spatial resolution (m)	Precipitation	Surface temp.	Soil moisture	Water storage	Snow water	Land cover	Topography
AMSR-E	2002	5,400–56,000	✓	✓	✓		✓		
ASTER (GDEM)	1999	15, 30, 90		✓	✓		✓	✓	✓
AVHRR	1991–2003	1,100	✓	✓					
GRACE	2002	300,000				✓			
ENVISAT-RA2	2002	≈1,000		✓				✓	✓
Landsat ETM+	1999	30, 60		✓				✓	
MODIS	1999	250, 500, 1,000		✓				✓	
OrbView-2	1997	1,100		✓				✓	✓
OrbView-3	2003	1, 4		✓					
RADARSAT-1	1995	8–100			✓				
SRTM	2000	30, 90							✓

With the continued availability of adequate and safe water supplies being central to national security and international harmony, changes in the water cycle are of concern for society (Gleick, 1993; Levy, 1995). Any perturbation plays a central role in the sustainability of the world's water resources. In order to effectively monitor change in these systems, it is necessary to have datasets that are both global and long-term. Remote sensing has yet to provide the reliability of *in situ* records, but is now used to parameterize the majority of Earth's water reservoirs and associated fluxes. To this end, the Integrated Global Water Cycle Observations (IGWCO) is a key theme within the Global Earth Observation System of Systems (GEOSS), an international programme discussed in more detail in Section 3.5.2 of this book.

7.1   Remote sensing of inland water quality

Suspended and dissolved substances in surface water change its colour. Clear water is blue, water rich in aquatic humus is yellow and the colour of turbid water depends on the mixture of the constituents and can vary from dark blue-green via bright green and brown to red. The colour is determined by the light scattered out of the water and that reflected at its surface. Since light that originates from below the water surface shows characteristic influences of the diverse colours of components in the water, optical remote sensing can be used to appraise water quality.

Clear water reflects very little solar irradiance, so its overall reflectance is generally low. Reflectance is maximum at the blue end of the spectrum and decreases as wavelength increases (Figure 7.1). Turbid water is capable of reflecting significant amounts of sunlight and has a higher overall reflectance. Algae-laden waters contain a high concentration of chlorophyll. By virtue of the strong absorption of this pigment at 450 nm and 670 nm, the greater the chlorophyll concentration, the lower the reflectance in the short wavelengths (blue light) and the higher the green light reflectance. As demonstrated by Figure 7.1, the spectra of clear and algae-laden water fall into concert beyond 750 nm. This is because, at these longer wavelengths, the absorption of light by water is more dependent on the water molecule itself and less on its constituents.

With the near-exponential attenuation of electromagnetic energy by water with increasing wavelength, the useable spectrum for water quality work is restricted to the visible range and, as demonstrated by Figure 7.1, maximum differentiation exists in the blue and green spectrum. For this reason, tried and tested techniques for monitoring chlorophyll content on land using the visible and infrared spectrum (e.g. NDVI – Chapter 5) are redundant over water. Inland waters are optically more complex than oceanic for several reasons:

- Many inland water bodies contain, in addition to chlorophyll, colour-producing agents such as suspended sediments and dissolved organic matter. Because of these characteristics, they are referred to as 'Case 2'. This concept

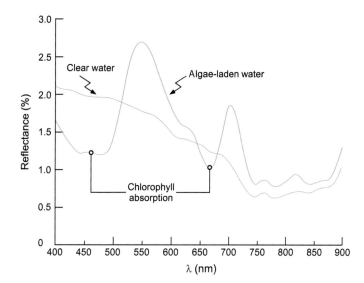

Figure 7.1 Reflectance characteristics of clear (blue) and algae-laden (green) water.

of characterizing water-types based on their optical properties is covered further in Chapter 8 (Section 8.4.2).

- These suspended sediments and dissolved organic matter also contribute to the overall reflectance of a water body, making it difficult to determine chlorophyll concentration in remotely sensed data. To determine chlorophyll concentration in inland water bodies, the absorption and scattering characteristics of suspended sediments and dissolved organic matter must also be accounted for. Absorption diminishes light intensity and scattering changes its angular distribution. The intensity of reflected light increases with the amount of scattering and decreases by absorption.

- Inland waters are often shallow enough that the lake bed contributes to the overall reflectance of the overlying water column. In such cases, the relative contribution of the substrate and water body has to be decoupled to obtain viable information about the characteristics (chlorophyll, turbidity, etc.) of the water. This is the inverse of the problem of mapping the seabed as covered in Section 8.4.3, but similar solutions apply.

- Case 2 inland waters can be highly turbid to the point that light only penetrates a few metres into the water column. Consequently, the technology only reports on the status of the surface or a very small fraction of the overall water body. This limitation becomes particularly pronounced when the water column is stratified, such that the upper portion is not representative of the bulk properties.

- Rivers and lakes are temporally very dynamic and demand repeated monitoring on short timescales at high spatial resolution. This is a tall order for satellite systems which, if they possess high spatial resolution, are limited in their repeat cycle by narrow swath-widths.

Table 7.2 Water quality parameters of lakes and rivers and useful wavelengths for their monitoring.

Parameter	Wavelength (nm)
Turbidity	705
Total suspended sediment	700
Volatile suspended solids	705
Chlorophyll-a, -b, -c	684 & 700

Because the spectral region useful for discerning water quality parameters is relatively narrow, remote sensors with numerous narrow spectral bands pooled into these wavelengths are more capable than instruments on many satellites, which have only a handful of rather broad bands in the visible region. As documented in Chapter 10, satellites that do have near hyperspectral capability, such as SeaWIFS, MODIS, and MERIS, have spatial resolutions of hundreds to thousands of metres and are hence better suited to work in the open ocean than inland water bodies.

In confined inland waters, the problem of terrestrial contamination in pixels also becomes prohibitive. Satellites with resolutions of metres to tens of metres, such as QuickBird, IKONOS, SPOT, and Landsat, have attractive pixel sizes but are spectrally limited in the blue and green portions of the spectrum. This disjoint has promoted the use of hyperspectral airborne sensors, which boast both high spectral and spatial resolution, as the cutting-edge technology for monitoring inland water quality (e.g. Hoogenboom *et al.,* 1998a; 1998b; Koponen *et al.,* 2002). Parameters related to water colour are routinely used to study environmental processes such as the primary production of biomass and the distribution of polluted suspended matter (Table 7.2).

7.2 Remote sensing sediment load and pollution of inland waters

In rivers and lakes, an accurate measure of suspended sediment is particularly important as it moderates fluxes of heavy metals and pollutants, such as polychlorinated biphenyls (PCBs), that become chemically bound to sediment particles (Dekker *et al.,* 2002). If the suspended sediment is organic in origin, it can also serve to control total primary production of the water body. Furthermore, suspended matter deposited in reservoirs reduces their storage capacity, minimizes flood control and also reduces the light penetration in water, thus minimizing the fish production (Li & Li, 2004). Over time, human-induced stresses such as pollution, sediment accumulation and the introduction of exotic biology all serve to disrupt the ecological balance of lakes and rivers.

Just as suspended sediments in coastal waters alter the colour of the water body, sediment load in rivers and lakes serves to change the colour to yellows, browns and reds. As previously noted, satellites with powerful hyperspectral capability for remotely discriminating concentrations of both sediments and chlorophyll in the oceans possess too coarse a spatial resolution to be of use in anything but the largest lakes.

However, this is not to say that satellites with pixels of metres to tens of metres and only a handful of water-penetrating bands are completely redundant. Provided that *in situ* measurements are available for calibration, regression analysis can be used to derive useful water quality parameters such as Secchi disc depth, chlorophyll-a and total suspended sediment concentration. This strategy proceeds under the assumption that the reflectance spectrum of water is, to a degree, dependent on these parameters. A model is developed through multiple regressions of field-sampled water quality measurements versus satellite-derived reflectance values. The technique thus levers the vast spatial coverage delivered by the remote sensing instrument to extrapolate a small number of field measurements to a large area.

Such regression-based models have been successfully developed for instruments such as Landsat and SPOT (Dekker *et al.*, 2001a; 2002; Kloiber *et al.*, 2002a), IKONOS (Sawaya *et al.*, 2003; Ekercin, 2007) and Hyperion (Giardino *et al.*, 2007). These advances are timely because, for rivers and lakes, ground-based monitoring is poorly poised to capture the dynamic patterns of water quality that vary over short timescales but vast areas.

Satellite monitoring is, however, not without its limitations, and a combination of these mean that remote sensing cannot exclusively replace on-water monitoring efforts. Disadvantages include:

- Satellite-based measurements often require more ground information than is available (Chipman *et al.*, 2004).
- The ≈2-week repeat cycle of most passive visible satellite systems is a limiting factor for monitoring temporally dynamic phenomena.
- The period appropriate for monitoring water clarity is typically short and remote sensing depends on clear skies – a cloudy overpass may mean no data for that month. Thus remote imagery cannot be used reliably to monitor short-term trends in water clarity. Satellite systems are, however, well suited to provide broad spatial coverage and 'coarse' long-term estimates of condition.
- Water quality parameters such as chlorophyll vary greatly on short timescales, such that well-fitting predictive relationships have to be based on ground measurements acquired within a few hours of the satellite overpass. This imposes restrictions on the use of historical image time-series (Li & Li, 2004).
- Since the remote differentiation of water quality patterns is conducted at the limits of sensitivity for sensors such as Landsat, SPOT, and IKONOS (which were primarily designed for terrestrial mapping), precise correction for atmospheric influence is demanded. This may require measurement of atmospheric status coincident with the time of satellite overpass, further restricting the use of historical image-sets.

- Even with rigorous ground calibration and attention to atmospheric correction, popular satellites such as Landsat, SPOT, and IKONOS do not provide adequate spectral resolution. In many cases, studies that have documented the ability to quantify multiple bio-optical properties in inland waters have relied upon airborne hyperspectral sensors with very fine spectral resolution (Hoogenboom *et al.*, 1998ab; Koponen *et al.*, 2002; Li & Li, 2004).
- Remote sensing estimates of water quality may not be precise enough to satisfy federal assessments.
- Particularly at the high spatial resolution such as provided by IKONOS, areas of a lake or river that are shadowed deliver poor retrievals of water constituents (Sawaya *et al.*, 2003).

Unlike suspended sediment and chlorophyll that have a pronounced spectral signature in the visible spectrum, many pollutants are invisible, persisting at very low but still extremely harmful concentrations. With the rapid growth of cities (Chapter 6), and the massive quantities of effluent and sewage that have been introduced into rivers in ever increasing quantities since the industrial revolution, the need for effective water quality measurements of pollutant load has become acute.

Pollutants without a spectral signature can be monitored by indirect methods, such as imaging the effects of water contamination on the vegetation of river banks through time using indices like NDVI (Suhong *et al.*, 2004). As we discussed in Chapter 5 (Section 5.2), such an approach demands only that the sensor should have bands in the visible red and infrared spectrum and an appropriate spatial resolution.

A second indirect method is to examine the effect that the pollutant has on the water body, a task made particularly easy if the contaminant is an agent for eutrophication. This is a common phenomenon in both marine and freshwater ecosystems, where nutrient enrichment by compounds containing high concentrations of phosphate, ammonia, and nitrate cause a massive, and often unsustainable, increase in phytoplankton biomass (Figure 7.2). Such 'organic pollution' can result from many factors, the most common being the excessive input of fertilizers from agriculture farming; food for aquaculture; untreated and/or treated sewage; or industrial wastewater inputs. Impacts of eutrophication range from fish kills to direct damage to human health through blooms of harmful algae (covered in detail in Chapter 10). Blooms of phytoplankton degrade the water quality by accelerating the growth of organic matter and decomposition, as well as decreasing the light availability.

Monitoring of eutrophication can be achieved remotely using the same techniques as previously described that utilize the visible spectrum to calibrate known reflectance characteristics to water quality parameters (Zilioli *et al.*, 1994; Oyama *et al.*, 2009). As always, to maximize the usefulness of satellite data for the monitoring of eutrophication, it is often essential to obtain *in situ* data almost simultaneously with the satellite acquisition.

Remote sensing can further be used to predict and quantify the intensity of runoff for a particular landscape and hence estimate the quantity of polluting runoff that will make its way into a river or lake. Relevant factors in this regard are the permeability of the ground, soil type, the topography of the watershed and its dominant vegetation types.

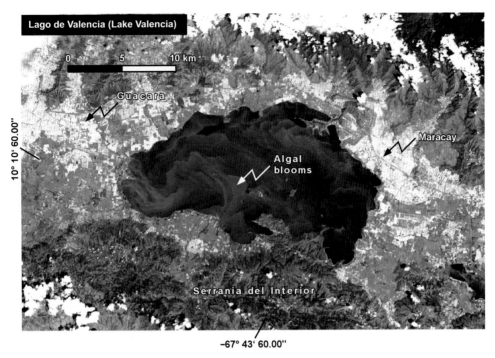

Figure 7.2 Lago de Valencia is the third largest lake in Venezuela. Despite a picturesque setting in the Maritime Andes, the lake suffers from algal blooms caused by continual influx of untreated wastewater from surrounding agricultural areas and towns such as Guacara and Maracay. Such a bloom has turned the lake waters a vivid green in this true-colour astronaut photograph (ISS010-E-5194), acquired on October 27, 2004 by the crew of the International Space Station. Credit: NASA.

7.3 Remote sensing non-coastal flooding

Altered patterns of precipitation are synonymous with global environmental change. While some areas of Earth, such as the southwestern North America and southern Europe, have become anomalously dry in the last decades (Breshears *et al.*, 2005; Piñol *et al.*, 1998), others, such as the tropics, appear to be experiencing increased rainfall (Gu *et al.*, 2007). In high-altitude regions such as the Himalayas, even when precipitation remains static, lake volumes are nonetheless increasing due to melting glaciers; also, rising temperatures are causing precipitation that would previously have fallen as snow to be more likely to fall as rain (Richardson & Reynolds, 2000).

Whether through rainfall or runoff, rivers swell under the burden of these massive quantities of water and the threat of flooding increases. On a geological timeframe, catastrophic flood events may be comparatively common, but human preference to site cities adjacent to major rivers, often on land within known historical flood plains, leads to ruinous events such as those that repeatedly play out in the Ganges river delta in Bangladesh.

Remote sensing non-coastal flooding differs from remote sensing water quality in that a simple partitioning of image pixels into water presence/absence may be sufficient to yield informative data. This lesser requirement widens the breadth of

Figure 7.3 A pair of Landsat 5 infrared (band 5) images depict the confluence of the Mississippi and Missouri rivers north of the town of St. Louis. The top image, acquired in 1991, shows the rivers at their normal level. Storm events in early 1993 caused massive flooding of the area (lower image). In both scenes, the strong absorbance of infrared light by water results in a near zero reflectance such that submerged areas appear black. Credit: NASA.

the electromagnetic spectrum such that the infrared and microwave wavelengths become useful despite the fact that, unlike the visible bands, they do not penetrate the water surface and hence are incapable of returning information about water quality. Figure 7.3 depicts the pronounced difference in spectral response of water versus land in the infrared spectrum in a pair of scenes captured prior to and shortly after the 1993 flooding of the Upper Mississippi River Basin – an event that caused between US$ 12 and 16 billion worth of damage (Hipple *et al.*, 2005).

Unlike the infrared spectrum, microwave remote sensing has the considerable advantage of providing all-weather imaging, a capability not to be underestimated in the cloudy skies that accompany extreme rain events. As shown in Figure 3.4 (Chapter 3), some microwave data allow measurements of flood water even obscured by vegetation (Ormsby *et al.*, 1985; Muller *et al.*, 1993).

Whereas the infrared is typically sensed with passive remote sensing instruments, microwave sensors with spatial resolutions amenable for monitoring rivers and lakes tend to be active. Current examples include the Advanced Synthetic Aperture Radar (ASAR) instrument onboard the ENVISAT satellite, which succeeded the Active Microwave Instrument (AMI) Synthetic Aperture Radar (SAR) instruments flown on the European satellites ERS-1 and ERS-2; Canada's next-generation RADARSAT-2; and the Japanese JERS-1 SAR.

Since microwaves display specular (mirror-like) reflection off water, the scattering is weak and submerged areas appear black on SAR images. By contrast, scattering from land is strong, so that a simple segmentation threshold may be sufficient to reliably separate emergent from submerged areas. However, active microwave radar sensors have the disadvantage that, because they are side-looking, shadows created by topography in mountainous areas may also return little to no scatter and hence could be misinterpreted as water. To preclude confusion, analysis of SAR data can be aided by the inclusion of high-resolution visible/infrared imagery and/or a digital elevation model. The SAR technology is discussed in detail in Chapter 6 (Section 6.2.4).

7.4 Bathymetry of inland waters

As will be discussed in Chapter 8, various algorithms have been developed for bathymetric mapping in the shallow coastal zone. In the multispectral case, these operate on the predictable but differential extinction of blue and green light by water. Hyperspectral remote sensing of bathymetry employs more advanced strategies such as optimization, spectral unmixing and look-up tables. Despite frequent application in the marine environment, it is only recently that the approach has been extended to investigate inland water systems (Fonstad & Marcus, 2005; Legleiter & Roberts, 2005; 2009).

As is the case for the coastal zone, any accurate solution must solve the classic problem of decoupling the remote sensing signal arising from the absorption of light by the water column from that of the underlying substrate. The limitations of the technique are the same as for marine application. Specifically, when depth varies on a sub-pixel scale, only one depth estimate can be assigned to each image pixel. The problem is particularly acute when the pixel size is large relative to the water body in question. Similarly, at the periphery of rivers and lakes, pixel radiance is a function of both the water body and its terrestrial bank (Legleiter & Roberts, 2005). Such mixed pixels may be difficult to detect and mask, leading to inaccurate depth estimates in the near shore.

Early application of bathymetric mapping in rivers was based on empirical solutions, whereby field measurements of depth were correlated with image pixel values (Lyon & Hutchinson, 1995; Lyon *et al.*, 1992; Winterbottom & Gilvear, 1997). These studies clearly demonstrated the proof of concept but did not yet delve into the underlying physical processes governing the interaction of light with the water column and substrate. Consequently, the depth products delivered were case-, scene- and sensor-specific and not yet mature enough for regional-scale application.

Later work conducted by Legleiter & Roberts (2005, 2009) built upon advances in the marine realm, and radiative transfer modelling (RTM) was used in the context of depth derivation for shallow stream channels. Unlike the earlier empirical algorithms, RTM is a physics-based approach and involves the quantification of the relative effects of the water column (as determined by its inherent optical properties; see Chapter 8, Section 8.4.2), the streambed substrate and the stream's surface upon the upwelling spectral radiance recorded by the remote sensing instrument. As is favoured by the marine community, the RTM modelling of river optics is commonly implemented using a computer model such as *Hydrolight* (Mobley & Sundman, 2001; www.hydrolight.info). Since the technique has a sound physical basis, it can be implemented at a large scale across multiple satellite scenes, including application to historical image datasets, thereby allowing changes in riverbed morphology to be quantified through time.

LiDAR, a predominantly airborne technology that is considered frequently in this book, is useful for mapping bathymetry of inland waters (Kinzel *et al.,* 2007; Hilldale & Raff, 2008; Marcus & Fonstad, 2008). Although, like the previously covered photogrammetric techniques, LiDAR is subject to water transparency limitations, it is attractive in being able to map large areas at relatively low cost – and, unlike boat-based methods, it is not limited to navigable rivers (i.e. those with deep and quiet waters). The fact that inland waters are typically shallow presents similar challenges to when bathymetric LiDAR is used in its more usual role, which is in the coastal zone. Furthermore, since many inland water bodies are small, it is highly desirable to map bathymetry in concert with the surrounding terrestrial terrain. By doing so, bathymetry can be considered in the overall context of the surrounding catchment and watershed. This is similarly useful in the coastal zone and will be discussed in that context in Section 9.4 of Chapter 9.

7.5 Mapping watersheds at the regional scale

Regional-scale watershed maps are a precursor to the analysis of many aspects of global environmental change, ranging from conservation planning to quantifying the global freshwater hydrological cycle. As outlined by Lehner *et al.* (2008), there exist many hydrographical maps of well-known river basins of individual nations, but there is a lack of seamless high-quality data on large scales, such as on continents or the entire globe.

A watershed describes the drainage basin or catchment area for a parcel of land. Watersheds drain into other watersheds in a hierarchical form, larger ones breaking into smaller ones or sub-watersheds, with the topography determining where the water flows. To reconstruct the geometry of a watershed system and its connections, it is vital to have a sound understanding of both the geomorphology and land cover of the area in question. Section 5.5 of this book describes various global vegetation and land cover mapping programmes relevant in this regard.

At a small scale, insight into geomorphology is relatively easily achieved using a digital elevation model (DEM) derived, for example, from an airborne LiDAR survey (see Sections 6.2.3 and 9.4). At the regional to global scale, such datasets do not exist and, until recently, assimilation of a DEM with sufficient accuracy for analysis of watersheds was not possible.

Three Earth-scale topography remote sensing programmes are relevant in regard to the reconstruction of global watershed maps:

- NASA's Shuttle Radar Topography Mission (SRTM) (Farr & Kobrick, 2000). The SRTM obtained elevation data on a near-global scale to generate a

Death Valley, California
September 24th 2003

0   5   10 km

Furnace Creek
Ranch

Figure 7.4 A 3-D view of Death Valley (California, USA) created by draping a 2003 simulated-natural-colour ASTER scene over GDEM topography. Credit: NASA.

high-resolution digital topographical database of Earth's surface between 60° N and 60° S latitudes. It consisted of a specially modified radar system that flew onboard the Space Shuttle Endeavour during an 11-day mission in February 2000. The resolution of the cells of the source data is one arc second (approx. 30 metres), but 1″ data have only been released over United States territory; for the rest of the world, only three arc second data (approx. 90 metres) are available. The SRTM product is available for download through the Mission's website (www2.jpl.nasa.gov/srtm).

- The ASTER Global Digital Elevation Model (GDEM) is a joint venture between NASA and Japan's Ministry of Economy, Trade and Industry (METI). Released in June 2009, the GDEM was created by stereo-correlating the 1.3 million scene ASTER VNIR archive, covering the Earth's land surface between 83° N and 83° S latitudes. The GDEM is produced with 30 metre postings and is formatted in 1° × 1° tiles as GeoTIFF files. Each GDEM file is accompanied by a quality assessment file, either giving the number of ASTER scenes used to calculate a pixel's value or indicating the source of external DEM data used to fill the ASTER voids. The GDEM is free of charge and available for downloading (http://asterweb.jpl.nasa.gov/gdem.asp). Figure 7.4 shows an ASTER simulated-natural-colour image draped over digital topography from the GDEM.

- The pairing of identical synthetic aperture radar satellites TerraSAR-X and TanDEM-X. Operated as a public/private partnership between the German space agency (DLR) and EADS Astrium, TerraSAR-X was launched in 2007, with TanDEM-X lofted in 2010 to sit in a tight helical orbit around its more established sibling. The two spacecraft come within 200 m of one another as they wind themselves along their respective orbits. Infoterra GbmH, a German subsidiary of Astrium, has exclusive rights to commercialize the data arising from the TerraSAR-X/TanDEM-X venture (by comparison, SRTM and GDEM are free). It is planned that, by 2013, the pair will have measured the variation in height across 150 million km^2 of the globe to an accuracy of better than two metres. With the use of only TerraSAR-X, a DEM of $10 m^2$ spatial resolution can be created. The pairing with TanDEM-X will increase the fidelity to around $2 m^2$, an order finer than delivered by SRTM or GDEM. The challenge facing the operators of TerraSAR-X/TanDEM-X over the next three years is that the pair produce vast quantities of data. To be used, these all need to be downlinked to Earth via ground stations. At the time of writing, the present infrastructure is reaching its limits and will need considerable investment to attain the necessary downlink capacity. Like the SRTM program, these German SAR satellites are unhindered by cloud, fog, and haze. The GDEM, which is based on stereo-ASTER imagery, demands a clear view of the Earth.

Since it was released only recently, it is too early to evaluate the full utility of the ASTER GDEM, but given time it will no doubt be used a great deal in the context of understanding Earth's geomorphology. Most importantly, the GDEM extends 20° of latitude further north and south than the older SRTM elevation data. Taking an alternative view, the GDEM covers >98 per cent of the globe as opposed to the ≈85 per cent achieved using the SRTM. The GDEM also serves to fill in areas, such as very steep terrains and in some deserts, where the Shuttle mission failed to acquire meaningful data.

Since the SRTM elevation data has been available for nearly a decade, it represents a well-established technology for regional-scale watershed mapping. The most ambitious such project is HydroSHEDS (Hydrological data and maps based on SHuttle Elevation Derivatives at multiple Scales (Lehner *et al.*, 2006), developed by the Conservation Science Program of the World Wildlife Fund. This provides georeferenced datasets (vector and raster), including river networks, watershed boundaries, drainage directions and flow accumulations. The data are available through the project's website (http://hydrosheds.cr.usgs.gov/) at three resolutions: 3, 15, and 30 arc seconds (approximately 90 metres, 500 metres, and 1,000 metres at the equator).

As outlined by Lehner *et al.* (2008), a problem confounding SRTM data is that regions exist with data gaps (voids). These omissions are particularly prevalent over large water bodies and in areas where radar-specific problems prevented the production of reliable elevation data (e.g. radar shadow in mountainous areas). In such cases, interpolation or ancillary data are relied upon to fill the voids with reliable elevation values. This step is critical if a watershed model such as HydroSHEDS is to emulate continuous flow across a catchment area. Figure 7.5 shows a section of the HydroSHEDS database for a small section of the Amazon basin.

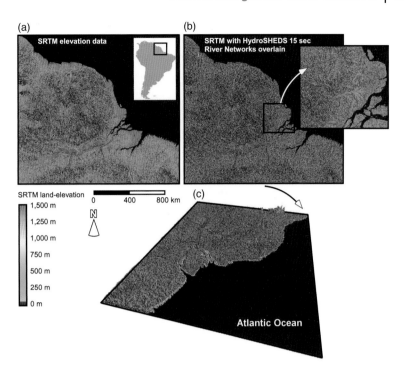

Figure 7.5 SRTM topography data (a) overlain by the HydroSHEDS river network for a 1,670 × 1,670 km segment of the Amazon Basin (b). (c) shows the dataset in 3-D. The resolution of the HydroSHEDS dataset spans scales of thousands of km² to 1 km². Data retrieved from http://hydrosheds.cr.usgs. gov/. For general information see www.worldwildlife.org/ hydrosheds and Lehner *et al.* (2008). Credit: WWF & USGS.

7.6 Remote sensing of land surface moisture

Terrestrial water storage (TWS) consists of groundwater; soil moisture and permafrost; surface water; snow and ice; and wet biomass. TWS variability tends to be dominated by snow and ice in polar and alpine regions, by soil moisture in mid-latitudes, and by surface water in wet, tropical regions such as the Amazon (Rodell *et al.*, 2004).

The mid-latitudes are particularly vulnerable to drought, where prolonged dry weather leads to a radical reduction in soil moisture. Southern Europe, the Iberian Peninsula in particular, has suffered recurrent droughts in the last decades, with rain falling during few but intense events (Martin-Vide, 2004). Because of limited access to irrigation, agricultural production in the region is highly vulnerable to this erratic climate. To the south, Section 5.6 considers the case of the Sahel, a mid-latitude semi-arid belt that straddles Africa from the Atlantic to the Red Sea. The Sahel is also particularly susceptible to drought, with a chronic rainfall shortage persisting from the late 1960s to the early 1990s, at which point there was a switch to anomalously high rainfall (Anyamba & Tucker, 2005).

The intensity and distribution of precipitation in both southern Europe and northern Africa are likely controlled by the same factors – the interaction of North Atlantic Oscillation and the El Niño Southern Oscillation (Rodó *et al.*, 1997). With global climate change predicted to alter these driving forces, mid-latitude climate will remain unpredictable and the need to monitor terrestrial water storage in the region will be of high priority. As it controls the prevalence and type of vegetation, the effect of soil moisture availability is a key modulator of the terrestrial carbon cycle, as well as an important integrating parameter in the processes responsible for the exchange of water and energy between the land and atmosphere. Aridification of the mid-latitudes is therefore of global significance.

The Earth naturally emits energy in the microwave region of the electromagnetic spectrum. The intensity of the microwave emission from a medium is referred to as its emissivity or brightness. There exists a trend of decreasing microwave emissivity with increasing soil moisture, such that the emissivity of dry soil is relatively high as compared to damp. This difference provides the basis for microwave sensing of soil moisture.

In the past, soil moisture measurements were typically performed from aircraft due the to poor resolution of passive satellite microwave radiometers. However, satellite-based observation has gradually matured into a leading technique for monitoring surface moisture at regional scales (Van de Griend & Owe, 1994; Njoku & Entekhabi, 1996). As discussed in section 10.5 of this book in the context of ocean salinity, the technique has the advantage of providing spatially integrated information over a large extent while providing the ability to repeat observations over a given area at regular time intervals. Microwave measurements have the further benefit of being largely unaffected by cloud cover and variable surface solar illumination. A disadvantage is that accurate soil moisture estimates are limited to regions that have either bare soil or only low to moderate amounts of vegetation cover. There are three primary effects that augment the relationship between soil moisture and soil emissivity and hence limit the accuracy with which soil moisture can be estimated:

- Long wavelength microwave radiation emanates from a greater soil depth than do the short wavelengths, and hence portrays the soil moisture from deeper below the surface. It should be noted that the penetration of microwaves into soil does not exceed a few tenths of a wavelength. With useable wavelengths ranging from ≈5–30 cm, the depth of penetration is a matter of centimetres only.
- Vegetation both absorbs and scatters the emitted soil radiation while also emitting its own microwave radiation. Longer wavelength radiation penetrates vegetative cover to the highest degree.
- Increasing the surface roughness of the soil generally increases its emissivity, because the surface area from which the emission is emanating is increased. Roughness is, however, wavelength-dependent, so a smooth soil surface at a 15 cm wavelength may appear rough to a 5 cm wavelength.

Early microwave sensors which have proved to be useful for retrieving soil moisture have included NASA's Nimbus-7 Scanning Multichannel Microwave Radiometer (SMMR) (Gloersen & Barath, 1977; Van de Griend & Owe, 1994), which operated from 1978 to 1987; and the Special Sensor Microwave/Imager

(SSM/I) (Hollinger *et al.*, 1990), which has provided global passive microwave products from late 1978 to the present. The latter instrument is flown on board the United States Air Force Defence Meteorological Satellite Program (DMSP), a mission considered in more detail in Chapter 6 (Section 6.5).

Launched in 1997, the Tropical Rainfall Measuring Mission (TRMM) is a joint space mission between NASA and the Japan Aerospace Exploration Agency (JAXA) which also caries a passive microwave sensor, the TMI, that can be used to retrieve soil moisture (Narayan & Lakshmi, 2006). The AMSR-E (Advanced Microwave Scanning Radiometer) lofted in 2002 aboard NASA's Aqua satellite also shows promise as a source of soil moisture data (Gruhier *et al.*, 2008). These satellites, which carry passive microwave sensors, are also considered in Chapters 10 and 12, where the technology is discussed as it pertains to deriving sea surface temperature, salinity, and the prevalence of polar ice.

7.7 Remote sensing of groundwater

The term 'groundwater' is considered to describe subsurface waters that originate from below the water table. That is, they will flow directly into a well. Soil moisture (previous section) is commonly termed 'subsurface water' and differs from groundwater as it occurs above the water table (Younger, 2007). Of all the liquid fresh water on Earth, groundwater is the largest reservoir, accounting for slightly less than 99 per cent of the total budget. The remaining 1 per cent is split between moisture in the atmosphere and rivers, lakes and freshwater wetlands. Groundwater can be within the matrix of sedimentary rocks, occupying the minute pore spaces between sediment grains; housed within rock fractures; or held within vast underground caverns.

While situated below the water table, the water is hidden from the gaze of remote sensing instruments. However, once it makes it to within a few centimetres of the surface, and as shown in the previous section, its microwave signature can be detected from orbit. Groundwater also discharges to the surface through springs that in turn feed rivers, lakes and wetlands. In some cases, fresh groundwater will also discharge through the seabed into the ocean (Figure 7.6).

In arid environments, the surface discharge of groundwater may be detected to lie beneath areas of unusually dense vegetation. In temperate regions, where vegetation is commonly dense, the speciation of plant cover may provide important clues to the occurrence of groundwater. These can be efficiently sensed using an NDVI index based on the reflectance properties of chlorophyll in the visible and infrared wavelengths (Chapter 5).

A curious property of discharging groundwater is that its temperature is relatively constant all year round and usually approaches the yearly average temperature for the region (Moore, 1982). This is in contrast to waters that occupy streams, rivers, and lakes which, being in contact with the atmosphere, largely mirror the diurnal and seasonal changes in temperature experienced at

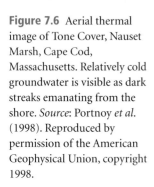

Figure 7.6 Aerial thermal image of Tone Cover, Nauset Marsh, Cape Cod, Massachusetts. Relatively cold groundwater is visible as dark streaks emanating from the shore. *Source*: Portnoy *et al.* (1998). Reproduced by permission of the American Geophysical Union, copyright 1998.

the Earth's surface. The potential for discriminating springs and seeps from their thermal signature is thus greatest during the summer and winter months (Becker, 2006). It is this difference in temperature that forms the basis for remote sensing groundwater discharge using signatures in the thermal infrared (TIR) portion of the electromagnetic spectrum (Rundquist *et al.,* 1985; Moore, 1996; Torgersen *et al.,* 2001; Handcock *et al.,* 2006). The technique obviously demands a rich time-series of data so that the thermal stability of the groundwater discharge can be detected against the seasonally fluctuating temperature of surface waters.

Because of the fine spatial scale demanded to resolve groundwater springs and seeps, airborne scanners such as NASA's Thermal Infrared Multispectral Scanner (TIMS) have been extensively utilized (Rundquist *et al.,* 1985; Banks, 1996). Airborne sensors have the advantage over satellites that they can be mobilized during weather conditions that maximize temperature differences between the discharging groundwater and other targets. Satellites such as IKONOS and QuickBird offer a metre-scale resolution that rival airborne systems, but they lack a thermal infrared channel.

As shown in Table 7.1, there exists only one satellite mission that can estimate the quantity of groundwater stored beneath the few centimetres that can be probed using passive microwaves. The GRACE (Gravity Recover and Climate Experiment) mission, lofted from Russia in 2002, consists of two identical satellites orbiting in unison at the same altitude but with a separation of 220 km.

The instruments are noteworthy as they are among the few in Earth orbit that do not look down. Instead, each of the pair monitors the relative position of the other, using microwave ranging instruments. As they overpass a positive gravity anomaly, the leading spacecraft approaches and speeds up due to higher angular

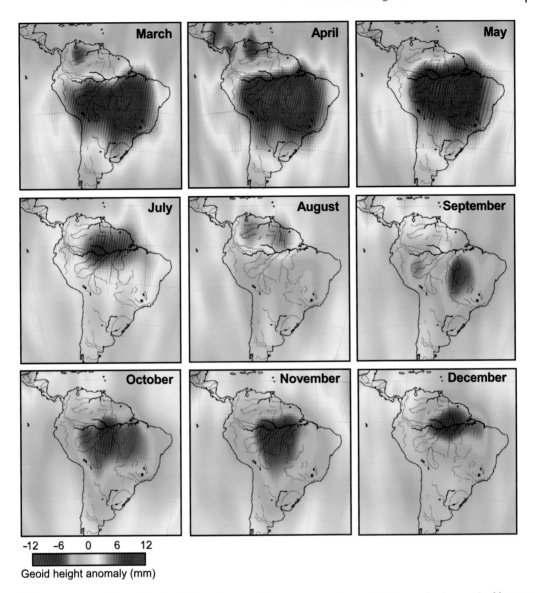

-12   -6   0   6   12
Geoid height anomaly (mm)

Figure 7.7 A time-series of monthly GRACE gravity anomaly measurements over the Amazon basin, acquired between March and December 2003. Areas of orange, red and pink on the maps show areas of higher than average gravity (and therefore more water), while green, blue, and purple are areas with lower than average monthly gravity. Credit: NASA.

momentum. This causes the pair to increase separation. The first spacecraft then passes the anomaly and slows down again; meanwhile, the second spacecraft accelerates and then decelerates over the same point. Recording spacecraft separation in this way enables gravity to be mapped, with cycles of growing/shrinking separation indicating the size and strength of anomalies. Using this method, GRACE will accurately map variations in the Earth's gravity field over the 5-year predicted lifetime of the mission.

The quantity of groundwater below Earth's surface influences its gravity field. Most importantly, as groundwater is consumed or recharged, water-induced gravity anomalies change through time. GRACE has the power to track these changes with reasonable accuracy (Yirdawa *et al.*, 2008; Chen *et al.*, 2009) and offers the opportunity to remotely construct a simultaneous global assessment of changes in water storage.

The technology is exciting in that it offers an alternative method of assessment from digging or drilling a large number of expensive monitoring wells into the water table. The latter technique is obviously spatially unrepresentative and prohibitively costly in remote regions. Figure 7.7 demonstrates seasonal changes in Earth's gravity field over the Amazon basin induced by combined fluctuations of stored groundwater and the volume of the Amazon and Orinoco rivers. A change in Earth's gravity field (the geoid) of 1 mm, as measured by GRACE, is roughly equivalent to the change that an additional 4 cm of water would produce.

7.8   Key concepts

1 Surface waters encompass streams, rivers, and lakes. Subsurface water defines water stored below Earth's surface but above the water table. Soil moisture is the dominant reservoir here. Groundwater differs from subsurface water in that it defines waters stored below the water table. All three repositories can be assessed with reasonable accuracy via remote sensing, though groundwater is by far the most challenging.

2 Both suspended sediments and some pollutants can be assessed spectrally by their influence on water quality. Sediments change the colour of water and can be quantified in the short-wavelength visible spectrum using regression-based optical models. Pollutants may alter riverbed vegetation and hence be sensed indirectly, or they may induce changes in phytoplankton, such as observed with eutrophication.

3 Since water has a pronounced signature in both the infrared and microwave portion of the electromagnetic spectrum, flooded land can be easily discerned at these wavelengths. Sensors utilizing the microwave spectrum have the advantage of all-weather capability and, in some cases, the ability to recognize the presence of water beneath vegetation.

4 As is the case for the coastal zone, the predictable but differential absorption of the visible spectrum by water can be used as a basis to derive bathymetry for rivers and lakes. The most advanced strategies employ a technique termed 'radiative transfer modelling'. The accuracy of the results is, however not in the same league, as that obtained from either LiDAR or acoustic techniques.

5 The flow of water from land to sea is dictated by the geometry of the watershed. Since water always flows downhill, this can be mapped provided

that a high-resolution digital elevation model (DEM) can be constructed. Over small areas, technology such as airborne LiDAR can serve this purpose. At regional to global scale, only two programmes are capable of providing sufficiently accurate DEMs. These are the Shuttle Radar Topography Mission (SRTM) and the ASTER Global Digital Elevation Model (GDEM).

6 The Earth naturally emits energy in the microwave spectrum and the moisture content of soil serves to augment the quantity of energy emitted. Satellite passive microwave remote sensing is capable of measuring these subtle differences of emissivity and can be used to quantify soil moisture to depths of several centimetres over vast areas.

7 If groundwater is discharged at the Earth's surface, it can be detected on the basis of its thermal infrared signature, which is largely invariant through the seasons, or else from the enhanced vegetation growth in the area surrounding the seep. Groundwater that remains below the water table has a subtle effect on Earth's gravity field. The GRACE satellite mission is able to detect these gravity anomalies and return an assessment of groundwater volume at the global scale.

Coral reefs, carbon and climate

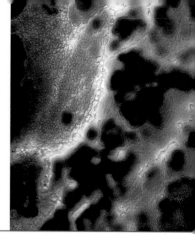

8.1 Introduction

Just as it has for the polar ice caps, the history of coral reefs has been punctuated with great change. During a span of more than 600 million years, reef ecosystems have enjoyed episodes of incredible production and diversification, building edifices of carbonate rock that exist today in the form of great mountain ranges such as the Dolomite Alps, or the exquisitely preserved Devonian patch reefs of Australia's Canning Basin (Figure 8.1). Reefs have also had to persevere through the bad times, dodging the eye of extinction on numerous occasions (Figure 8.2).

Coral reefs are not by any means made of corals alone. In fact, many other calcareous organisms, both animal and plant, may contribute more to the volume of a reef than do the corals (Blanchon et al., 1997; Wood, 1999; Braithwaite et al., 2000; Perry et al., 2009). Corals, however, are the architects and builders of the reef (Bosscher & Schlager, 1992). Living coral polyps establish themselves on a shallow platform and begin to develop as colonies. These grow horizontally and vertically, producing forms characteristic of their species. As they get larger, they grow closer together and the spaces between them become in-filled by many kinds of smaller calcareous organisms, both entire skeletons and fragments broken up by wave action and boring animals and plants. Small pieces of mollusc shells, sea urchins, foraminifera and calcareous plants are among the most important detrital ingredients of the filler.

Both the corals and other calcareous inhabitants are built primarily from calcium carbonate ($CaCO_3$), with small amounts of magnesium salts and other trace minerals. As communities of reef organisms assemble vast carbonate edifices, they act both as sinks and as sources for inorganic carbon and, most importantly, carbon dioxide. Through production of a carbonate skeleton, reef building thus sequesters vast quantities of inorganic carbon from the atmosphere. Currently, coral reefs act as a sink for ≈ 110 million tons of carbon each year. However, though

Remote Sensing and Global Environmental Change, First Edition. Samuel Purkis and Victor Klemas.
© 2011 Samuel Purkis and Victor Klemas. Published 2011 by Blackwell Publishing Ltd.

Figure 8.1 Ancient coral patch reefs now exposed on land have been described extensively. Shown are the Devonian reef complexes of the Canning Basin, Western Australia. Frasnian outcrops are exposed in the Bugle Gap area, where limestone ranges represent exhumed platforms and valleys coincide with basinal deposits. As a result, the present-day topography mirrors the depositional paleotopography of the Devonian sea floor ≈370 million years ago. Outcrops like these allow easy ground-truthing and precise quantification of the anatomy of reefs that existed in ancient seas. Credit: Erwin Adams.

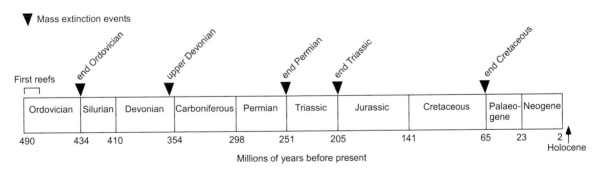

Figure 8.2 Timeline of extinction events for Earth's reefs. There have been five mass extinctions in the last 500 million years that have yielded periods of several million years where Earth has been devoid of reefs. It is open to debate whether we are presently witnessing the sixth extinction event with the current reef crisis.

this number may appear huge, it is low by comparison to the amount of carbon dioxide released by our own activities. Considered as such, reefs are only able to mop up the equivalent of two per cent of the present output of anthropogenic CO_2 (Kinsey and Hopley, 1991).

Over geological timescales, reefs have had great impacts on ocean chemistry, carbonate deposition, atmospheric carbon and global climate change (Ridgwell & Zeebeb, 2005). In the short term, however, and faced with our current warming predicament, reefs can do little to alleviate the upward trend of atmospheric CO_2. In fact, global changes in ocean chemistry, atmospheric composition and climate are all actually acting to jeopardize the future of coral reefs (Glynn, 1996; Hughes *et al.*, 2003). In addition, we are placing ever greater demands on this delicate ecosystem through burgeoning tourism and coastal development projects, unsustainable fishing and land use change. If current climate predictions are accurate and our pressure on this resource continues unabated, the survival of corals in even the short term is far from guaranteed (Hoegh-Guldberg *et al.*, 2007). If it happens, then extinction of reef builders may take millions of years to repair because these organisms are k-strategists with very long life cycles and, therefore, slow evolution.

Being sensitive to perturbations in Earth's chemistry, we can use coral reefs as a sensitive 'dip-stick' for climate change. Unlike the slow and sustained march of the ice caps that require enormous exchanges of energy over centuries to initiate a recordable growth or recession, coral polyps react to changes in their environment on timescales of days in a 'boom or bust' manner. In periods of 'boom', carbonate is deposited at scales of centimetres per year. When 'bust', production dwindles to zero. Just as tree rings can be counted, the rate of deposition of carbonate leads to bands of different density that can be used to reconstruct the climate under which the coral skeleton was deposited (Fairbanks & Dodge, 1979; Swart & Grottoli, 2004; Smith *et al.*, 2006). As shown in Figure 8.3, each calendar year is characterized by a light band (high-density skeleton corresponding with August to October growth) and a dark band (low-density skeleton corresponding with November to July). Persistent dense (light) bands are representative of years in which colony extension rate was reduced by stress. Such interpretations yield valuable information about the coral itself and about the environment in which it lived.

There is a curious irony to the present fate of Earth's reefs. On the one hand, patterns in fossil reef growth allow us to reconstruct the planets climate in unprecedented detail (e.g. Figure 8.3) and this information is crucial in enabling us to view our present climate in an historical context. On the other hand, however, we have also come to appreciate ancient fossil reefs, with their thick and well-developed carbonate rocks, for another important reason. Many buried fossil reefs contain vast petroleum reserves. On a conservative estimate, over one-third of all known oil fields sit in ancient reef structures. It is the ease of extraction of hydrocarbons from these porous rocks that has led to today's total reliance on oil to power our society.

Extracting and burning hydrocarbons releases carbon dioxide into the atmosphere. Even with our present rate of hydrocarbons use, atmospheric CO_2 is predicted to rise to concentrations sufficient to threaten the very existence of today's coral reefs. Their porous nature has come as a mixed blessing to reefs.

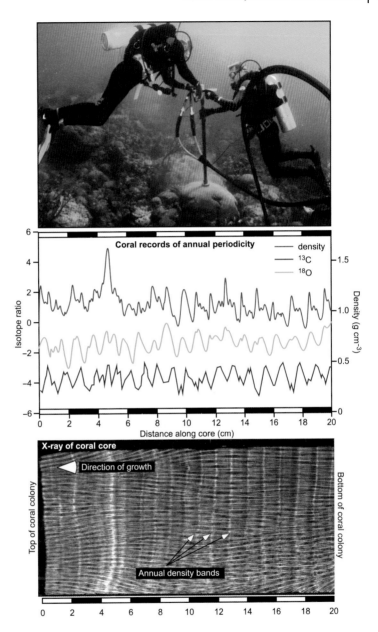

Figure 8.3 The upper pane of this figure shows divers drilling a core from a coral colony. The middle pane shows variation in isotopic signature of carbon and oxygen, used to reconstruct temperature, from the *Montastarea faveolata* coral X-rayed in the lower pane. Yearly density bands are clearly visible in this X-radiograph. Credit: K.P. Helmle, with isotope data provided by P.K. Swart and photograph by D. Gilliam.

8.2 The status of the world's reefs

As with so many natural ecosystems, the coral reefs we observe today are a shadow of their former selves. Many (perhaps the majority) are dead or dying, with remote and healthy examples gravely threatened (Glynn, 1996; Hughes *et al.*, 2003; Hoegh-Guldberg *et al.*, 2007; 2009). Coral reefs react quickly to new stressors because they thrive in a narrow range of environmental conditions and are very sensitive to small changes in temperature, light, water quality and hydrodynamics. Even if these stress factors are sub-lethal, the coral colony has little spare energy for sex. Population sizes fall and so does growth rate. This decline under stress is exacerbated by poor survival rates of young colonies. Numerous demographic studies show a general decline in mortality rates for older, larger colonies (Connell, 1973; Tanner *et al.*, 1996). Recent work has even established a strong link between the prevalence of coral disease and ocean warming (Bruno *et al.*, 2007; Bruckner & Bruckner 2006; Muller *et al.*, 2008; Rosenberg *et al.*, 2009).

The decline of reefs is closely linked to human activity, because the tropical coastlines that host them are often heavily populated. Change in coastal waters is likely to be greater today than that experienced in 100 million years of reef evolutionary history. Among the documented impacts on corals, operating on a variety of scales, are global climate change (e.g. increases in sea surface temperature, sea level, CO_2 saturation and the frequency and intensity of storms and associated surge and swells); regional shifts in water quality; and more localized impacts due to increased loading of sediment, over fishing, contaminants, and nutrients reaching coastal environments. These factors have had demonstrable impacts on the health of coral-dominated coastal ecosystems worldwide (Jackson *et al.*, 2001; Hughes *et al.*, 2007; Pratchett *et al.*, 2008; Sale, 2008; Randall & Szmant, 2009; Riegl *et al.*, 2009).

With so many deleterious stressors on the ecosystem, reefs are subject to an endless period of convalescence after the most recent cataclysm. Some reefs impacted during the 1998 El Niño-Southern Oscillation (ENSO) event have not yet recovered, potentially resulting in decreased fish abundance, phase- and strategy-shifts in benthic community structures, diversity loss and decreased overall productivity. Worryingly, the potential for recovery of many sites, even those that are very isolated, is not guaranteed (Graham *et al.*, 2006). Statistics vary according to source, but estimates suggest that 20 per cent of the world's coral reefs are already lost, 24 per cent are under imminent risk of collapse and another 26 per cent are in grave danger of irreparable damage (Byrant *et al.*, 1998; Wilkinson, 2006; Figure 8.4).

8.3 Remote sensing of coral reefs

Before tackling the important issue of scale in reef remote sensing, it is necessary to clarify some definitions. Large-scale means large areas, small-scale means small areas. Scaling up means moving from small to large. But what is 'small' and what is 'large' in the context of a reef?

In this sense, there is a move by the community to adopt terms such as 'colony-scale', 'community-scale', 'reef-scale', 'regional-scale' or 'global-scale'. Remote sensing can provide information across these expanses of extent, but reef-scale is typically the most accessible. The very small and the very large are the most problematic to resolve, requiring a synergy of many different remote sensing instruments.

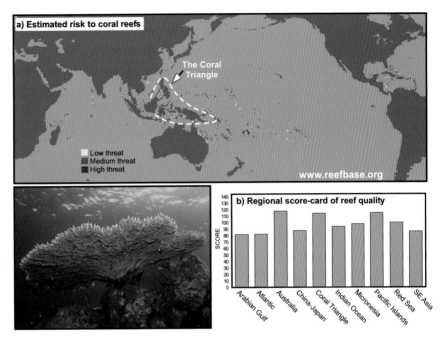

Figure 8.4 Coral reefs around the world are threatened by a variety of natural and man-made factors: (a) From ReefBase online GIS (www.reefbase. org); (b) ReefCheck data from Wilkinson (2006): the lower the index, the more degraded are the reefs of the area. Photo courtesy of A. Hagan.

Reefs themselves resist efforts to be quantified at an extent commensurate with our technological capabilities. Reef ecosystems occur over broad spatial scales (tens to hundreds of kilometres) and are characterized by a continuum from fragmented patches of centimetre diameter to continuous tracts of similar benthic character stretching for kilometres.

To effectively address the issue of reef degradation, it is necessary to develop an understanding of the type, geographical extent and condition of reef resources on a global scale. However, such understanding is difficult to obtain when dealing with a marine resource. Unlike when we are dealing with terrestrial ecosystems, we cannot readily observe submerged ecosystems. Their spectacular topography and diversity are hidden from casual view and may only be detected and even become known with systematic detailed surveys. On land, regular field observations are often sufficient to provide a good general understanding of wildlife activity patters, areas of importance and general abundance of the most common species. However, within the marine environment, even the most basic determination of common species present usually requires a rather comprehensive and expensive systematic inventory to determine. Reefs are particularly difficult to monitor routinely by virtue of their great size, and the fact that many of the richest examples sit in the remotest reaches of our oceans.

As we will see throughout this chapter, benthic classification from remote imagery is almost always less accurate than traditional field methods (such as diver survey) and less capable than comparable studies conducted over land. However, the rapid acquisition and processing of remote sensing imagery allows for mapping and monitoring large reefal areas with much less effort and expense than traditional mapping techniques.

The capability of mapping reefs with remote sensing has historically been built around technologies developed for other – usually terrestrial – applications. Perhaps the greatest advances in recent years have been made using airborne hyperspectral sensors that offer the spatial and spectral capability to discern the subtle spectral states of a reef that can be used as a proxy for coral health. The majority of airborne hyperspectral radiometers are flexible in that they can be 'tuned' to the demands of a specific project. To monitor crop health in Ohio, a suite of spectral bands can be positioned through the visible and into the infrared spectrum. The same instrument can be re-configured in minutes (and in flight) to pool all of its spectral capability in the shorter water-penetrating wavelengths, as demanded by a reef mapping campaign.

The power of remote sensing is not restricted to imaging reefs themselves. Indirect observation can be used to sense the environment around the reef, providing valuable information with which to predict the likelihood of impacts. In this category falls sensing of the ocean temperature (e.g. the NOAA Virtual Stations covered in Chapter 10); wave height; sea level; turbidity; chlorophyll and coloured dissolved organic matter concentrations; the atmosphere (aerosols, rain, solar insolation, or cloud cover – see Chapter 11); or nearby landmasses (vegetation cover, watershed structure, or urban growth – see Chapters 5, 6 and 7). These environmental variables describe the boundary conditions around the reefs and can be related to processes occurring on the reefs themselves (Andréfouët & Riegl, 2004).

8.4 Light, corals and water

8.4.1 Light and the water surface

Coral reefs, by virtue of the fact that they are submerged, are not well predisposed to monitoring using satellite data. Water changes both the magnitude and distribution of energy with respect to wavelength. The effect becomes more acute as wavelength and water depth increases.

Before tackling the effect of the water body on radiant energy, in this section we will consider the way in which light interacts with the water surface (more properly termed the air-water interface). Many photons downwelling on a water body do not penetrate down into the water column due to strong interactions at the surface. This component of the light field is reflected from the surface back into the atmosphere and is an unwanted signal in remote sensing imagery. The signal of interest is the fraction of light which passes into the water column, persists to interact with the seabed and is then reflected out across the air-water interface to be detected by a sensor.

However, photons that have interacted with the water column or seabed must then approach the air-water interface from below and may again interact strongly with the surface. Unable to exit, these are reflected back into the water body (Figure 8.5) and are lost in the remote sensing case. As happens with downwelling radiant energy, the upwelling passage of light from the water body through the

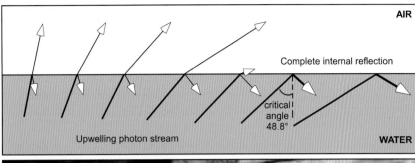

Figure 8.5 Total internal reflection causes this Red Sea reef top to be reflected on the water surface. In water, this critical angle is 48.8°, and beyond this no light is able to pass through the surface. Photo courtesy of B. Riegl.

water surface will change the magnitude and directional properties of the light, but not its spectral distribution (i.e. colour). This effect is again relevant to how much energy propagates into the atmosphere and can be intercepted by an over-flying remote sensing instrument.

As light travels from air to water (or from water to air), it is refracted due to differences in the optical density of the two mediums (the refractive index of air can be assumed to be 1; water has an index of 1.33). The process of refraction is described by Snell's law and can most simply be summarized as a change in speed (and thus direction) of a light field as it passes from one medium to another. The implications of refraction at the surface are that, for a flat sea surface, the whole of the hemispherical irradiance from the atmosphere which passes across the interface is compressed into a cone of underwater light with a half angle of 48.8° (Dekker *et al.*, 2001).

This phenomenon also has implications for (upwelling) reflected radiance. Arising as a consequence of Snell's law is a phenomenon known as 'total internal reflection'. If the ratio of the refractive indices of two media is less than 1, there will be some critical angle of incidence above which the angle of refraction is always greater than 90°. For water, this critical angle is 48.8°. Once this threshold is exceeded, no incident light will be transmitted through the sea surface to the atmosphere and all upwelling energy will be reflected back down into the water body (Figure 8.5).

While Snell's Law describes angles of refraction, Fresnel equations (pronounced FRA-nel) describe the behaviour of light when moving between media of differing refractive indices. The reflection of light that the equations predict is known as Fresnel reflection. The Fresnel reflectance varies with respect to zenith angle from ≈0.02 for normal incidence to ≈0.03 at 40°, and then increases strongly with increasing angle size in a well-behaved and well-known manner. In practice, even the lightest of winds ruffle the sea surface and uniform sky conditions are rarely encountered – except in heavily overcast days, which are not of relevance to reef remote sensing applications since the sea surface cannot then be viewed from above using the visible spectrum.

Sunglint is the specular reflection of the sun off water. As with all specular reflection, the angle of the incident ray of light equals the angle of the reflected ray. Thus, if the angle between the sun and the water surface is equal to that through which a remote sensing instrument's field of view looks upon the surface, the sensor will image a mirror-like reflection of the sun.

Full sunglint in images typically results in bright silver to white coloration of the water surface, which can obscure and generally overwhelm the water-leaving radiance to a point that retrieval of seabed information is impossible. In reality, the sea surface is not perfectly flat, since wind roughens it to a myriad of differently angled wave facets. A proportion of these will meet the angle of specular refection, resulting in a broken glitter pattern, as typically seen on the ocean when the sun is low to the horizon (Figure 8.6a).

Under conditions that promote glint, the degree to which the seabed can be imaged is heavily compromised, so every effort must be made to minimize the effect. The degree of glint can be reduced drastically using an appropriate choice of sensor geometry, setting the time of image acquisition to provide a solar zenith angle between 40° and 55° and close to nadir imaging. When remote sensing is operating from an aircraft, careful flight planning and a degree of luck with the sea state must be relied upon to minimize the problem of glint.

However, though not ideal for sea floor characterization, glint-compromised images can reveal numerous details of water circulation that are otherwise invisible, with patterns of circulation, wind and tidal action exposed by their differing surface roughnesses resulting in a different hue and brightness of the glint (Figure 8.6b).

Despite the best laid plans, a degree of sunglint in an image is typically unavoidable. Satellite data are particularly vulnerable due to having a wide field of view – plus, the user has little control on acquisition conditions other than cloud cover. For seabed mapping, this is a nuisance, though, through an extraordinary strategy, Philpot (2007) made the best of a bad deal, proving that glint could be harnessed to estimate atmospheric transmission in an image scene.

In cases where a degree of water-leaving radiance is present in glint affected pixels, there does exist a means for empirically tackling the problem. Most commonly the near-infrared band, which exhibits maximum absorption and minimal water-leaving radiance over clear waters, can be used to characterize the spatial distribution and intensity of the glint contamination. The result from this is subtracted from the visible bands and the glint effect can be reduced and, in some cases, completely removed (e.g. Hochberg *et al.*, 2003; Hedley *et al.*, 2005; Kay *et al.*, 2009; Kutser *et al.*, 2009; Figure 8.7).

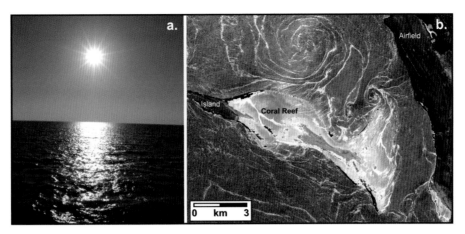

Figure 8.6 (a) The bright reflection from the water surface is an example of sunglint – a direct specular mirror-like reflection. Courtesy A. Hagan. (b) This QuickBird satellite image from the Red Sea shows eddy and sinuous surface water patterns on the leeward side of a coral reef caused by wind-driven currents. The image was acquired during storm conditions and the surface patterns are revealed by the presence of sunglint. Satellite image: DigitalGlobe.

Figure 8.7 True-colour IKONOS satellite image before (inset) and after deglinting using the algorithm of Hochberg *et al.* (2003) operating on the infrared band. This scene depicts the Cordelia Shoal, a reef system off the Island of Roatán (Honduras). This area is of particular interest because, at the time of writing and as shown in the photographs, it still boasts dense thickets of the branching coral *Acropora cervicornis*, a species once prevalent in the Caribbean but now severely degraded (Keck *et al.*, 2005). Satellite image: GeoEye.

Correcting imagery for glint requires a near-infrared waveband near 900 nm, while correcting imagery for atmospheric effects requires two wavebands, often 750 nm and 865 nm. Even though these long wavelengths offer little to no water penetration, they are nevertheless for this reason demanded for seabed mapping. In cases of very shallow water (<1 m), the assumption that all the water leaving signal in the infrared spectrum originates from the sea surface may be invalid, and shallow water pixels may be over-corrected by glint removal as described.

8.4.2   Light and the water body

Water serves to attenuate light strongly (Figure 8.8). Since photons that have interacted with the reef and have subsequently arrived at an orbiting sensor have made a two-way journey through the water column, the effect of submergence dominates the received signal, even if the water depth is comparatively minor. A photon of light propagating through water may become absorbed or scattered by molecules of the water itself, or by molecules of dissolved or suspended materials within it. The combination of absorption and scattering is termed attenuation.

The degree of attenuation will depend upon the clarity/turbidity of the water and is wavelength dependant. Attenuation increases approximately exponentially with wavelength, such that short wavelength (blue) light is attenuated to a far lesser degree than the longer wavelength (red) light. As any SCUBA diver knows, this is the reason that the ocean progressively loses colour with depth descended.

The optical properties of pure water are well constrained in the visible (Smith & Baker, 1981; Pope & Fry, 1997), but the optical properties of the ocean rarely approximate to such conditions, even in the highly transparent nutrient-deficient waters associated with coral reefs. Depending on their attenuation characteristics, Jerlov (1976) classified water bodies into set water types – types I, II and III for clear oceanic waters, with type I being the clearest, and types 1,3,5,7 and 9 for coastal waters (Figure 8.9).

Introduced by Morel & Prieur (1977) and refined by Gordon & Morel (1983), a partition between oceanic and coastal waters was made on the basis of their turbidity and pigment content. Case 1 waters were classed as those where chlorophyll concentration is high relative to scattering; and Case 2 where the presence of inorganic particles (and therefore scatter) are higher than the concentration of phytoplankton. For coral environments, the majority of primary production occurs in the benthos and, since the water column remains oligotrophic (low in nutrients and thus unable to support much plant life), the optical properties tend to be dominated by inorganic particles and, for the most, part approximate to Case 2 conditions. Even in the rare cases where Case 1 conditions prevail, shallow reef waters are typically still classed as Case 2, since the bottom reflectance significantly contributes to the water leaving radiance signal (Dekker *et al.*, 2001).

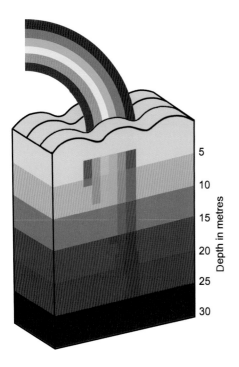

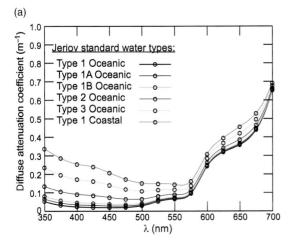

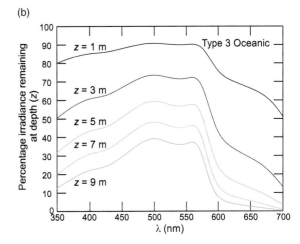

Figure 8.8 Where colour vanishes. Sunlight is rapidly attenuated by water molecules, plankton and suspended detritus. Blue light penetrates to the greatest depth and persists for several tens of metres into the water body. Red light, conversely, is extinguished within a matter of a few metres of the surface.

Figure 8.9 (a) Jerlov's water types. Attenuation of light by water is wavelength-dependent and increases approximately exponentially beyond 575 nm. (b) The absorption of light by a Jerlov Oceanic Type 3 water. By a depth of 9 m, light in wavelengths 600 nm–700 nm is less than ten per cent of that at the surface. Such attenuation limits corals to inhabit relatively shallow and well-lit waters, where the ambient illumination is sufficient to satisfy their photosynthetic needs.

Termed irradiance, several hundred watts per square metre of visible light are incident on the ocean's surface for typical daytime conditions. Irradiance below the water's surface can change greatly in magnitude in a matter of seconds if a cloud passes in front of the sun, or if a gust of wind changes the sea surface from glassy smooth to rippled. However, observation shows that certain ratios of radiometric quantities are relatively insensitive to environmental factors such as sea and sky state. Likewise, the rate of change with depth of a radiometric quantity is often well-behaved. As introduced by Preisendorfer (1961, 1976), these ratios and depth derivatives are termed Apparent Optical Properties (AOPs) and include the coefficients of attenuation for the water body and reflectance.

Conversely, factors that are dependent on the directional structure of the ambient light field, such as absorption and scattering, are termed Inherent Optical Properties (IOPs). These, importantly, affect the ocean surface colour. If the both the light field above the ocean's surface and the IOPs are well constrained, then there exists – in theory at least – the opportunity to predict the light field anywhere within the water body. This is the realm of radiative transfer models, which will be discussed in the following section.

8.4.3   Reflectance models for optically shallow waters

In pursuit of developing a meaningful way to decouple the contribution of the sea floor and water column in a reflectance signature received by a remote sensing instrument, it has been proven useful to model the light field analytically within shallow waters above coral reefs. Such simulation allows for the computation of the quality, quantity and distribution of water-leaving radiance as would be produced under a variety of scenarios that would be difficult (or impossible) to test *in situ*. Such simulations utilize radiative transfer (RT) theory and are termed radiative transfer models. Commonly a 'forward' approach to modelling is adopted, whereby spectral radiance distributions (spectra) are computed from known inherent optical properties of the water column.

Early forward models operated using the Monte Carlo method for simulating the propagation of photons through water (Kattawar & Plass, 1972; Kirk, 1981). This approach is conceptually easy to understand but was historically avoided because of the high computational burden associated with tracing the fates of millions of virtual photon packets, according to statistical probabilities, through the atmosphere and water column. To lessen the required computations, the fate of each photon is typically determined by six simple interactions (Mobley, 1994):

1 Loss of photons by conversion of radiant energy to non-radiant energy (absorption);
2 Loss of photons by scattering to other directions without change in wavelength (elastic scattering);
3 Loss of photons by scattering with change in wavelength (inelastic scattering);
4 Gain of photons by conversion of non-radiant energy into radiant energy (emission);

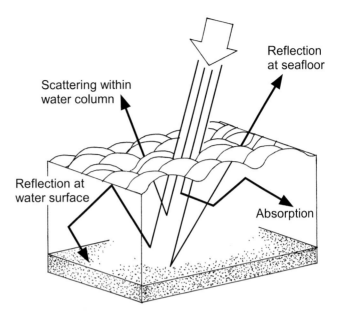

Reflection
at seafloor

Scattering within
water column

Reflection at
water surface

Absorption

Figure 8.10 A schematic diagram depicting the dominant fates of photons in optically shallow water. Only the signal that has reflected off of the sea floor contains viable information that can be used by a remote sensing instrument to interpret benthic character.

5 Gain of photons by scattering from other directions without change in wavelength (elastic scattering);
6 Gain of photons by scattering with change in wavelength (inelastic scattering).

Each simulated photon path is randomly distinct from the others, as determined by the probabilities of absorption and scattering in the water and air and by the probabilities of transmission, reflection or absorption at the water surface or seabed.

As photons enter at the sea surface, they scatter within the water, reflect internally at the sea surface or sea floor, and eventually either escape into the atmosphere or are absorbed by the water or sea floor (Figure 8.10). Only a minor proportion of those input into the system will exit the water surface and adopt a trajectory that will be intercepted by an over-flying sensor.

A second approach to forward modelling is based on the Invariant Imbedding Technique developed by Preisendorfer (1976) and Mobley (1994). The most popular commercially available 'forward-type' model is the well-regarded *HYDROLIGHT* package (Mobley & Sundman, 2001; www.hydrolight.info). The public domain software *WASI* also offers a fast and user-friendly tool for forward and inverse modelling of optical data (Gege, 2004). Within such models, parameters such as the concentration of phytoplankton, suspended matter, gelbstoff absorption, bottom reflectance, bottom depth, surface wind speed, solar zenith angle and viewing angle can all be varied in order to parameterize the dependence of the water-leaving light field on such factors. Typical output is a large number of spectra resulting from multiple simulations.

Forward models can be 'inverted' to solve for water or seabed optical properties from the water leaving radiance distribution (as recorded by a remote sensor). Such inverse schemes are now relatively refined and are typically empirical-solved (Gordon & Morel, 1983; Dekker, 1993), or are tackled through non-linear optimization (Lee *et al.*, 1999; Durand *et al.*, 2000; Hedley & Mumby, 2003). In the latter case, *in situ* simulated and remotely sensed data are integrated into a shallow-water reflectance model to simultaneously derive both seabed reflectance and bathymetry. Such an approach is limited to the exhaustive datasets offered by a hyperspectral instrument and is thus generally confined to airborne platforms (typically AVIRIS, HYMAP, and CASI).

The presently available hyperspectral satellite EO-1 HYPERION shows promise (Kutser *et al.*, 2003), although it is limited by a relatively coarse spatial resolution of $30\,m^2$ and a swath-width of only $7.5\,km$. Future spaceborne sensors will undoubtedly offer similar spectral capability but with a more attractive pixel size and repeat cycle. The potential of spaceborne hyperspectral sensors will be discussed later in the chapter.

These modelling approaches are conceptually powerful since they are derived from the physical principles of the interaction of light and water in optically shallow environments. Provided sufficient spectral detail exists, such 'inverse' models can be transferred between both airborne and spaceborne platforms. Unfortunately, as pointed out by Andréfouët & Riegl (2004), there is still a long way to go before these deterministic strategies will be fully operational and can be harnessed by many users. Factors limiting the practicality of the approach are, among others: the necessity for powerful computers; data calibration issues; the extreme complexity of radiative transfer processes in natural waters; and the heterogeneity of the real coral reef world and the waters in which the reefs lie.

8.4.4 Reflectance signatures of reef substrata

The reflectance spectra of reefal substrata (corals, seagrasses, macro- and micro-algae, sand, mud, etc.) can be considered analogous to a fingerprint. Subtle differences in spectrum shape can be used to identify and discriminate between habitats. This approach can be considered an extension of the principles developed in Chapter 2 (Section 2.5), for mapping terrestrial land cover. The shape of each spectrum is dictated primarily by absorption and scattering, and, to a lesser degree, by fluorescent emission of the substrate. The majority of reef substrata have a pronounced chlorophyll signature (strong absorption at $450\,nm$ and $670\,nm$; Figure 8.11) by virtue of their own chlorophyll pigments (seagrass and algae), photosynthetic symbionts (corals) or by the ubiquitous presence of benthic unicellular micro-algae that inhabit the surfaces and interstitial cavities of unconsolidated sediments (carbonate sands and muds).

The best chance of discriminating between reef substrata lays in the detection of slight differences in spectral form that arise from different chlorophyll assemblages. However, since the overlying water column filters all radiation beyond $650\,nm$

Figure 8.11 Generic reflectance of healthy coral tissue in comparison to normalized absorption spectra of two common photosynthetic pigments. Chlorophyll-a absorbs in the blue and in the red, while carotenoids absorb in the green. The form of the coral spectra is clearly modified by the peaks of chlorophyll absorption. It is the different (and often exotic) assemblages of chlorophyll pigments contained within algae, seagrasses and corals that give rise to their characteristic reflectance spectra.

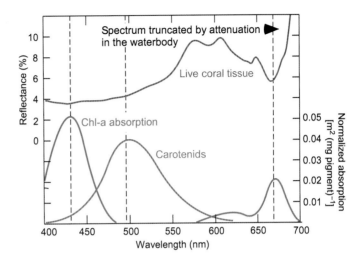

(Figure 8.11), it is not possible to utilize the red-edge signature to discriminate chlorophyll-containing from non-chlorophyll-containing materials (e.g. Chapter 5 high reflectance at near-infrared – Figure 5.1). This constraint imposes limitations to how the spectral separation of optically similar targets can be achieved, and it demands that unique solutions need to be found in the marine realm.

As with terrestrial studies, it is routine to support the remote mapping of reefs with field optical measurements to measure the light field above and within the water column. Such measurements are required:

1 to parameterize spectral models, including those incorporating radiative transfer equations; and
2 to aid in the identification of subtle spectral signatures of reef substrata that may be used to drive a reliable classifier (Hochberg & Atkinson, 2000; Hochberg et al., 2006; Lesser & Mobley, 2007).

These data are collected using portable and often submersible spectroradiometers (Figures 8.12, 8.13, 8.14).

Optical measurements conducted in the field to characterize the substrates and water bodies associated with coral reefs typically concentrate on the AOPs of the system and in particular reflectance ($R_{rs(z=a)}$, $R_{rs(z=w)}$, R_b, and R_w). Quantification of IOPs requires more advanced optical instruments, and IOPs are typically measured in the laboratory – although this can be achieved *in situ* from a ship (Bricaud et al., 1995) or by a diver carrying specialized equipment (Zaneveld et al., 2001; Zaneveld & Boss, 2003).

Being an intrinsic property and virtually independent of the illumination, atmospheric and water column conditions at the time of measurement (i.e. an AOP), reflectance is the quantity used to describe the interaction of light with a surface. The spectral irradiance reflectance (R), as defined by Mobley (1994), is the ratio of spectral upwelling (E_u) to downwelling (E_d) plane irradiances. For reef

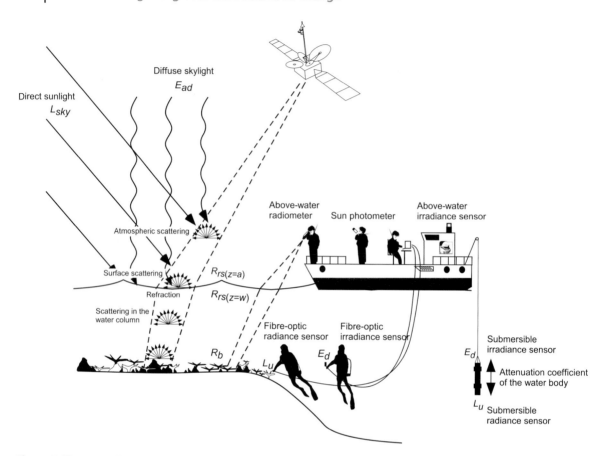

Figure 8.12 A typical measurement strategy used to quantify the components of the radiative transfer sequence between satellite and reef substrate using a suite of optical tools. Radiometric terms are given in Table 8.1.

mapping, it is common to utilise remote-sensing reflectance (R_{rs}), which is similar to irradiance reflectance, but the diffuse upwelling component (E_u) is replaced by radiance measured over a solid angle, termed the water-leaving radiance (L).

Reflectance is a relatively easy parameter to determine in the field with standard instrumentation, either using a twin channel instrument capable of measuring the up- and downwelling components concurrently with use of a cosine corrector, or else evaluated in rapid succession with E_d determined against a Spectralon lambertian reflectance panel (e.g. Chapter 4: Figure 4.10). Since spectral reflectance offers the primary criteria with which to differentiate between different benthic substrates in a remote sensing case, the most informative measurements are those that contain minimal influence from the intervening water column.

The parameter strived for is termed substrate reflectance (R_b), which describes the reflectance of (wet) substrate material with no water cover. Absolute measurement of R_b is unattainable *in situ* owing to shadowing by the instrument itself as the substrate is approached and the small-scale horizontal variability of the target. To obtain R_b spectra, it would be possible to remove the target substrate from the water and

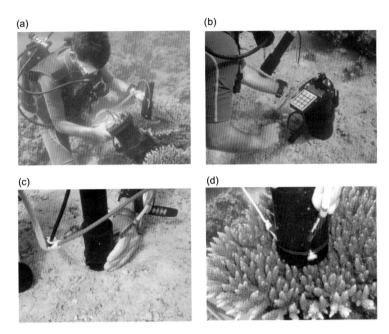

Figure 8.13 Measurement of the radiative properties of reef substrata by the author (Purkis) using a Physical Sciences Inc. DiveSpec submersible spectrometer.

conduct spectral measurements in the laboratory under controlled conditions. Such destructive interference is, however, generally impractical if a large number of substrates are to be quantified. Conversely, a strategy of evaluating a signal approximating to substrate reflectance is commonly adopted by measuring reflectance in close proximity to the substrate, so as to preclude water column effects, yet not so close as to be compromised by instrument self-shading (Figure 8.13).

Of the several multispectral imaging systems that have been deployed on reefs, only one active system provides results at centimetre scale. Mazel *et al.* (2003) deployed a narrow-beam in-water line-scanning multispectral fluorescence imaging system at night on Floridian and Bahamian reefs. This technique is in its infancy, yet, by exploiting differences in three fluorescence bands, the main benthic functional groups of the reef were determinable. It is not beyond the realm of possibility that the fluorescence signatures of reef biota may also be exploited using spectrally calibrated LiDAR beams, such as are already available on the CHARTS fusion package that will be discussed in Chapter 14 (Section 14.1.1), as well as the terrestrial full-waveform LiDARs considered in Chapter 6.

8.5 Passive optical sensing

Passive optical remote sensing is well poised for the task of reef monitoring in that it offers a good temporal resolution, coupled with affordable imagery spanning vast areas. Satellite and aircraft sensors are emerging as a leading technology in reef research. Because of the fact that long-wavelength electromagnetic energy is

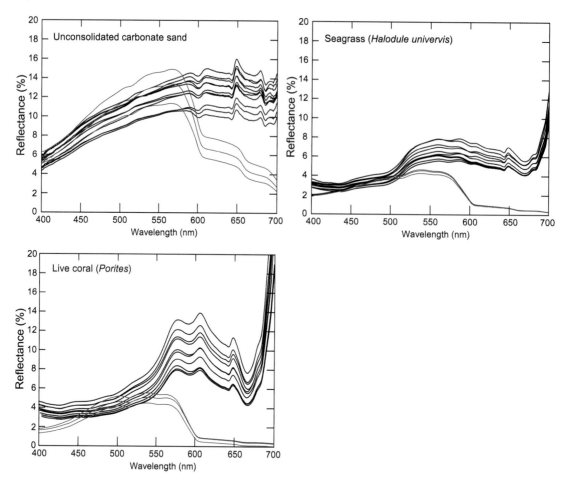

Figure 8.14 Concurrently measured reflectance spectra of sand, seagrass and live coral, evaluated beneath the water surface at a depth of 4 m (solid lines) and just above the water surface (purple lines). The spectra evaluated above the water surface lack any meaningful information at wavelengths longer than 650 nm, due to the rapid extinction of light by water. The plots demonstrate the narrow breadth of visible wavelengths useful for remote sensing applications.

rapidly attenuated by water, passive optical sensors with spectral bands positioned in the visible blue-green spectral region can be used most successfully to monitor coral habitats. Although they require that the water column is sufficiently transparent, and that the coral reef substrate is within the depth where a sufficient amount of light reaches the bottom and is reflected back out of the water body, passive remote sensing instruments can be used to map the seabed.

Numerous studies in clear waters, such as offered by the Bahamas and the Red Sea (Purkis *et al.*, 2002; Lesser & Mobley, 2007), indicate a penetration of blue-green wavelengths in clear water up to 20 m, which is also the critical depth for reef growth (Bosscher & Schlager, 1992). As such, the depth limitations of passive sensing do not diminish the utility of the technology, provided the water is clear, which is a typical condition of reef environments. The skill of the reef mapper lies in the

Table 8.1 Radiometric terms of light and water.

Term	Description	Term	Description
λ	Wavelength [nm]	E_{ad}	Downwelling irradiance in air [W/m^2/nm]
z	Water depth [m]	E_d	Downwelling irradiance [W/m^2/nm]
k	Attenuation of the water body [m^{-1}]	E'_d	Normalized downwelling irradiance [W/m^2/nm]
AOT_{550}	Aerosol optical thickness ($\lambda = 550$ nm) [–]	ρ_{panel}	Reflectance of the lambertian reflectance panel [–]
L	Spectral radiance at the sensor aperture [W/m^2/sr/nm]	ρ_{sky}	Fresnel reflection coefficient for the wind-roughened sea surface [–]
L_{sky}	Downward radiance of skylight [W/m^2/sr/nm]	$R_{rs(z=a)}$	Remote sensing reflectance just above the water/air boundary [sr^{-1}]
L_t	Radiance emanating from the water surface [W/m^2/sr/nm]	$R_{rs(z=w)}$	Remote sensing reflectance just beneath the water/air boundary [sr^{-1}]
L_w	Water-leaving radiance [W/m^2/sr/nm]	R_b	Substrate reflectance [sr^{-1}]
L_u	Under water upwelling radiance [W/m^2/sr/nm]	R_w	Reflectance of optically deep water [sr^{-1}]
L_p	Radiance from the lambertian reflectance panel [W/m^2/sr/nm]		

interpretation of a weak and water-attenuated signal into a reliable assessment of benthic character.

Due to having to look through a layer of water, even in these 'penetrating' bands, the magnitude of the optical signal that contains any information about the sea floor is low by comparison to that arising from the atmosphere. For a typical sea floor submerged under ten metres of water, the atmosphere produces approximately 80–90 per cent of the radiance seen by a satellite and 10–20 per cent of the signal comes from the water itself, with an even smaller portion coming from the seabed. For the case of the relatively broad spectral bands of a satellite system, the signal is typically little stronger than the background noise of the instrument, even when water depth is only several metres. The situation is further complicated by the fact that many coral reef substrata are optically very similar (typically green or brown in colour), so differentiation using their spectral signature is difficult.

In its defence, satellite imagery is relatively inexpensive and, with the metre-scale resolution offered by sensors such as QuickBird, WorldView-2 and IKONOS, the mapping of reefscapes is possible so long as at least some on-ground knowledge exists. Hyperspectral imagery, with its many narrow contiguous spectral bands, has been more successful in discerning the composition of the seabed on the basis of spectral signatures. The Compact Airborne Spectrographic Imager (CASI) has been the tool of choice for many years

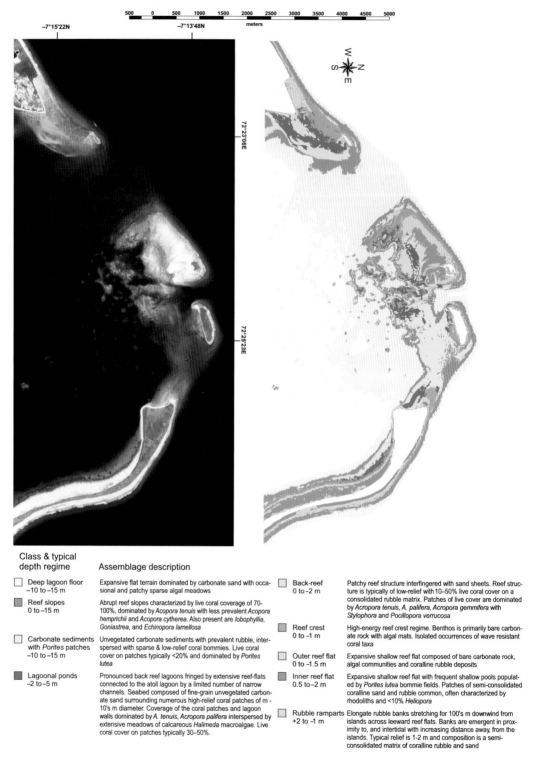

Class & typical depth regime

Assemblage description

☐ Deep lagoon floor
−10 to −15 m

Expansive flat terrain dominated by carbonate sand with occasional and patchy sparse algal meadows

▨ Reef slopes
0 to −15 m

Abrupt reef slopes characterized by live coral coverage of 70-100%, dominated by *Acopora tenuis* with less prevalent *Acopora hemprichii* and *Acopora cytherea*. Also present are *lobophyllia*, *Goniastrea*, and *Echinopora lamellosa*.

☐ Carbonate sediments with *Porites* patches
−10 to −15 m

Unvegetated carbonate sediments with prevalent rubble, interspersed with sparse & low-relief coral bommies. Live coral cover on patches typically <20% and dominated by *Porites lutea*

▨ Lagoonal ponds
−2 to −5 m

Pronounced back reef lagoons fringed by extensive reef-flats connected to the atoll lagoon by a limited number of narrow channels. Seabed composed of fine-grain unvegetated carbonate sand surrounding numerous high-relief coral patches of m - 10's m diameter. Coverage of the coral patches and lagoon walls dominated by *A. tenuis*, *Acropora palifera* interspersed by extensive meadows of calcareous *Halimeda* macroalgae. Live coral cover on patches typically 30–50%.

☐ Back-reef
0 to −2 m

Patchy reef structure interfingered with sand sheets. Reef structure is typically of low-relief with 10–50% live coral cover on a consolidated rubble matrix. Patches of live cover are dominated by *Acropora tenuis*, *A. palifera*, *Acropora gemmifera* with *Stylophora* and *Pocillopora verrucosa*.

☐ Reef crest
0 to −1 m

High-energy reef crest regime. Benthos is primarily bare carbonate rock with algal mats. Isolated occurrences of wave resistant coral taxa

☐ Outer reef flat
0 to −1.5 m

Expansive shallow reef flat composed of bare carbonate rock, algal communities and coralline rubble deposits

☐ Inner reef flat
0.5 to −2 m

Expansive shallow reef flat with frequent shallow pools populated by *Porites lutea* bommie fields. Patches of semi-consolidated coralline sand and rubble common, often characterized by <10% *Heliopora*

☐ Rubble ramparts
+2 to −1 m

Elongate rubble banks stretching for 100's m downwind from islands across leeward reef flats. Banks are emergent in proximity to, and intertidal with increasing distance away, from the islands. Typical relief is 1-2 m and composition is a semi-consolidated matrix of coralline rubble and sand

Figure 8.15 Classification of an IKONOS scene from Diego Garcia (Chagos Archipelago – British Indian Ocean Territory). Nine assemblages describing seabed cover are resolvable. The Chagos archipelago is one of the largest of the few remaining remote and pristine reef sites. Satellite image: GeoEye.

0   5   10 **Kilometers**

Figure 8.16 Upper left is a Cessna 208 seaplane used in cooperation with the Khaled bin Sultan Living Oceans Foundation to fly the CASI-500 hyperspectral instrument (upper right – pictured with Herb Ripley of Hyperspectral Imaging Ltd). The lower pane shows a mosaic of CASI lines obtained over the coral reefs of the Farasan archipelago (Red Sea – Saudi). Credit: Khaled bin Sultan Living Oceans Foundation.

(Mumby *et al.*, 1997; Bertels *et al.*, 2008). The instrument is versatile and mounted easily in an aircraft.

Figure 8.16 shows the CASI instrument and an example of the type of small aircraft on which it is easily mounted. Slight radiometric differences mark the boundaries between adjacent flight lines in the imagery shown in this figure. Compared with satellite data, the narrow swath-width of an airborne sensor increases the post-processing burden required to form a seamless image mosaic that can be used for mapping.

As shown in Figure 8.17, however, the spectral fidelity of CASI is far superior to that of a multispectral satellite. QuickBird, for example, has three broad spectral bands in the visible wavelengths and one in the infrared. Each band is on the order of 100 nm wide. In contrast, CASI boasts 22 near-contiguous bands, each only 10–20 nm wide. Furthermore, the CASI bands can be positioned by the operator to suit the remote sensing application. In the band-set depicted in the figure, they have been pooled into the visible portion of the spectrum to maximize water penetration.

The market share of CASI is now being challenged by several other sensor manufactures, most notably Ocean PHILLS (Ocean Portable Hyperspectral Imager for Low-Light Spectroscopy) and Specim's AISA family of airborne hyperspectral sensors (Goodman & Ustin, 2007). Active instruments avoid many of the problems associated with passive sensors arising from the detection of a very low seabed signal. The two most widely used active remote sensing systems are bathymetric LiDAR and vessel-based acoustic (sonar) systems.

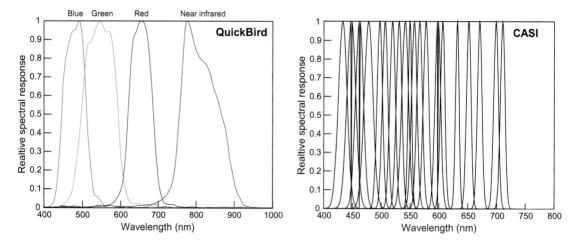

Figure 8.17 Left pane is the relative spectral response of DigitalGlobe's QuickBird satellite. The superior spectral resolution of CASI (right pane) facilitates far improved discrimination of seabed type on the basis of spectral quality.

8.6 Sensor-down versus reef-up sensing

Coral reef mapping via remote sensing follows two approaches: 'sensor-down' or 'reef-up'. Sensor-down is simple in being purely image-based. The approach utilizes a classifier to partition image pixels into classes. Guided by on-ground knowledge, these are subsequently named as assemblages (seagrass, algae, coral rubble, etc.). The classification commonly acts upon one or more spectral bands, though aspects of image derivatives such as texture and object-orientated indices (e.g. edge-strength) have also proven useful (Kohler *et al.*, 2006; Purkis *et al.*, 2006).

As favoured by NOAA's reef mapping initiative, manual digitization of imagery can be considered sensor-down in its most straightforward guise. Though technically simple, human interpretation is a tried and tested technique and has been adopted in the NOAA case because automated classification is deemed unable to offer sufficient accuracy across the vast swaths of US tropical sea floor that fall under their responsibility to map. Visual interpretation remains in favour for large-scale mapping, as automated spectral methods are unable to recognize complexities in a reefscape that are obvious to a trained interpreter. This is a pertinent reminder of the present limitations for satellite reef mapping at regional to global scales.

Coined in a 2000 paper by Hochberg & Atkinson, the so-called 'reef-up' method relies on analyses of field reflectance spectra to identify the wavelengths of characteristic features that facilitate spectral separation of seabed types. Bands of interest are teased out of the field spectra using derivatives (Hochberg & Atkinson, 2000; Hamylton, 2009). In a similar vein, field spectra can be pooled into an exhaustive 'spectral library', which in turn is harnessed as a training set in a classification. The approach is powerful in that the classifier is trained solely using *in situ* optics and thus is image-independent. Flexibility is also maintained, as the spectral library used for training can be resampled to the bandwidths and sensitivity of any instrument used in future flight campaigns.

Collecting ground-truth data to validate reef maps is time-consuming and expensive. The use of a spectral library eliminates the need to collect ground data that will be used to guide a supervised classification, allowing more time to be spent on collecting information to be used for accuracy assessment. Unfortunately, the technique has yet to be employed at a truly regional scale for reefs.

Reef-up strategies are better suited to airborne hyperspectral remote sensing, since the few broad-wavelength bands offered by satellites do not readily allow substrate mapping by their reflectance spectra. This does not mean that field spectra are redundant for use in conjunction with multispectral satellites, though, and they have been shown to have significant utility, providing that a reasonable degree of optical closure can be attained through robust correction of water column and atmospheric effects (Purkis, 2005).

8.7 Spectral unmixing

Individual coral colonies rarely achieve diameters exceeding several metres. In contrast, communities of colonies frequently attain extents of many hundred of metres, which coalesce to semi-continuous structures that can be hundreds to thousands of kilometres in length (fringing and barrier reefs). Remote sensing the location of reefs is therefore achieved easily with fairly coarse-resolution instruments that offer pixel sizes of tens to hundreds of metres. Satellites within the Landsat programme have been utilized for this purpose for three decades (Lyzenga, 1981; Zainal *et al.*, 1993; Ahmad & Neil, 1994; Dustan *et al.*, 2001; Purkis & Pasterkamp, 2004; Palandro *et al.*, 2008).

Remote imaging of individual colonies is more problematic and poses a suite of difficulties associated with the identification of sub-pixel targets in spectrally mixed pixels. Metre-scale commercial satellites such as IKONOS, QuickBird, and WorldView-2 are too coarse for this purpose, and presently only aircraft-mounted instruments can be used to resolve reef architecture at the colony level.

In terrestrial settings, linear spectral unmixing algorithms have been very successful in reconstructing the proportion of cover by different end-member land cover types (Adams *et al.*, 1986; Foody & Cox, 1994). Such approaches proceed under the assumption that that the reflectance of a pixel is the sum of the end-member spectra scaled in linear proportion to the cover of each end-member within the pixel. As shown in Figure 8.18, unmixing strategies in shallow water systems are possible (Hochberg & Atkinson, 2003; Hedley *et al.*, 2004; Conger *et al.*, 2006; Goodman & Ustin, 2007) but, due to the optical complexity of the light-field in these environments, these approaches have yet to be implemented routinely at reef-scale.

The primary barrier to unmixing reef pixels is finding a way to devise an algorithm that is sufficiently sensitive to tease out multiple, optically very similar, target substrates, while remaining adaptive to a spatially inhomogeneous distribution of

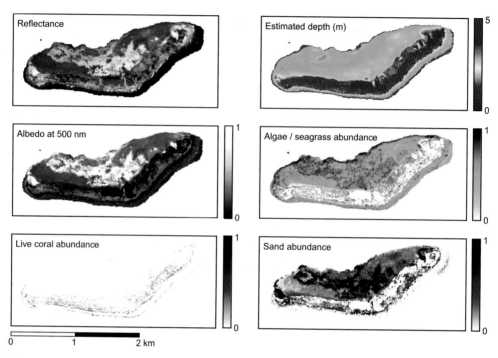

Figure 8.18 Example output derived from hyperspectral imagery of Enrique Reef, Puerto Rico (following techniques described in Goodman & Ustin, 2007; Guild *et al.*, 2008). From top left to bottom right: true-colour surface remote sensing reflectance; model-derived water depth in metres; bottom albedo at 550 nm; and estimated habitat composition depicted as a function of the abundances of algae, seagrass, coral, and sand. *Source data*: AVIRIS (500 × 225 pixels at 3.7 m spatial resolution, covering 1.5 km2; 10 nm spectral resolution from 400–800 nm). Credit: J. Goodman.

water quality. Furthermore, the nature of benthic reflectances and their mixing, as they propagate to the surface through the water column, is almost certainly non-linear in the majority of cases. However, having said this, such techniques will become ever more feasible as further hyperspectral data become available with the launch of future hyperspectral satellites.

Counter to the limitations of sub-pixel unmixing based on the spectral component of an image, there exist spatial techniques to determine the composition of a pixel at a finer scale than the resolution at which the data were collected. Such strategies are very much at the experimental stage and utilize the existence of 'scale-invariance' in the geometry of reef habitats. They allow the extrapolation of area-frequency relations observed at the coarse scale to elucidate the occurrence of habitats at the fine scale. Investigations in this realm reveal that reefs are fractal within certain thresholds of scale (Purkis *et al.*, 2005; 2007).

It is presently unclear whether the observed scaling behaviour can be attributed to the biological or sedimentological functioning of the reef depositional environment or, indeed, an interplay between the two. Purkis & Kohler (2008) and Purkis *et al.* (2010) discuss possible mechanisms and their likely contribution to scaling functions.

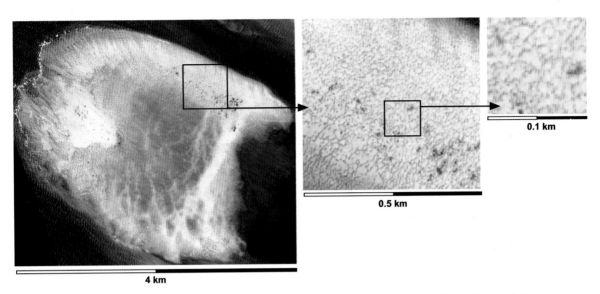

Figure 8.19 Enhanced QuickBird image of an offshore carbonate shoal, Abu Dhabi (United Arab Emirates). The reticulated 'honeycomb' structure that these coral frameworks adopt is persistent across a series of length scales (Purkis *et al.*, 2010). Such 'fractal' behaviour can possibly be harnessed to predict patchiness at spatial scales beyond the resolution of the imagery. Satellite image: DigitalGlobe.

8.8 Image-derived bathymetry

Providing that the sea floor is shallow and well lit, we have seen that remote sensing is a powerful technology to map reef features over vast areas. The product of such work is, however, two-dimensional and fails to capture the vertical complexity of the reef system. Nonetheless, this depth component is critical for interpreting the geomorphology of the reef structure, which in turn may offer significant insight into factors such as the distribution of fish and invertebrate life (Andréfouët *et al.*, 2005; Purkis *et al.*, 2008; Mellin *et al.*, 2009) as well as the geological history of the system (Boss, 1996; Wilkinson & Drummond, 2004; Purkis & Kohler, 2008; Brock *et al.*, 2008; Purkis *et al.*, 2010).

In rare cases, ancillary information on water depth for an area of interest may be available. This could, for example, have been acquired from vessel soundings or from an airborne bathymetric LiDAR survey. In such cases, the two-dimensional product from the remote sensing analysis can be fused with the separate bathymetry survey to yield a three-dimensional reef dataset (Figure 8.20). However, such data are not available for most reefs because of the difficulty in obtaining soundings over vast expanses of shallow water.

There exists the possibility, however, of extracting bathymetric dataset directly from multispectral/hyperspectral satellite or airborne images such as Landsat, IKONOS or CASI (Green *et al.*, 2000; Andréfouët *et al.*, 2005). These optical

Figure 8.20 True-colour QuickBird image draped over an optically derived DEM for a fringing reef system in the central Red Sea. The sea floor geomorphology is clearly resolved. Data courtesy of the Khaled bin Sultan Living Oceans Foundation. Satellite image: DigitalGlobe.

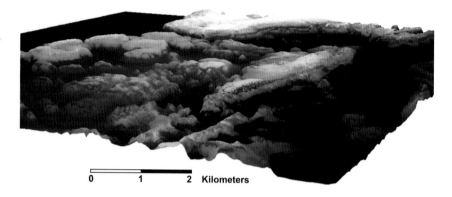

Figure 8.20 True-colour QuickBird image draped over an optically derived DEM for a fringing reef system in the central Red Sea. The sea floor geomorphology is clearly resolved. Data courtesy of the Khaled bin Sultan Living Oceans Foundation. Satellite image: DigitalGlobe.

bathymetric algorithms are sufficiently reliable to now be implemented commercially. Resulting from research conducted in the 1980s, hydrographical charts of French overseas territory (SHOM-SPT 1990) now include bathymetry derived using Satellite Pour l'Observation de la Terre (SPOT) satellite data and a multi-regression algorithm. NOAA is updating bathymetric maps of the Northwestern Hawaiian Islands using a revised ratio-algorithm applied to IKONOS data (Stumpf *et al.*, 2003).

By virtue of its limited spectral resolution, multispectral imagery is less well poised than hyperspectral to provide a bathymetry estimate. However, the strategy proposed by Stumpf (2003) is very satisfactory providing that a sufficient number of independent soundings are available with which to tune the algorithm's coefficients. This empirical ratio approach capitalizes on the differential attenuation of blue and green light by water. Lyzenga *et al.* (2006) achieved a similar result using a radiative transfer model. With the additional number of bands that hyperspectral imagery provides, water depth can be retrieved using techniques such as optimization (Lee *et al.*, 1999; 2001), spectral unmixing (Hedley & Mumby, 2003, Hedley *et al.*, 2004) or look-up-table methodology (Louchard *et al.*, 2003; Mobley *et al.*, 2005; Lesser & Mobley, 2007), to name but a few. The optical derivation of water depth is also covered in Chapter 7 in the context of rivers.

8.9 LiDAR

Bathymetric LiDAR instruments are exclusively aircraft-mounted and operate on the transmission of green laser light (typically ≈500 nm) and the recording of the amount of energy back-scattered from the seabed. LiDAR instruments measure the distance to the surface of the water and the distance to the bottom of the water body. The difference between these two measurements is the depth of the water at that spot. Coupled with robust GPS information, a repeated series of measurements along designated transects provides a highly accurate characterization of bathymetry. The cost of conducting aircraft-based LiDAR surveys varies greatly with the mapping density desired and can range from approximately \$375 per km^2

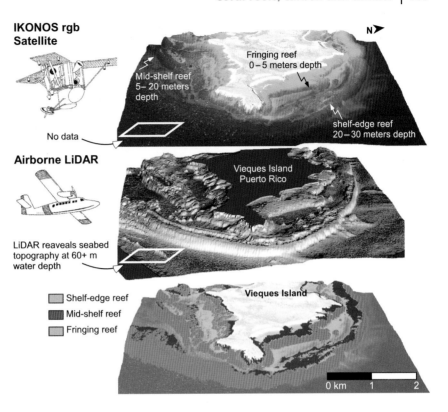

IKONOS rgb Satellite

Fringing reef
0 – 5 meters depth

Mid-shelf reef
5 – 20 meters
depth

shelf-edge reef
20 – 30 meters depth

No data

Airborne LiDAR

Vieques Island
Puerto Rico

LiDAR reaveals seabed
topography at 60+ m
water depth

Shelf-edge reef

Mid-shelf reef

Fringing reef

Vieques Island

0 km   1   2

Figure 8.21 Combined reef mapping using IKONOS multispectral satellite data (top pane) and airborne bathymetric LiDAR soundings (middle pane). While the LiDAR lacks any spectral capability, it is capable of resolving the seabed to depths exceeding 60 metres (white rectangle). The resulting map product (lower pane) is both highly accurate and in 3-D. The data were acquired over the eastern point of the Island of Vieques (Puerto Rico). Satellite image: GeoEye.

at 5×5 m resolution to \$2,000 per km² at 2×2 m resolution sensing (Rohmann & Monaco, 2005). The technology and operation of LiDAR sensors will be considered in greater detail in Chapter 9.

While not a pixel-based imaging technology, bathymetric LiDAR soundings can be interpolated to a raster and viewed as a single-band image (Figure 8.21: middle pane). Processed in this way, the data reveal a picture of seabed topography and can be used as the basis for mapping, much like a satellite image (Storlazzi *et al.*, 2003; Brock *et al.*, 2004, 2006; Collin *et al.*, 2008; Walker *et al.*, 2008; Nayegandhi & Brock, 2008; Purkis & Kohler, 2008; Costa *et al.*, 2009). The advantage of LiDAR over passive visible imagery is that it typically offers a greater depth of penetration into deep or turbid water, while maintaining a high spatial resolution (several m² and easily comparable to the capability of QuickBird or IKONOS; Figure 8.21). That is not say that the laser is not rapidly attenuated but, by virtue of its intensity, it travels deeper into the water than diffuse sunlight and commonly to two to three times the Secchi Depth (Wang & Philpot, 2007), which could be as deep at 60 m in clear reefal waters (Figure 8.21: white rectangle). This is far superior to the depth penetration of passive optical systems, which are generally limited to no better than 1.5 Secchi depths.

The majority of bathymetric LiDARs do not return a spectrally calibrated signal, and hence the resulting interpolated image cannot be used in a classification

Figure 8.22 Sea floor reflectance image of a section of the Florida reef tract (offshore Dania Beach) obtained at 532 nm from the SHOALS-3000 LiDAR instrument, a component of the CHARTS instrument array. In addition to water depth, SHOALS is able to provide two additional products that have been unavailable with previous LiDAR instruments: sea floor reflectance and water column attenuation. Credit: Optech International.

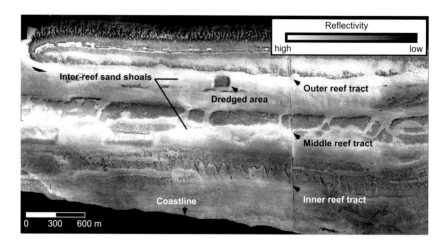

that keys from substrate albedo (as would be the case for a multi- or hyper-spectral scene). It is therefore common practice for the data to be interpolated manually, with a user digitizing reef features (Walker *et al.*, 2008), or else a classifier is devised that operates on the uncalibrated laser backscatter values (Collin *et al.*, 2008). In either case, good ground control is demanded in the form of seabed descriptions, photographs or video.

Some of the more advanced bathymetric LiDAR instruments, such as the Optech SHOALS-3000 (Figure 8.22) and the ADS Mk II Airborne System operated by Tenix LADS (Figure 14.1) are calibrated and return a reflectance 'image' of the seabed at the wavelength of the laser beam (Tuell *et al.*, 2004; Kopilevich *et al.*, 2005; Tuell *et al.*, 2005). The interpolated image, while still only a single band, can be considered spectral, so offers the opportunity to map the seabed on the basis of its greenness (bathymetric LiDAR lasers are green).

This technology will be covered in Chapter 14 (Section 14.1.1). Calibrated LiDARs are also covered in detail in Chapter 6 in the context of mapping land cover in cities.

8.10 Sonar

Powerful as optical remote sensing is, the technology is limited to relatively shallow clear waters. While this is the preferred home of corals, extensive coral reefs do exist in shallow waters too turbid for airborne or satellite imagery (Riegl *et al.*, 2007), and many deeper frameworks exist that have diverse and extensive coral cover (Riegl & Piller, 2003). These areas, sometimes termed 'coral carpets', may be especially important when inclement conditions damage shallow water resources but spare corals at depth. These survivors may then provide a battery of larvae that will recolonize the upper reef, allowing the ecosystem to survive to fight another day. The severity of heat-induced bleaching has been observed to be differential with depth (Berkelmans & Oliver, 1999; Sheppard *et al.*, 2008). Relatively insensitive to turbidity and capable

Table 8.2 The utility of four sonar-based technologies for reef characterization.

Sonar	Attributes
Single-beam echo sounders	A cost-effective platform for bathymetry and sediment classification mapping, but requires interpolation to produce spatially continuous coverage from point data. Single-beam sonar data are increasingly used for benthic habitat characterization and have found particular application in the mapping of vegetative cover in variable water depths (Riegl & Purkis, 2005; Foster *et al.*, 2009).
Side-scan sonar	Side-scan sonar can provide high-resolution imagery of the sea floor (Goff *et al.*, 2000; Collier & Humber, 2007) but cannot provide bathymetric depth information by itself, which is of key importance in classifying benthic habitats for management and ecological purposes. Survey systems that provide both side-scan sonar and concurrent depth sensory readings are very expensive. Because side-scan sonar instruments must be towed below the surface, the exact positioning of the data can also be difficult to control and determine.
Multi-beam echo sounders	Multi-beam sonar instruments are desirable for marine benthic mapping because they can provide both spatially accurate bathymetry and benthic characteristic information over a continuous swath of sea floor (Kostylev *et al.*, 2001; Wilson *et al.*, 2007). Multi-beam sonars have transducers that send and receive a large number of concurrent, highly accurate and precisely located signals in an array across the swath of the instrument beneath the vessel. The return signals are analyzed for the time required for return and return echo strength. The time required for the sound wave to reflect off the sea floor and return is the indicator of depth and provides the bathymetry data. The strength of the echo return (termed 'backscatter') is a function of the incident angle of the sound pulse, the roughness or surface characteristics of the sea floor (e.g. coral, sand, or seagrass) and the composition or density of the bottom (e.g. rock or mud). Multi-beam sonar instruments are generally hull-mounted on the vessel for maximum stability. One problem with multi-beam sonar is that swath-width is a function of depth and, under varying topographical conditions, sample width and backscatter signal characteristics along the vessel track can vary significantly.
Interferometric sonar	A unique capability of interferometric systems is that they combine acoustic bathymetry with simultaneous side-scan sonar data acquisition. The result is a co-registered, well-positioned bathymetry and 'backscatter' imagery that can be used to better identify and classify habitat characteristics. The technique is experimental and costly and has been trialled by NOAA in the Florida Keys (Rohmann, personal communication).

of penetrating hundreds of metres of water column, acoustic vessel-based sensors can probe these areas that are inaccessible to optical technologies.

Sonar (SOund Navigation And Ranging) is a technique based on the principle of emitting sound energy and analyzing the echoes returned from sea floor features. There are four types of sonar systems typically used in reef habitat or bathymetry mapping work (Table 8.2).

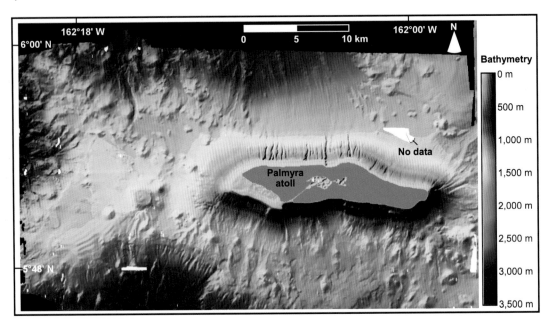

Figure 8.23 A NOAA multi-beam bathymetric survey of Palmyra Atoll (US Pacific). Spatial resolution of the dataset is 40 m^2 and they were collected by the NOAA Coral Reef Ecosystem Division. Multi-beam is an active ship-mounted remote sensing technology utilizing acoustic energy and is thus unlimited by water depth. Reef geomorphology is clearly resolved. Credit: NOAA.

Side-scan sonars and multi-beam swath systems (Figure 8.23) represent the state of the art, but are expensive (on a par with a hyperspectral airborne campaign) and thus limited to big-budget projects. For example, multi-beam sonar data can range from ≈$1,500–$5,000 per km^2, depending on water depth (Rohmann & Monaco, 2005). For surveys of less than 100 km^2 with water depths of less than 50 m, airborne LiDAR is deemed to be more cost-effective than multi-beam sonar and faster to acquire.

As highlighted by Costa *et al.* (2009), these higher efficiencies for LiDAR are due to the system's distinct acquisition geometry, wider swath-widths, and faster survey speeds. In particular, the average acquisition speed is much faster for LiDAR systems, approximately 140 knots, while the average speed of a survey ship is approximately 8 knots. Also, for LiDAR, swath-width is primarily determined by scan angle and is nearly independent of depth and aircraft altitude (Stephenson & Sinclair, 2006). Conversely, the relationship between swath-width and water depth is proportional for multi-beam systems (i.e. the shallower the water, the narrower the swath and the smaller the area that can be mapped on a single survey line).

Less capable but many orders of magnitude cheaper than LiDAR, multi-beam, or side-scan, single-beam echo-sounders can be used to discriminate seabed character from small boats (Figure 8.24). These systems have comparatively modest data storage requirements and potentially quick post-processing turnaround times, and they are easily deployed. The common approach to acoustic seabed

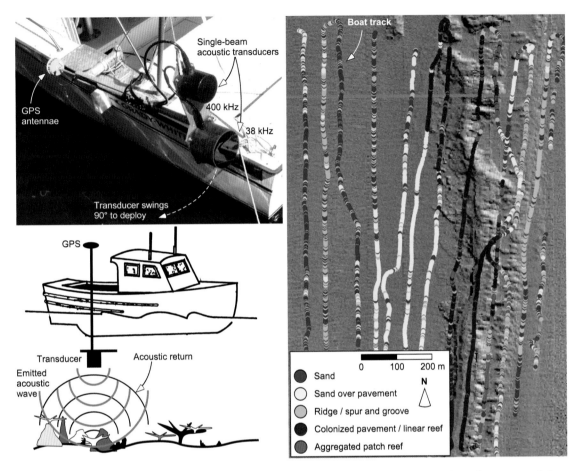

Figure 8.24 Single-beam acoustic ground discriminating sonar characterizes return echo waveforms by energy and shape parameters. By matching acoustic signatures with observed bottom types, discrete predictions of seabed character can be produced along the track of the vessel. This sonar can be used for mapping but, since, acoustic observations are only made beneath the transducer, interpolation is required between track lines. Credit: G. Foster.

mapping has been to use sediment classification as a surrogate for benthic habitat, using either the shape of the first returning acoustic echo received from the sea floor by the instrument or else incorporating information from multiple echo returns. The former case is adopted by QTC View echo-sounders (Moyer *et al.*, 2005; Freitas *et al.*, 2006), while multi-echo logging is used by instruments manufactured by RoxAnn (Greenstreet *et al.*, 1997; Hamilton *et al.*, 1999), ECHOplus (Riegl *et al.*, 2007) and BioSonics (Foster *et al.*, 2009). Particular success has been achieved using single-beam acoustic tools to map sea floor vegetation (Freitas *et al.*, 2008; Riegl *et al.*, 2005; Gleason *et al.*, 2006; Foster *et al.*, 2009). Here, acoustic energy interacts strongly with tiny air bubbles in the leaves of algae and seagrass, allowing clear discrimination from surrounding substrata.

The interaction of the acoustic beam with the seabed is very complex and depends on a number of environmental (e.g. seabed composition, surficial roughness, epi- and infaunal biota) and mechanical (e.g. frequency, pulse duration, beam width) factors. The first part of an echo return is primarily energy reflected back to the receiver as a specular return, i.e. a mirror-like reflection of the sound waves making normal-incident contact with the seabed, while the trailing edge of the echo is primarily incoherent backscatter, formed by the outer beam making less than normal-incident contact with the seabed. These two sections of the echo return generally relate to the hardness and roughness of the seabed, respectively, but also encode a myriad of information relating to the structural and biological characters of the benthos. Thus, an acoustic wave returns very different information than does a stream of upwelling photons.

Since the acoustic wave emanates in a cone from the transducer, the greater the water depth, the larger the area on the sea floor that is insonified. In areas with high benthic heterogeneity, this can lead to confusion because the acoustic wave is interacting with a dissimilar area of seabed and is hence modulated differently. Resolution is also a factor of depth; for a given beam width, the likelihood of insonifying a 'pure' class will diminish with increasing depth, due to the increasing diameter of the spreading signal (Foster *et al.*, 2009).

Future challenges lie with better translating the acoustic signal into meaningful biologic information, as well as integrating acoustic seabed inventories with those produced using optical sensors (e.g. Riegl & Purkis, 2005). Through very different mechanisms, both LiDAR and sonar are capable of resolving the geomorphology of the seabed in three dimensions, an important property that can be a good predictor for the distribution of reef fish and coral abundance. In this respect, remote sensing can convey the extraordinary utility of assessing the likely abundance of reef fish communities (Kuffner *et al.*, 2007; Mellin *et al.*, 2007; Pittman *et al.*, 2007; Purkis *et al.*, 2008). Though powerful for studies of limited scale, these acoustic tools are prohibitively expensive to map reef formations across entire ocean basins – a calling that only satellites can answer to.

8.11   Sub-bottom acoustic profiling

While the remote technologies considered thus far in this chapter sense the surface of the seabed or overlying water column, acoustic instruments can be used to view the internal structure of reefs. This is useful, as a reef will always strive to grow to sea level. In the absence of tectonic uplift or subsidence, the ability to look inside a reef and observe the level to which it grew can reveal the height of the sea during past geological periods. In recent geological history, sea level dropped to $\approx 140\,m$ below present due to vast quantities of water being locked up in northern hemisphere glaciers during the last ice age (approximately 20,000 years ago). What is still a topic of debate is whether the sea rose in a smooth curve up to its present position as the ice melted, or if the rise was punctuated by instances of very rapid rise, interspersed with periods of near stasis. This is relevant as we consider the

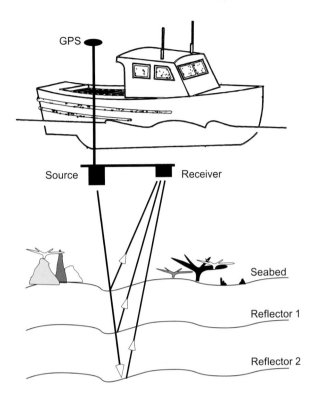

Figure 8.25 Operation of a sub-bottom acoustic profiler. Sound produced at the source reflects off the seabed, reflector 1 and reflector 2, which are areas of rapid density change. The receiver captures the reflected sound waves and forms an image of the two-dimensional cross-section through the seabed (i.e. Figure 8.26).

likely sea level trajectory under present global warming (Chappell & Polach, 1991; Blanchon *et al.*, 2009; Chapter 1, Figure 1.1).

Sub-bottom profiling systems operate using a low-frequency, high-intensity acoustic wave that is emitted from a transducer held beneath the water surface. The transducer may be hull-mounted or towed on a platform behind the vessel (e.g. Figure 9.12). The sound wave is emitted vertically downwards and, by virtue of its low frequency, it travels through the water column with little attenuation. Upon reaching the seabed, it passes through the sediment/water interface and continues to propagate into the subsurface (Figure 8.25). This wave will be reflected by boundaries between buried layers that have different acoustical properties and hence cause an 'impedance contrast'. A receiver, also mounted below the water, is used to record the travel time, comparing the time that the signal was emitted with the times that these acoustic returns are received. Post-processing of the reflected signal provides information as to the depth and type of layers within the body of the reef.

The sub-bottom profile depicted in Figure 8.26 demonstrates how the surface of the modern (Holocene) reef (green line), as well as the position of the now-buried Pleistocene reef (magenta line) that was actively growing 80,000–150,000 years before the present, can be resolved. The profile also reveals the flanks of the reef, showing that they rise from a flat seabed of approximately 400 m depth. The pronounced break in the angle of the slope at 85 m likely depicts a change in the rate of sea level rise following the last ice age.

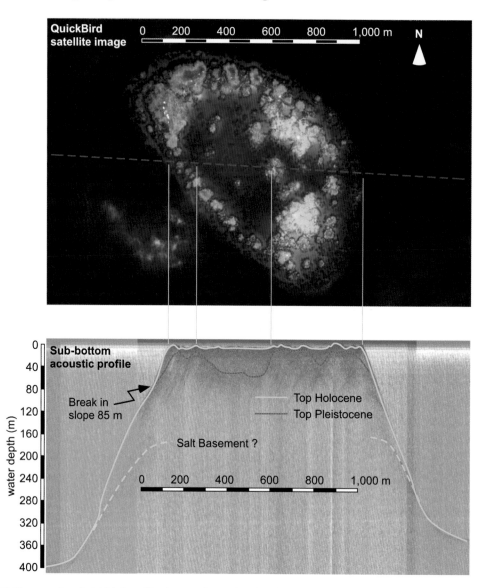

Figure 8.26 Top pane is a QuickBird satellite image of an isolated reef in the Saudi Arabian Red Sea. The broken red line is a track acquired by a sub-bottom acoustic profiler (bottom pane). Data courtesy of the Khaled bin Sultan Living Oceans Foundation. Satellite image: DigitalGlobe.

The frequency of sub-bottom instruments typically falls in the range of 2 to 30 kHz, an order lower than a bathymetric echo-sounder (200 kHz), which explicitly does not intend the signal to penetrate the seabed. The lower the frequency of the sub-bottom, the greater the depth of penetration, but the coarser the vertical resolution with which sediment layers are resolved. For this reason, many systems operate at multiple frequencies.

A limitation of the technology is that the depth of acoustic reflector layers can only be calculated once the speed of sound through the sediment matrix is known. As with all remote sensing technologies, this requires ground-truth, which is conducted by drilling a core through the reef and retrieving samples whose acoustic properties can be measured in the laboratory. Such drilling is, however, expensive, considering that it frequently must be conducted far from land, either from a boat or using a drill mounted on the seabed (i.e. Figure 8.3). Furthermore, sub-bottom instruments can penetrate to depths exceeding 60 m and the core must consequently be rather long and invariably expensive to obtain. Sub-bottom profiling is, however a powerful technology for reef characterization, and it complements planimetric images such as those obtained from satellites (Figure 8.26).

8.12   Radar applications

Operating in the microwave portion of the electromagnetic spectrum (1 mm–30 cm), Synthetic Aperture Radar (SAR) (covered in detail in Chapter 6, Section 6.2.4) does not penetrate into the water body; as seen in Chapter 7, this property makes microwave sensing useful for flood detection and tracking. Instead, the microwave energy reflects off of the upper micro-layer of the surface. At first glance, SAR does not, therefore, seem an intuitive instrument with which to study coral reefs. However, while it cannot discern seabed type, radar can be used to extract seabed geomorphology and bathymetry. These properties are calculated solely from properties of the micron-thick skin of the water surface.

Although the methodology is well refined, the technique has rarely been applied for studying reefs. One of the few articles that consider the technology is provided by Jones *et al.* (2006), who demonstrate utility for mapping slicks of coral spawn on the sea surface in the Timor Sea. The signature of the spawn, Jones notes, may easily be confused with polluting oil, so care must be taken to discriminate between natural events and hydrocarbon slicks emanating from the petroleum industry. This extraordinary application demonstrates the diverse capabilities of remote sensing for reef research.

The caveat to retrieving water depth from SAR is that the image must be acquired under favourable meteorological and hydrodynamic conditions, i.e. moderate winds of 3–10 m/sec and significant currents of about 0.5 m/sec (Alpers & Hennings, 1984, Vogelzang, 1997). These criteria are necessary to ensure that the water body is in motion over the seabed. This flow (typically tidally driven) interacts with the bottom topography, causing modulation of the surface current velocity, which in turn give rise to local variations in surface wave patterns. An over-flying SAR senses such variations through modulations in the backscattered radar signal (e.g. Lyzenga, 1991).

As with the previously discussed optical derivation of water depth, some reference soundings are required to tune the extraction model. Suitable conditions for SAR imaging can be elusive and have, to date, mitigated widespread use of the technology for resolving reef geomorphology. In cases where conducive conditions are found, the potential of SAR is, however, heightened by its ability to image the sea surface through cloud cover at any time of the day or night, coupled with data far cheaper than optical imagery, yet with comparable pixel size. For example, in its 'spotlight' mode, the TerraSAR-X satellite, lofted from Kazakhstan in 2007, can acquire images of an area measuring $5\,km \times 10\,km$ with a resolution as fine as $1\,m^2$.

8.13   Class assemblages and the minimum mapping unit

Because of the limitations of today's passive-optical remote sensing technologies, reliable sub-pixel unmixing is unattainable in the majority of mapping projects. Instead, a classification is adopted that encompasses the biotic diversity of an area through the definition of assemblage classes. These classes must adequately describe the geomorphology of the reef, while offering a sufficient summary of the overlying biota so as to be ecologically relevant. The classification scheme must also be able to adapt to deliver a number of classes commensurate with the spatial and spectral resolution of the sensor being used.

As highlighted by Mumby & Harborne (1999), most reef mapping to date has been conducted on an *ad hoc* basis with little consistency in terminology. This has prevented the integration of maps from different sources into a greater archive. The team also noted that this lack of consistency is detrimental to coral reef science and management at regional scales and requires urgent standardization. This problem has undoubtedly arisen from the fact that reef remote sensing has undergone a long period of technology and algorithm development, and only recently has being employed in government-led regional to international mapping efforts.

The last few years have seen this picture changing, with the initialization of multi-million dollar mapping projects covering thousands of square kilometres. One of these is that of NOAA, which at the time of writing is well on its way to publishing coral reef maps for all US territories. To accomplish this task, they have adopted a flexible system termed the 'Coral Ecosystem Classification Scheme' (Monaco *et al.*, 2001; Rohmann *et al.*, 2005). The scheme was developed to provide a 'common language' to compare and contrast digital maps derived from various remote sensing platforms (envisioned to be IKONOS, colour aerial photography and vessel-based active acoustic instruments). Importantly, a hierarchical structure is taken to enable users to add habitat categories if and when they are required. Such flexibility is paramount when dealing with large and diverse ecosystems such as reefs.

In the NOAA scheme, each habitat polygon is assigned three attributes: zone (e.g. fore-reef, reef-flat, lagoon, etc.); geomorphological structure (mud, sand, spur-and-groove, etc); and biological cover (seagrass, coral, etc). By allowing the use of combinations of descriptors, such a 'multi-layered' classification scheme requires relatively few headings to describe a multitude of seabed types.

Beyond the system adopted by NOAA, there are two further classification schemes that can be employed for large-scale reef mapping:

- **CMECS** is a hierarchical standard ecological classification system intended to be universally applicable for coastal and marine ecosystems. It has been developed by NatureServe (www.natureserve.org) and is currently being evaluated and considered for adoption as a United States national federal standard. The CMECS framework is applicable on spatial scales of less than one square metre to thousands of square kilometres and can be used in littoral, benthic and pelagic zones of estuarine, coastal and open ocean systems. The hierarchical framework currently contains six nested levels (Regime, Formation, Zone, Macro-habitat, Habitat, and Biotope), each containing clearly defined classes and units. Conceptually, the hierarchy can be divided into two types of data requirements. The upper levels (i.e. Regime to Zone) can generally be mapped from remote imagery and bathymetric data. In contrast, the lower levels (i.e. Macro-habitat to Biotope) exist at local spatial scales, and underwater photographic or direct observational data are usually required to populate them.
- **SCHEME** (Florida's System for Classification of Habitats in Estuarine and Marine Environments) is a system for classification of marine and estuarine benthic habitats which has been adopted by Floridian government agencies but also has utility to be applied in diverse reef systems. Classification is again hierarchical in structure, with five levels (Class and four possible Subclass designations) and two lists of modifiers (General and Taxonomic) applied to each of the appropriate levels. SCHEME is both compatible with NOAA's 13-level classification system and NatureServe's CMECS system.

The minimum mapping unit (MMU) is the smallest feature (e.g. individual patch reef, seagrass patch or other distinguishable feature or aggregate of features) that will be delineated on a map being produced from a given source of imagery (e.g. aerial photography, satellite imagery or LiDAR) and mapping protocol (e.g. electronic image analysis or visual interpretation). Deciding on the MMU to be used is a balance between providing maps with sufficient detail to meet the requirements of the people using them and the time needed and cost to produce the product.

Generally, the size of the MMU selected will be a trade-off between the desire to map small features (e.g. individual coral heads or patch reefs) that may be important to habitat interpretation versus the time required to identify and classify all features of this size visible in the data. The smaller the MMU adopted, the more individual features there will be to map and the more expensive the project will be.

The ability to assess the thematic accuracy (i.e. how many of the benthic habitat features are correctly classified) will also be a factor in setting the MMU size. The smaller the MMU, the more field habitat observations will need to be collected to adequately assess map thematic accuracy. The NOAA Coral Ecosystem Mapping Team has used an MMU of approximately 1 acre ($\approx 4,000\,m^2$) to map the benthic habitats in Puerto Rico, US Virgin Islands, Hawaii, American Samoa, Guam, and the

Northern Mariana Islands with visual interpretation of satellite imagery. A MMU of approximately $100\,m^2$ (0.0247 acres) was used when mapping the benthic habitats of the Northwest Hawaiian Islands with semi-automated image analysis. For regional-scale mapping of the entire Saudi Arabian Red Sea, the Living Oceans Foundation and the National Coral Reef Institute opted for a MMU of $4\,m^2$, even though the area covered was comparable to the NOAA project (Purkis *et al.*, 2010).

8.14 Change detection

Changes in coverage from live stands of coral to benthic algae indicate a degradation in the health of the reef system. It is necessary to monitor the extent of such successions, termed phase-shifts, to understand whether the perturbation is local and likely part of the natural reef cycle or, more worrying, reef-wide and indicative of an impending collapse of an entire ecosystem. It follows that if a remote sensing system is to be able to assess the biological state of a coral reef, it must at least be capable of distinguishing living coral from dead coral covered by an overgrowth of benthic algae.

The most common cause of reef-wide coral death is bleaching (Baker *et al.*, 2008; Baird *et al.*, 2009). This was first documented in 1911 in the Florida Keys and became a well known phenomenon during the 1980–90s when the large-scale occurrence of bleaching became evident. As demonstrated in both 1998 and 2002, the most severe bleaching events typically accompany coupled ocean-atmosphere phenomena, such as the El Niño-Southern Oscillation (ENSO) (Glynn, 1993; 1996). Bleaching results in the catastrophic loss of coral cover in some locations and has changed coral community structure in many others, with a potentially critical influence on the maintenance of biodiversity in the marine tropics (Riegl *et al.*, 2008).

Models of coral reef dynamics that incorporate future accelerated bleaching are not optimistic (Done, 1999, Hoegh-Guldberg, 1999; Sheppard, 2003; Donner, 2009). Bleaching has facilitated or initiated increases in many coral diseases, the breakdown of reef framework by bioeroders and the loss of critical habitat for associated reef fish and other biota (Pratchett *et al.*, 2008; Wilson *et al.*, 2009). Likewise, increasing CO_2 concentrations associated with ocean acidification are predicted to reduce the rates at which corals deposit their calcium carbonate skeletons by 20–60 per cent (Langdon *et al.*, 2000; Kleypas & Langdon, 2006; Manzello *et al.*, 2008), leading to vastly increased bioerosion and reef destruction (Guinotte & Fabry, 2008). Further discussion on this topic can be found in Chapter 11 (Section 11.6).

Bleaching involves the forced expulsion of a coral's symbiotic algae (*zooxanthellae*), which carry photosynthetic pigment and give the colony its distinctive colour (Figure 8.27). With the *zooxanthellae* providing up to 90–95 per cent of the energy requirements of the coral host, the expulsion of these valuable symbionts induces death by slow starvation. Bleaching may be induced by a variety of causes that stress the coral, and typically is related to factors such as water

Figure 8.27 Scanning electron microscope images of the coral *Montastraea cavernosa*. The spherical features housed within the layer of tissue named the mesoglea are the symbiotic *zooxanthellae* algae. These symbiotic algae are expelled during bleaching, causing the colony to loose its colour. Credit: A. Renegar.

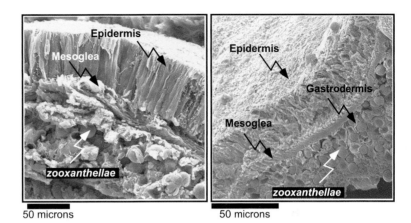

temperatures above or below the thresholds that the colony can tolerate, intense illumination by ultraviolet radiation, sediment, chemical or salinity stress, and disease (Rosenberg *et al.*, 2009). Once the algae have vacated, the denuded carbonate skeleton of the coral becomes visible, rendering the colony bright white in colour (hence the term bleaching).

At this time, the bleached coral framework has a much higher reflectance in the visible spectrum than that of live coral tissue or macro-algae. Similar to the bright signature of coral sand, the shape of the spectrum is relatively flat, with monotonous increase towards greater wavelengths (Figure 8.28). This distinct spectral signature of bleached coral is reasonably easy to detect, even with the impoverished spectral resolution of a multispectral satellite (Figure 8.29).

However, the bare coral skeleton is rapidly colonized by turfing algae (Diaz-Pulido *et al.*, 2009; Norstrom *et al.*, 2009) that have a spectral signature similar to the live coral (Figure 8.28). This presents a relatively narrow window of time, often less than a couple of weeks, during which the bleached reef remains spectrally different from its healthy live state. This window is typically much shorter than the repeat cycle of the commercial satellites that offer the necessary metre-scale pixels required to resolve such an event (IKONOS, QuickBird, WorldView-2), and too brief to allow a comprehensive airborne survey to be mobilized.

Remote detection of bleaching therefore remains a hit-and-miss affair, rather than a robust monitoring tool. Nonetheless, coral bleaching has been directly observed using aerial photographs (Andréfouët *et al.*, 2002), IKONOS (Elvidge *et al.*, 2004; Rowlands *et al.*, 2008), Landsat (Yamano & Tamura, 2004) and even side-scan sonar (Collier & Humber, 2007).

The challenge of change detection lies in the separation of meaningful changes of interest from incidental image-to-image change due to differing illumination conditions, seasonal changes, clouds, water column differences or spatial mis-registration. Furthermore, it must be appreciated that there is a natural broad continuum of pigment state (colour) in corals, as healthy colonies naturally go through seasonal changes in *zooxanthellae* densities (Fitt *et al.*, 2000). These uncertainties have, to date, precluded the routine use of satellite data as a means of bleaching detection.

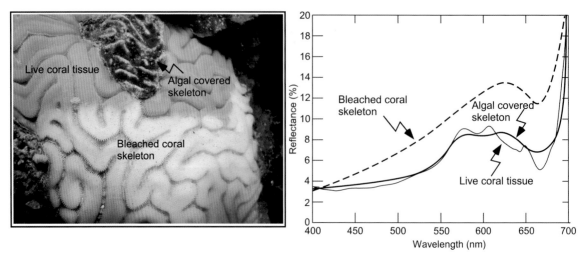

Figure 8.28 A coral colony demonstrating the three-stage bleaching process with associated spectral signatures. Live coral tissue has a very similar spectral signature to that of a bleached colony overgrown by turfing algae. The bright white of the denuded carbonate skeleton is spectrally distinct but short-lived.

Figure 8.29 Detection of bleaching using IKONOS data on a patch reef system offshore Roatan, Honduras. As the photographs portray, the study site supported healthy thickets of *Acropora cervicornis* coral in 2004 that bleached severely in 2005. Pixels flagged as containing bleached coral following analysis of a time-series of imagery are highlighted in magenta. Satellite image: GeoEye.

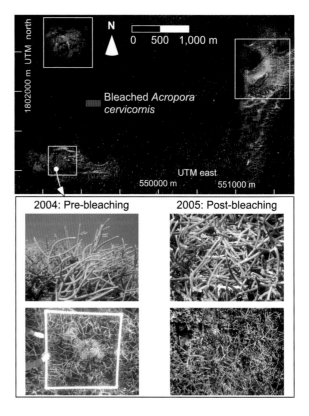

8.15   Key concepts

1 The architects of coral reefs are stony corals which possess a skeleton that is primarily composed of calcium carbonate. Shallow-water reefs are formed in a zone extending approximately 30°N to 30°S of the equator. Being manufactured from calcium carbonate over geological timescales, reefs sequester vast quantities of CO_2 from the atmosphere and have a significant role to play in Earth's carbon cycle.

2 Being extremely sensitive to subtle changes in both ocean chemistry and temperature, corals act as an early indicator of climate change. The future survival of reefs is threatened by a multitude of factors, but the most serious are thought to be: i) an increase in seawater temperature that raises the frequency of mass coral bleaching events; and ii) the decrease in the pH of the oceans as a result of the increased quantity of carbon dioxide in the atmosphere. The 1998 ENSO event had a detrimental impact on an unprecedented expanse of the world's coral reefs. Many still show no sign of recovery even after more than ten years.

3 Since reefs are situated underwater, microwave sensors have very limited use for monitoring. Instead, instruments operating in the short wavelength visible spectrum (blue and green light) penetrate the water to a high degree and hence are of great utility. The principal goal of reef remote sensing is to quantify the character of the sea floor. Multispectral satellites such as IKONOS and QuickBird are capable of differentiating between areas composed of sand, seagrass, algae and corals. In theory, at least, hyperspectral sensors such as CASI are better capable of separating live coral from dead, provided the water column is not too deep (several metres at maximum).

4 The water column overlying the reef alters both the magnitude and distribution of light with respect to wavelength. The effect becomes more acute as wavelength and water depth increase. The optical signal that exits the water surface, having interacted with the sea floor, is a fraction of the strength of the light that originally downwelled onto the surface. By virtue of this weak and distorted signal, remote sensing instruments have to operate at the limits of their sensitivity in order to make meaningful observations of the sea floor. Furthermore, advanced processing using algorithms, such as spectral unmixing and/or radiative transfer theory, may have to be used.

5 Maps of coral reef habitats are typically distributed in a GIS format. There are several regional-scale mapping programmes in operation, and of note is the NOAA initiative in the United States and its territories.

6 Change detection in coral reef environments using time-separated images is challenging because of the subtle and short-lived spectral differences that typify bleaching events. Hyperspectral capability and a short revisit-time maximize the chance of an ecologically relevant observation.

9

Coastal impact of storm surges and sea level rise

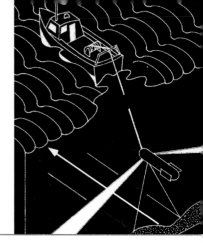

There is a scientific consensus that average temperatures are increasing worldwide and that global mean sea levels have been rising at a rate of about 2 mm per year (Figures 1.1 and 1.2). The sea level is rising because water expands as it is warmed and because water from melting glaciers and ice sheets on land is added to the oceans. Many scientists believe that, due to melting glaciers and expanding ocean water, the sea level rise will accelerate in the future (IPCC, WMO/UNEP, 2007). Since 1993, satellite observations have permitted more precise calculations of global sea level rise, and this has now been estimated to be closer to 3 mm per year over the period 1993 to 2003. In the mid-Atlantic region from New York to North Carolina, tide-gauge observations indicate that relative sea level rise rates were even higher than the global mean and ranged between 2.4 and 4.4 mm per year.

Coastal areas such as barrier islands, beaches, wetlands and estuarine systems are highly sensitive to sea level changes. For instance, rapid sea level rise can cause segmentation of barrier islands or disintegration of wetlands, especially if the sediment supply cannot keep up with the rising sea.

The substantial sea level rise predicted for the next 50 to 100 years will impact coastal economic development, beach erosion control strategies, salinity of estuaries and aquifers, coastal drainage and sewage systems and coastal wetlands (NOAA, 1999). For example, the susceptibility of the Louisiana coast to sea level rise can be seen in Figure 9.1, which shows that large wetland areas might be lost by the time the sea level has risen 1 m. Coastal wetlands, including tidal salt marshes, tidal freshwater marshes and mangrove swamps, are generally within fractions of a metre of sea level and could thus be lost if the sea level rises significantly. In the face of rising sea level, the lowest nation in the world, the Maldives, is currently investigating the feasibility of acquiring a new homeland in preparation for the inundation of the archipelago that will likely occur within a century. The islands rise little more than 1 m above present sea level and even a modest increase in the level of the ocean would render the nation uninhabitable for the future descendants of the 300,000 islanders.

Remote Sensing and Global Environmental Change, First Edition. Samuel Purkis and Victor Klemas.
© 2011 Samuel Purkis and Victor Klemas. Published 2011 by Blackwell Publishing Ltd.

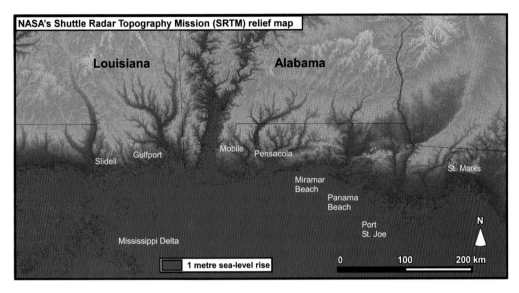

Figure 9.1 Gulf Coast susceptibility to one metre of sea level rise, as depicted by a digital elevation model derived from NASA SRTM data. Source for this dataset is the Global Land Cover Facility. Credit: Landsat 7 project and EROS Data Center.

Global warming, enhanced by human activity, is raising the temperature of the world's oceans as well as increasing their evaporation rate. These two factors are increasing the strength and frequency of cyclones, including Atlantic hurricanes. As demonstrated by the strong hurricanes in 2004 and 2005, increases in ocean temperature can turn more tropical disturbances into hurricanes or can pump up an existing storm's power and add to its rainfall (Trendberth, 2007). The number of hurricanes in any given year is also influenced by the seasonal ocean patterns known as El Niño and La Niña. For instance, El Niño calmed Atlantic cyclone activity in 2006.

Over the next 15 years, the USA's coastal population is projected to increase by about 25 million people, reaching 166 million people by the year 2015. Coastal storms account for 71 per cent of recent annual US disaster losses. Coastal erosion due to storm surges and sea level rise will claim roughly 1,500 homes each year for several decades (Gregg, 2007). Each event costs roughly $500 million, not including direct damages from erosion of the coastline. With 14 events in a year, losses would total $7 billion annually (Island Press, 2000).

Good overviews of hurricane impacts on coastal ecosystems are provided in recent journal articles (Greening *et al.*, 2006; Mallin & Corbett, 2006; Sallenger *et al.*, 2006). An important conclusion resulting from these articles is that many ecological components of estuaries and coastal systems were initially severely altered by the hurricanes, yet were quite resilient. For example, hydrodynamic effects were strong but short-lived, lasting only days (Rankey *et al.*, 2004). Water quality and phytoplankton productivity showed an initial response to winds and increased rainfall in the watershed, but effects were again relatively short-lived, lasting only months. Even fish assemblages recovered within weeks, after displaying severe initial hurricane impacts

(Greening *et al.*, 2006). In sharp contrast, the long-term effects on man-made infrastructure, including roads and social systems such as towns, were devastating.

9.1　Predicting and monitoring coastal flooding

Many of the world's cities and towns are in areas vulnerable to rising sea levels, and millions of people are at risk of being swamped by flooding and intense storms. This became abundantly clear when Hurricane Katrina made landfall near New Orleans in 2005 (NOAA, 2006; 2008). Before landfall, emergency managers and other local officials needed to know the predicted trajectory, strength, lateral extent and progress of a storm, plus an estimate of expected storm surge, wave height and wind velocity (Klemas, 2009). The use of remote sensing in storm forecasting and tracking is discussed in detail in Chapter 11.

　　Most coastal communities in North America have developed flood-risk maps, which are available to regional planners to assist in developing long-term protection and adaptation strategies. A typical model for determining the risk to coastal populations exposed to storms and sea level rise is shown in Figure 9.2. Using local information, like subsidence data, one must first determine the relative local sea level rise (SLR) from global SLR data. Next, the raised flood levels must be calculated from predicted storm surge and flood curves. Knowing the local coastal topography, one can then determine the size of the flood hazard zone. As shown in

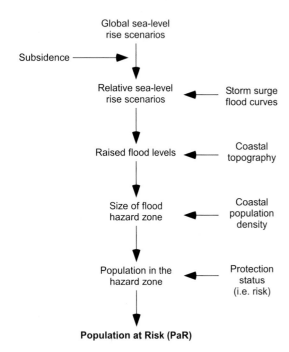

Figure 9.2　The flood model algorithm as used in GVA2. Modified with permission from Nicholls & Hoozemans (2005) and Springer Science and Business Media.

Figure 9.2, from the coastal population density and their protection status, one can estimate the population in the hazard zone and the population at risk (PaR), respectively. The results can then be used to plan emergency evacuation paths and other storm- and SLR-related emergency procedures (Nicholls *et al.*, 1999; Nicholls & Hoozemans, 2005).

Storms producing large waves and storm surges can cause severe erosion of sandy beaches and can inundate wetlands and coastal developments (Mather *et al.*, 1967; Klemas, 2009). Individual storm characteristics play a major role in the type and temporal extent of impacts. For instance, the storm's direction and speed of approach, point of landfall and intensity all influence the magnitude of storm surge and resultant flooding and environmental damage.

Storms with significant rainfall have longer-term effects on downstream systems than faster-moving storms with higher winds. Also, areas with higher population densities and related land use activity are more susceptible to environmental damage and long-term effects from storms. Environmental impacts from hurricanes include excessive nutrient loading, algal blooms, elevated oxygen demand resulting in hypoxia and anoxia, fish kills, large scale releases of pollutants and debris, and the spread of pathogens (Mallin & Corbett, 2006).

One important model used by the NOAA National Hurricane Center to estimate storm surge heights and winds resulting from predicted or hypothetical hurricanes is the SLOSH (Sea, Lake and Overland Surges from Hurricanes) model. This takes into account the pressure, size, forward speed, track and winds of a hurricane. SLOSH is used to evaluate the threat from storm surges, and emergency managers use this data to determine which areas must be evacuated. SLOSH model results are combined with road network and traffic flow information, rainfall amounts, river flow, and wind-driven waves to identify at-risk areas (NOAA/NHC, 2008).

The dynamic SLOSH model computes the water height over a geographical area or basin. The calculations are applied to a specific locale's shoreline, incorporating unique bay and river configurations, water depths, bridges, roads and other physical features. Computations have been run for a number of basins covering most of the Atlantic and Gulf Coasts of the US and the offshore islands. The typical SLOSH grid contains over 500 points located on lines extending radially from a common basin centre. The distance between grid points ranges from 0.5 km near the centre (where surge water heights are of more interest) to 7.7 km in the deep water at the edge of the grid. Bathymetric and topographical map data are used to determine a water depth or terrain height for each grid point.

The model consists of a set of equations derived from the Newtonian equations of motion and the continuity equation applied to a rotating fluid with a free surface. The equations are integrated from the sea floor to the sea surface. The coastline is represented as a physical boundary within the model domain. Subgrid-scale water features (cuts, chokes, sills and channels) and vertical obstructions (levees, roads, spoil banks, etc.) can be parameterized within the model. Astronomical tides, rainfall, river flow, and wind-driven waves have not been incorporated into the model.

Digital Globe QuickBird true-colour satellite image acquired Aug. 31st, 2005 Eastern New Orleans

N

Surekote Road levee break

0   50   100 m

Figure 9.3 QuickBird satellite image of the Surekote Road levee break, eastern New Orleans, acquired August 31, 2005, two days after Hurricane Katrina struck. This breach allowed water to flow from Lake Pontchartrain into New Orleans. The resolution of the data is 2.4 m². Satellite image: DigitalGlobe.

The primary use of the SLOSH model is to define flood-prone areas for evacuation planning. The flood areas are determined by compositing the model surge values from 200–300 hypothetical hurricanes, and separate composite flood maps are produced for each of the five Saffir-Simpson hurricane categories. The SLOSH model can also be run using forecast track and intensity data for an actual storm as it makes landfall (NOAA/NHC, 2008).

After Hurricane Katrina made landfall, high-resolution satellites such as IKONOS and QuickBird (Chapter 3, Table 3.6) and airborne sensors were used to document the details of the flooding and destruction, including actual breaks in the levees protecting the city. As discussed in Chapter 6, the levees that failed were most likely weakened by subsidence in the years running up to Katrina. Sections of levees designed to channel canals through New Orleans crumbled under the battering waves and storm surge of the hurricane. The breaks allowed water to flow from Lake Pontchartrain into New Orleans, inundating the city (Klemas, 2009).

The actual levee breaks, with water rushing into the city, could clearly be seen in high-resolution satellite imagery. On August 31, 2005, the QuickBird satellite

Figure 9.4 QuickBird satellite images of New Orleans prior to and post Hurricane Katrina. Flooding caused by the breach of the 17th Street Canal is clearly visible on August 31, 2005. Satellite image: DigitalGlobe.

captured images of levee breaks, as shown in Figure 9.3. The image shows the 100-metre-long breach in the levee along the Industrial Canal in East New Orleans. The enlarged lower image shows water pouring through a break in the levee at Surekote Road. The streets on the opposite side of the Industrial Canal are also flooded because of similar breaches in canals to the west.

Satellites, aircraft and helicopters also provided valuable imagery showing damaged bridges, highways, port facilities, oil rigs and other coastal infrastructure (DigitalGlobe, 2003; NASA, 2005). To assess the hurricane damage to the levee system, the Army Corps of Engineers also used topographical data obtained with a helicopter-mounted LiDAR sensor over the hurricane protection levee system in Louisiana. This information was very valuable for planning specific repairs and general reconstruction efforts. Pre- and post-Hurricane Katrina images of New Orleans obtained by the QuickBird satellite are shown in Figure 9.4. The flooding condition of city blocks, and even each building, can be clearly recognized in the images.

Most coastal wetlands can cope with gradual sea level rise (Reed, 2005). However, when one combines sea level rise with coastal storms such as Hurricane Katrina, wetlands can be devastated and have little chance of full recovery in the future (NOAA, 2006; 2008). Katrina's strong winds, storm surges and heavy rainfall damaged many ecosystems along the Gulf coast. In southeastern Louisiana, Katrina transformed nearly 200 km^2 of marsh into open water. Most of the loss east of the Mississippi river was attributed to the effects of Katrina's storm surge. Vegetation was ripped out and sand washed in, scouring and damaging mangrove roots and

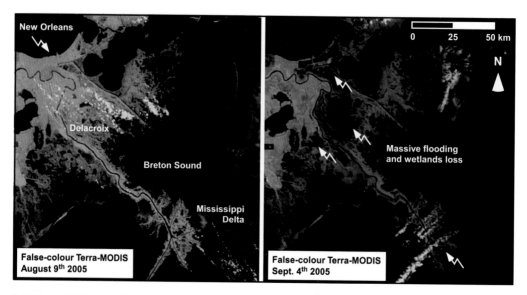

Figure 9.5 Terra-MODIS satellite images of wetland losses in the Mississippi delta before and after the August 29, 2005 landfall of Hurricane Katrina. Extensive flooding in the later of these time-separated false-colour images is black against the vivid green of the wetlands vegetation. Credit: NASA.

harming the animals that live there. Large influxes of eroded sediment reduced habitat for coastal birds, mammals and invertebrate species. Barrier islands were submerged and eroded. Entire seagrass beds, which are critical for fish, sea turtles and marine mammals, were uprooted and destroyed during the storm. Coral reef beds were scoured, torn and flattened, causing population reductions in animals such as sea urchins, snails and fish. Katrina inundated marshes and swamps with saltwater and polluted runoff from urban areas and oil refineries, impacting amphibians and reptiles due to their sensitivity to toxins and other pollutants. Large areas of wetlands were lost, some of them permanently (Barras, 2006; Provencher, 2007).

Time-series of medium-resolution (10 m to 30 m) images from Landsat TM, MODIS, aircraft and other platforms have been used not only to observe the immediate damage to wetlands, but also to monitor their recovery (Klemas, 2005; Lyon & McCarthy, 1995). For example, the resulting wetland losses caused by Hurricane Katrina were mapped by NASA, NOAA and USGS scientists over large areas using medium-resolution satellite imagery and geographical information systems.

Figure 9.5 shows two images taken at a resolution of 250 metres by NASA's Moderate Resolution Imaging Spectroradiometer (MODIS) on August 9 (left) and September 4 (right). The flooded areas are black against bright green vegetation in these false-colour MODIS images. As the right image demonstrates, extensive flooding is visible in the wetlands south of New Orleans after the hurricane passed, as compared to the left image taken before Katrina (NASA, 2005).

The satellite images demonstrate how coastal wetlands function to protect inland regions and coastal communities from storm surges unleashed by powerful hurricanes. The wetlands act as a sponge, soaking up water and diminishing the storm surge. If the wetlands had not been there, the storm surge could have penetrated much farther inland. By contrast, there were no wetlands to buffer New Orleans from Lake Ponchartrain, so the storm forced lake water to burst through the levees that separated it from the city (NASA, 2005).

Inland systems can also be impacted by coastal storms, since they are susceptible to saline incursions and flooding during storms. For instance, elevated soil salinity levels lingered for months after Hurricane Hugo hit South Carolina. This had adverse affects on upstream wetland and tree survival, as well as on nutrient cycling processes. (Blood *et al.*, 1991) Salinity and soil moisture can be obtained using airborne and satellite microwave radiometers (Burrage *et al.*, 2003; Parkinson, 2003), and the technology is covered in Chapters 7 and 10.

Only SAR can penetrate the clouds to observe coastal inundation conditions during the time of a hurricane's landfall. Radar can detect flooded coastal marshes, because they usually provide a weaker radar return than non-flooded ones. The marsh grasses may calm the water surface, accentuating specular reflection (Ramsey, 1995). The radar return from flooded forests is usually enhanced compared to returns from non-flooded forests. This enhancement is due to the double bounce mechanism, where the signal penetrating the canopy is reflected off the water surface and subsequently reflected back toward the sensor by a second reflection off a tree trunk (Hess *et al.*, 1990) (Figure 3.4).

9.2 Coastal currents and waves

Currents and waves strongly affect coastal ecosystems, especially in the near-shore, which is an extremely dynamic environment. Currents influence the drift and dispersion of various pollutants and, together with breaking waves, they can mobilize and transport sediments, resulting in erosion and morphological evolution of natural beaches. Changes in the underlying bathymetry, in turn, affect the wave and current patterns, resulting in a feedback mechanism between the hydrodynamics and morphology.

The ability to monitor these processes is necessary in order to understand and predict the changes that occur in the near-shore region. Arrays of current meters, acoustic Doppler velocimeters and pressure sensors are not very effective for determining surface currents and waves over large coastal regions, since these instruments measure currents at a point and are expensive when large numbers of sensors have to be deployed. Therefore, shore-based high-frequency (HF) and microwave Doppler radar systems are frequently used to map currents and determine swell-wave parameters over large coastal areas with considerable accuracy (Graber *et al.*, 1997; Paduan & Graber, 1997; Bathgate *et al.*, 2006; Chant *et al.*, 2008).

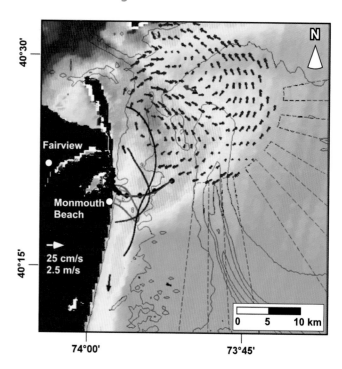

Figure 9.6 New York Bight surface currents as measured with shore-based HF radar. Modified with permission from Chant *et al.* (2008) and the American Geophysical Union.

The surface current measurements use the concept of Bragg scattering from a slightly rough sea surface, modulated by Doppler velocities of the surface currents. When a radar signal hits an ocean wave, it usually scatters in many directions. However, when the radar signal scatters off a wave that has a wavelength half that of the transmitted signal wavelength, Bragg scattering will return the signal directly to its source, resulting in a very strong received signal. Extraction of swell direction, height and period from HF radar data is based on the modulation imposed upon the short Bragg wavelets by the longer faster-moving swell.

Figure 9.6 shows low-pass-filtered surface currents in the New York Bight derived from Coastal Oceans Dynamics Application Radar (CODAR) and overlaid on an ocean colour image obtained by the Ocean Color Mapper (OCM). Drifter and dye trajectories are also shown. CODAR is a land-based high-frequency (HF) radar system specifically designed for measuring coastal currents. This figure demonstrates how integration of HF data with satellite imagery and drifter data can provide a more complete picture of sea surface dynamics.

HF radars can determine coastal currents and wave conditions over a range of up to 200 km (Cracknell & Hayes, 2007; Chant *et al.*, 2008). While HF radar provides accurate maps of surface currents and wave information for large coastal areas, their spatial resolution, which is about 1 km, is more suitable for measuring mesoscale features than small-scale currents. On the other hand, microwave X-band and S-band radars have resolutions of the order of 10 m, yet have a range of only a few kilometres.

9.3   Mapping beach topography

Topographical and hydrographical information are basic elements in studies of near-shore geomorphology, hydrology and sedimentary processes. In order to plan sustainable coastal development and implement effective beach erosion control and coastal ecosystem protection strategies, coastal managers need information on long-term and short-term changes taking place along the coast, including beach profiles, changes due to erosion, wetlands changes due to inundation, etc.

Before the advent of GPS and LiDAR, shoreline position analysis was based on historical aerial photographs and topographical sheets (Jensen, 2007; Morton & Miller, 2005). To map long-term changes of the shoreline due to beach erosion, time-series of historical aerial photographs and topographical maps have been used. Aerial photographs are available dating back to the 1930s, and topographical maps exist to extend the record of shoreline change to the mid- to-late 1800s. Such data are held by local, state and federal agencies, including the US Geological Survey and the USDA Soil Conservation Service. They also have various maps, including planimetric, topographical, quadrangle, thematic, ortho-photo, satellite and digital maps (Rasher & Weaver, 1990; Jensen, 2007).

To perform a shoreline position analysis, the shoreline can be divided into segments which are uniformly eroding or accreting. The change in the distance of the waterline can then be measured in reference to some stable feature, such as a coastal highway. The instantaneous water line in the image is not a temporally representative shoreline. The high water line, also referred to as the wet/dry line, is a commonly used indicator because it is visible in most images. Other indicators include the vegetation line, the bluff line or man-made shore vestments (Boak & Turner, 2005; Thieler & Danforth, 1994).

As shown in Figure 9.7, beach profiles can change rapidly with the seasons and after storms, in addition to exhibiting slower changes due to littoral drift and sea level rise. Long-term changes of shorelines due to littoral drift or sea level rise can be aggravated by man-made structures such as jetties, seawalls and groins. Figure 9.8 shows how a coastal structure forces the wave-induced littoral drift to deposit its sand on the right side and prevents the sand from replenishing the left side, which is being eroded by winter storms.

Topographical and depth data can now be effectively acquired at various spatial scales by airborne laser surveying using LiDAR (Light Detection and Ranging) (Krabill *et al.*, 2000; Ackermann, 1999; Guenther *et al.*, 1996). A laser transmitter/receiver mounted on an aircraft transmits a laser pulse that travels to the land surface or the air-water interface, where a portion of this energy reflects back to the receiver. The land topography is obtained from the LiDAR pulse travel-time. On water, some of the energy propagates through the water column and reflects off the sea bottom. The water depth is calculated from the time lapse between the surface return and the bottom return.

Global Positioning Systems (GPS), combined with LiDAR techniques, make it possible to obtain accurate topographical maps, including shoreline positions.

Figure 9.7 Changes in beach profiles between summer and winter due to changes in wave climate. During winter storms, the beach is eroded and the resulting seaward cross-shore sediment transport produces offshore bars.

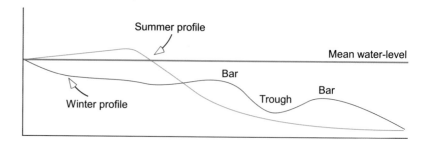

Figure 9.8 Shoreline change due to interruption of longshore transport of sand by a jetty. The updrift side fills up and the downdrift side erodes until the updrift side is filled and bypassing begins.

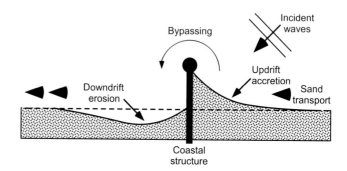

LiDAR surveys can produce a 10-centimetre vertical accuracy at spatial densities greater than one elevation measurement per square metre. This is important for various coastal research applications of LiDAR, including mapping change along barrier island beaches and other sandy coasts (Brock & Purkis, 2009). The ability of LiDAR to rapidly survey long, narrow strips of terrain is very valuable in this application, because beaches are elongate, highly dynamic sedimentary environments that undergo seasonal and long-term erosion or accretion and are also impacted by severe storms (Kempeneers *et al.*, 2009; Krabill *et al.*, 2000; Stockdon *et al.*, 2002; 2009).

A typical beach profile mapping procedure may include cross-shore profiles every 10 metres from available LiDAR data. Beach slope and location, elevation of the berm, dune base and dune crest can also be determined from these beach profiles. One can use a known vertical datum to remove the subjective nature of identifying the shoreline. The water line is then readily identified, because laser returns from the sea are noisy. All points that lie seaward from this line are deleted from the profile. A vertical range around the elevation datum is then chosen (e.g. 1.0 metre) and all points that do not fall within this range are removed from the profile. Finally, a linear regression is fit through the cluster of points to produce the horizontal position of the shoreline and the slope, using the elevation datum and regression analysis (Stockdon *et al.*, 2002; 2009).

A typical LiDAR aircraft mapping configuration includes a light aircraft equipped with a LiDAR instrument and GPS, which is operated in tandem with a GPS base station (Figure 9.9). In coastal applications, the aircraft flies along the coast at a height of about 500–1,000 metres, surveying a ground swath directly below the aircraft. The aircraft position throughout the flight is recorded by an onboard GPS receiver. The aircraft GPS signals are later combined with signals concurrently collected by a nearby GPS base station. Differential kinematic GPS post-processing determines the aircraft flight trajectory to within about 5 centimetres.

Although airborne laser mapping may be carried out at night, flight safety dictates that coastal LiDAR operations are normally confined to daylight hours and timed to coincide with low tide to maximize coverage of the beach face. LiDAR overflights should not be conducted during high wind conditions, since the rough water surface will scatter the LiDAR pulse and make it difficult to detect them (Irish & Lillycrop, 1999; Brock & Sallenger, 2000).

Mapping submerged aquatic vegetation (SAV) and coral reefs requires high-resolution (1–4 m) imagery (Mumby & Edwards, 2002; Purkis, 2005). Coral reef ecosystems usually exist in clear water and can be classified to show different forms

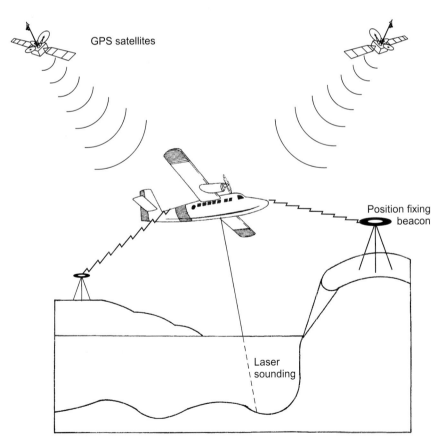

Figure 9.9 Illustration of a LiDAR topography operation.

of coral reef, dead coral, coral rubble, algal cover, sand lagoons and different densities of seagrasses, etc. However, SAV may grow in more turbid waters and thus is more difficult to map. High-resolution (e.g. IKONOS) multispectral imagers have been used in the past to map eelgrass and coral reefs. Hyperspectral imagers should improve the results significantly by being able to identify more estuarine and intertidal habitat classes (Maeder *et al.*, 2002; Garono *et al.*, 2004; Philpot *et al.*, 2004; Mishra *et al.*, 2006; Wang & Philpot, 2007). Remote sensing of coral reefs is covered in Chapter 8.

9.4 LiDAR bathymetry

In LiDAR bathymetry, a laser transmitter/receiver mounted on an aircraft transmits a pulse that travels to the air-water interface, where a portion of this energy reflects back to the receiver (Figure 9.10). The remaining energy propagates through the water column and reflects off the sea bottom. The water depth is calculated from the time lapse between the surface return and the bottom return. Each sounding is corrected for water level fluctuations, using either vertical aircraft positioning from GPS or by referencing the LiDAR measurements of water surface location with water level gauge measurements. Since laser energy is lost due to refraction, scattering and absorption at the water surface, the sea bottom and inside the water column, these effects limit the strength of the bottom return and limit the maximum detectable depth.

Examples of LiDAR applications include regional mapping of changes along sandy coasts due to storms or long-term sedimentary processes and in the analysis of shallow benthic environments (Guenther *et al.*, 1996; Sallenger *et al.*, 1999;

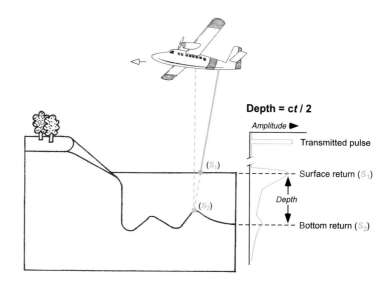

Figure 9.10 Principles of operation of a LiDAR bathymeter. The water depth can be calculated from the travel time difference (t) between the water surface (S_1) and bottom (S_2) pulse returns. Here c represents the velocity of the laser light pulse.

Gutierrez *et al.*, 1998; Bonisteel *et al.*, 2009; Brock & Purkis, 2009; Kempeneers *et al.*, 2009). In the coastal zone, there is considerable utility in being able to capture seamlessly and simultaneously topographical LiDAR above the water with bathymetric postings in the adjacent ocean. This objective is achievable but, as will be seen below, it demands the use of multiple lasers and/or advanced profiling technologies.

To maximize water penetration, bathymetric LiDARs employ a blue-green laser with a typical wavelength of 530 nm to range the distance to the seabed. With the near-exponential attenuation of electromagnetic energy by water with increasing wavelength, a pure blue laser with a wavelength shorter than 500 nm would offer greater penetration. However, this wavelength is not used because, first, blue light interacts much more strongly with the atmosphere than longer wavelengths; and second, creating a high-intensity blue laser is energetically less efficient than blue-green and consumes a disproportionately large amount of instrument power. This, combined with the fact that blue lasers suffer from temperature problems at high powers, explains why blue-green is the preferred wavelength for bathymetric LiDAR profilers.

Conversely, terrestrial topographical LiDARs typically utilize near-infrared (NIR) lasers with a wavelength of 1,064 nm. As is the case for the blue-green laser used for hydrography, this NIR wavelength is focused and easily absorbed by the eye, and hence the maximum power of the LiDAR system is limited by the need to make them eye-safe. Although less accurate, military instruments often utilize lasers with wavelengths as long as 1,550 nm, which hold the dual advantage of being eye-safe at much higher power levels and the beam is not visible using night-vision goggles. While bathymetric lasers are limited in their accuracy by water column absorption, terrestrial infrared lasers suffer from null or poor returns from certain materials and surfaces such as water, asphalt, tar, clouds and fog, all of which absorb NIR wavelengths.

Because they do not penetrate water, NIR topographical lasers cannot be used to assess bathymetry. However, blue-green hydrographical lasers do reflect off terrestrial targets and can be used to measure terrain. Traditionally, their accuracy and spatial resolution has been lower than provided by a dedicated NIR topographical instrument. However, dual-wavelength LiDAR provides both bathymetric and topographical LiDAR mapping capability by carrying both an NIR and a blue-green laser. The NIR laser is not redundant over water because, since it reflects off the air-water interface, it can be used to refine the surface position as well as to distinguish dry land from water using the signal polarization (Guenther, 2007).

In addition, specific LiDAR systems like the SHOALS which is covered in sections 8.9 and 14.1 of this book, record the red wavelength Raman signal (647 nm). The Raman signal comes from interactions between the blue-green laser and water molecules, causing part of the energy to be backscattered while changing wavelength (Guenther *et al.*, 1994). This is also useful to localize the air/water interface when experiencing incorrect surface detections due to land reflection or the presence of unexpected targets such as birds.

Figure 9.11 Coastal topography for a section of the Assateague Island National Seashore acquired using the airborne Experimental Advanced Airborne Research LiDAR (EAARL) (Bonisteel *et al.*, 2009). Operating in the blue-green portion of the electromagnetic spectrum, the EAARL is specifically designed to measure submerged topography and adjacent coastal land elevations seamlessly in a single scan of transmitted laser pulses. Assateague Island National Seashore consists of a 37-mile-long barrier island along the Atlantic coasts of Maryland and Virginia. Credit: USGS.

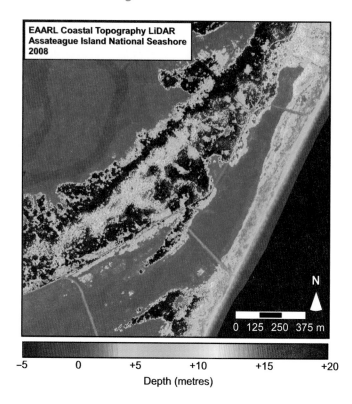

By employing a very high scan-rate, state-of-the-art systems such as the Experimental Advanced Airborne Research LiDAR (EAARL) have demonstrated the capability of measuring both topography and bathymetry from the return time of a single blue-green laser (Bonisteel *et al.*, 2009; McKean *et al.*, 2009; Nayegandhi *et al.*, 2009). Figure 9.11 shows such a bathymetric-topographical DEM of a section of the Assateague Island National Seashore, captured by the EAARL. This experimental advance signals a future move towards commercial implementation of dual-application but single-wavelength instruments.

While the EAARL and dual-wavelength LiDARs offer near seamless profiles between bathymetry and terrestrial terrain, neither bathymetric systems can acquire dependable bathymetric data in very shallow depths or over white water in the surf zone. When white-caps are present, the laser does not penetrate the water column. Even when the water surface is clear, if the depth is less than 2 m, it becomes difficult to separate the peak in the returning laser-waveform that corresponds to the water surface from that of the river or lake bed (Philpot, 2007). For coastal mapping, both problems are obviated by combining successive flights at low tide with a topographical LiDAR, and at high tide with a bathymetric LiDAR (Pastol *et al.*, 2007). Such a strategy is not possible for coastal areas that do not boast large tidal variations, or for non-tidal inland water bodies.

Laser depth sounding techniques have proven most effective in moderately clear, shallow waters. Typical flight parameters for airborne LiDARs used in

Table 9.1 LiDAR flight parameters (DGPS = differential GPS mode; KGPS = kinematic GPS mode).

Flying height	200–500 m (400 m typical)
Vertical accuracy	±15 cm
Horizontal accuracy	DGPS = 3 m; KGPS = 1 m
Max mapping depth	60 m (clear water)
Typical *kd* product	4
Coastal *k*	0.2–0.8 (d = 5–20 m)
Estuarine *k*	1.0–4.0 (d = 1–4 m)
Sounding density	3–15 m
Sun angle	18°–25° (to minimize glare)
Scan geometry	Circular (220 m swath typical)
Sea state	Low (0–1 Beaufort scale)
Water penetration	Green LiDAR (532 nm) used
Aircraft height	Infrared LiDAR (1,064 nm) used

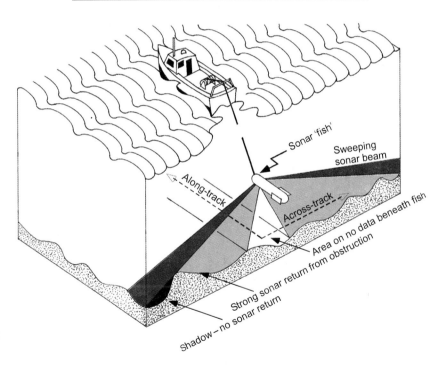

Figure 9.12 Side-scan sonar imaging system.

bathymetry are shown in Table 9.1. Optical water clarity is the most limiting factor for LiDAR depth detection, so it is important to conduct the LiDAR overflights during tidal and current conditions that minimize the water turbidity due to sediment re-suspension and river inflow.

The LiDAR system must have a *kd* factor large enough to accommodate the water depth and water turbidity at the study site (k = attenuation coefficient;

d = max. water depth). For instance, if a given LiDAR system has a $kd = 4$ and the turbid water has an attenuation coefficient of $k = 1$, the system will be effective only to depths of approximately 4 metres. Beyond that depth, one may have to use acoustic echo-sounding techniques or side-scanning sonar systems, as shown in Figure 9.12 (Brock & Sallenger, 2000). Typically, a LiDAR sensor may collect data down to depths of about three times the Secchi (visible) depth, but, in optically clear water, LiDAR sensors have successfully reached depths of nearly 60 m (Sinclair, 1999).

CASE STUDY: LiDAR application to modelling sea level rise at the Blackwater National Wildlife Refuge

A good case study for demonstrating the successful application of LiDAR is the Blackwater National Wildlife Refuge Restoration Project on the eastern shore of Chesapeake Bay in Maryland (Figure 9.13). The Blackwater Refuge, established in 1933, includes tidal marshes, freshwater ponds and forests, and is recognized as a wetland of international importance by the United Nations' Ramsar Convention.

The refuge has been featured prominently in studies of the impact of sea level rise on coastal wetlands. Most notably, it has been cited by the Intergovernmental Panel on Climate Change (IPCC) as a key example of wetland loss attributable to rising sea level due to global warming. Studies of aerial photos taken since 1938 show an expanding area of open water in the central area of the refuge, and this seems to parallel the record of sea level rise over the past 60 years. The US Fish and Wildlife Service (FWS) manages the refuge to support migratory waterfowl and to preserve endangered upland species. High marsh vegetation is critical to FWS waterfowl management, yet a broad area, once occupied by high marsh, has decreased with rising sea level.

Since 1938, 8,000 acres of marsh have been lost in the refuge – a rate of nearly 130 acres per year. The marsh is less than one metre above sea level and almost the whole of it has been breached and is being drowned. Several factors have contributed to the area's severe marsh loss, including wildlife damage (primarily geese and nutria), a rising sea level, severely altered hydrology and salinity and an increase in wave energy associated with greater stretches of open water. While most marshes build upwards through sediment deposition, Blackwater has no source of incoming sediment because the hydraulic structure is degraded and the sea level continues to rise (NOAA, 2005; Larsen, 2004).

Considering the most recent forecast of sea level rise, it becomes apparent that, without intervention, the entire Blackwater Refuge area will be submerged in the next century. Various engineering adjustments, such as channels, dams, housing developments and new roads have occurred, and an understanding of tidal characteristics will minimize the ecological impact of these changes and will eliminate such problems as further erosion or a rapid change in salinity that could harm marsh species. The USGS has developed an inundation model centred

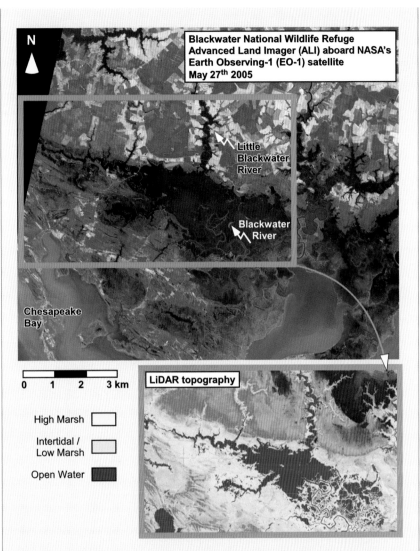

Figure 9.13 Top pane is a true-colour EO-1 image of the Blackwater National Wildlife Refuge in the Chesapeake Bay. EO-1 is a NASA satellite that flies in formation with the Landsat ETM+. It carries both the multispectral Advanced Land Imager with a spatial resolution of $10\,m^2$ (this image) and the hyperspectral Hyperion system ($30\,m^2$ spatial resolution). Inset is a detailed USGS LiDAR topo-map. Rising sea level has led to widespread degradation of the Chesapeake Bay marshes. Credit: USGS-NASA.

on the refuge and surrounding areas. Such models require a detailed topographical map upon which to superimpose future sea level positions.

As shown in Figure 9.13 (inset), LiDAR mapping of land and shallow water surfaces has helped solve this problem. The USGS has developed a detailed

LiDAR map of the refuge area at a 30 cm contour interval. With the model, the new map enables the present marsh vegetation zones to be identified as well as prediction of the location and area of future zones on a decade-by-decade basis over the next century, at increments of about 3 cm per decade of sea level rise.

The most recent runs of the model suggest that wetland habitat in the refuge could be sustained, but only for the next 50 years, through a combination of public and private preservation efforts, including easements and Federal land acquisitions. After 50 years, this area will become open water (Larsen *et al.*, 2004).

9.5　Key concepts

1　Sea levels are rising globally at a rate of about 2–3 mm per year. This can be attributed to a combination of thermal expansion of the oceans and input from melting ice sheets. The trend should be put in context against a near-continuous rise since the last ice age, when sea level stood more than 100 m lower than at present.

2　The substantial sea level rise predicted for this century will severely impact low-lying coastal zones – and wetlands, marshes and swamps in particular. Beyond simple inundation of land, salt incursion can irrevocably alter ecosystems that have historically been bathed in fresh water. These environments represent critical habitat for a multitude of species and are becoming populated by cities, industry and transport networks. Low island nations such as the Maldives are taking the unprecedented step of preparing to abandon their homeland with a view to re-establishing it on a continent that is not threatened by sea level rise.

3　The coastal zone becomes increasingly vulnerable to storm surge as sea levels are raised. Though still a point of contention, several climate models predict hurricane frequency and intensity will increase as global warming progresses. Hurricane Katrina and the devastation of the Louisiana coastline stand as testament to the impact a storm can have on an area that barely rises above sea level.

4　Most coastal communities in North America have produced flood-risk models and maps, which are available for developing long-term protection and adaptation strategies. These models use local information such as subsidence data and relative local sea level rise data, and raised flood levels are calculated from predicted storm surge and flood curves. The size of the flood hazard zone is then determined using the local topography The results can be used to plan emergency evacuation paths and other storm- and SLR-related procedures.

5 One model used by the NOAA National Hurricane Center to estimate storm surge heights and winds resulting from predicted or hypothetical hurricanes is the SLOSH (Sea, Lake and Overland Surges from Hurricanes) model. This model takes into account the pressure, size, forward speed, track and winds of a hurricane. SLOSH model results are used to evaluate the threat from storm surges and are combined with road network and traffic flow information, rainfall amounts, river flow and wind-driven waves to identify areas at risk.

6 Beach topography is a critical parameter to understanding the vulnerability of the coastal zone to storm surge. Aircraft-mounted LiDAR, a technology that utilizes precision laser ranging, represents the state of the art for mapping the profile of beaches. The same LiDAR technology is also used to create bathymetric charts in relatively clear coastal waters. For deep or turbid waters, acoustic depth sounders or side-scan sonar must be used.

7 Currents and waves strongly affect coastal ecosystems and influence the drift and dispersion of pollutants. Together with breaking waves, they also mobilize and transport sediments to cause beach erosion. Shore-based high-frequency (HF) and microwave Doppler radar systems can be used to map currents and determine swell-wave parameters over large coastal areas.

8 High-resolution satellite sensors and observers on small aircraft and helicopters can be used to evaluate the impact of storm surges and flooding in the coastal zone.

9 A good case study showing the successful application of LiDAR is the Blackwater National Restoration Project. A detailed LiDAR topographical map at a 30 cm contour interval was used with a model to predict the inundation due to sea level rise on a decade by decade basis over the next century, at increments of about 3 cm per decade.

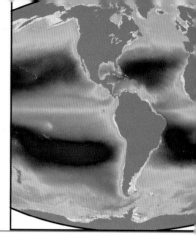

10 Observing the oceans

10.1 Introduction

Oceans influence the Earth's climate because they transfer and redistribute heat over the entire planet and absorb greenhouse gases. Oceans also cover about 70 per cent of the Earth's surface and are dynamic across a spectrum of temporal and spatial scales. They contain most of the Earth's water, support important marine ecosystems, help stabilize the atmosphere and climate, provide ship routes for transportation and are one of the main food suppliers for many coastal and island countries. Yet the oceans' size and depth make it difficult to monitor them with ships and buoys alone. Satellites, with a wide range of sensors, are proving to be highly effective for observing the physical and biological features of large ocean and coastal areas.

Ocean observations from space are critical to a wide range of applications, including weather and hurricane tracking and prediction, coral reef monitoring, climate change studies and fisheries management. Some of the key ocean features which can be mapped from satellites are shown in Table 10.1.

As the table shows, all EM wavelength regions are employed. For example, ocean colour, chlorophyll and productivity can be obtained with multispectral and hyperspectral imagers operating primarily in the visible part of the spectrum. Sea surface temperatures can be mapped with thermal infrared sensors (TIR), and ocean salinity with passive microwave radiometers. Physical ocean properties can be obtained using various radar instruments, such as winds from scatterometers, sea surface elevation and currents from altimeters, and sea surface slicks and waves from synthetic aperture radar (SAR). These measurements are used by oceanographers to study the ocean's general circulation, its large-scale, low-frequency variability, biological productivity, turbulent eddy energy, and air/sea interaction. Some of this information, such as sea surface temperature and elevation, can be used in global change models if the total available data covers long time periods (Robinson, 2004; Martin, 2004; Chuvieco, 2008).

Remote Sensing and Global Environmental Change, First Edition. Samuel Purkis and Victor Klemas.
© 2011 Samuel Purkis and Victor Klemas. Published 2011 by Blackwell Publishing Ltd.

Table 10.1 Spaceborne ocean sensing techniques. Modified from Pinet (2009), with permission from Jones & Bartlett Learning (www.jblearning.com).

Sensor type	Type of measurement	Oceanographical application
Visible and near-infrared radiometer and imager	Backscattered solar radiation from surface layer	Surface water turbidity; phytoplankton concentration
Thermal infrared radiometer and imager	Thermal emission from sea surface	Sea surface temperature; surface heat flux
Microwave radiometer and imager	Sea surface microwave emission	Sea surface temperature; surface heat flux; salinity; sea state; soil moisture
Radar altimeter (nadir-looking)	Return time of pulse; shape of return pulse	Ocean currents and tides; significant wave height
Radar scatterometer (side-looking radar)	Strength of return pulse from different directions	Surface wind speed and direction
Synthetic Aperture Radar (SAR) (high-resolution imaging radar)	Strength of return pulse from small area (Doppler shift of wave frequency with distance)	Swell patterns; internal wave patterns; oil slicks

10.2 Ocean colour, chlorophyll and productivity

The realization that passive optical imagery can be used to measure important oceanic variables occurred in the early 1970s. The launch of the Coastal Zone Colour Scanner (CZCS), carried aboard the NASA-operated Nimbus-7 satellite, provided the first implementation and confirmation of this concept. With a swath of 1,556 km, CZCS produced imagery at a spatial resolution of 825 m from 1978 until its demise in 1986.

The instrument was remarkable for two reasons. First, the mission was designed as a proof-of-concept exercise, expected to return data for a single year – not for nearly a decade, as turned out to be the case. Second, CZCS was the first orbital instrument dedicated to imaging near-surface ocean colour and temperature. Indeed, four of the six spectral bands were so sensitive as to saturate when used over land.

The primary objective of the mission was to quantify the large-scale lateral distribution of organic and inorganic materials in ocean waters. To this end, bands were positioned in the 400 nm to 670 nm range to give maximum water penetration and to correspond to the spectral features of chlorophyll and coloured dissolved organic matter (CDOM, also termed 'gelbstoff'). As shown in Table 10.2, the CZCS global datasets laid the scientific foundation for a new generation of satellite ocean colour measurements using systems such as the Sea-Viewing Wide-Field-of-View Sensor (SeaWiFS), Moderate Resolution Imaging Spectroradiometer (MODIS) and Medium Resolution Imaging Spectrometer (MERIS) (Martin, 2004; Ikeda & Dobson, 1995).

Table 10.2 Some satellite remote sensing systems used to measure ocean colour. Note that the MODIS instrument is carried aboard two platforms (Terra and Aqua). Modified from Jensen, John R., *Remote Sensing of the Environment: An Earth Resource Perspective*, 2nd edition, Copyright 2007. Printed and electronically reproduced by permission of Pearson Education, Inc.

Sensor	Agency	Satellite	Operating dates	Spatial resolution (m)	Number of bands	Spectral coverage (nm)
CZCS	NASA	Nimbus-7	1978–86	825	6	433–12,500
SeaWiFS	NASA	OrbView-2	Launch 1997	1,100	8	402–885
MODIS-Terra	NASA	Terra	Launch 1997	250/500/1,000	36	405–14,385
MODIS-Aqua	NASA	Aqua	Launch 2002	250/500/1,000	36	405–14,385
MERIS	ESA	Envisat-1	Launch 2002	300/1,200	15	412–1,050

The long lineage of dedicated ocean colour instruments is particularly important to our understanding of global change. First, primary production from phytoplankton lies at the base of the food chain for the vast majority of marine ecosystems and serves to draw down vast quantities of the carbon dioxide into the deep ocean. Second, rising sea temperatures are likely to alter patterns of production within the oceans. Observation of the global distribution of productivity, and its change through time, are thus seen as an important proxy for shifting climate.

Approximately one half of the biosphere's net primary production is attributed to photosynthesis by oceanic phytoplankton. In context, this amounts to more than a hundred million tons of carbon dioxide fixed into organic material per day (Behrenfeld *et al.*, 2006). This process is a vital link in the cycling of carbon between living and inorganic stocks, and it has a profound influence on ocean chemistry. Photosynthesis removes carbon dioxide dissolved in seawater to produce sugars and other simple organic molecules; oxygen is released as a by-product of the reaction. Without a sound understanding of the cycling of carbon in the open ocean, it is difficult to forecast the rate of rise of CO_2 in the atmosphere through continued anthropogenic emissions, and in turn to predict the likely magnitude of Earth warming that should be expected.

The reason satellites are at the forefront of this science is that the magnitude and variability of oceanic primary production are poorly known on a global scale, largely because of the high spatial and temporal variability of marine phytoplankton concentrations. For instance, wind-induced upwelling in coastal regions brings nutrients to the surface and creates patchy zones of high biological productivity, accompanied by high concentrations of chlorophyll and phytoplankton, which can be detected by colour sensors on satellites (Table 10.3). The waters off Peru and California are good examples, where long-term upwelling events influence the abundance of fish over periods of months (Robinson, 2004). Oceanographical vessels move too slowly to map dynamic, large-scale variations in productivity, making global coverage by shipborne instruments impossible. Only spaceborne observations can provide the rapid global coverage required for studies of ocean productivity worldwide.

Table 10.3 Gross primary productivity of ocean and terrestrial areas. Modified from Pinet (2009), with permission from Jones & Bartlett Learning (www.jblearning.com).

Quantity (gC/m²/yr)	Ocean area	Terrestrial area
< 50	Open Ocean	Deserts
50–100	Continental Shelves	Forests; grasslands; croplands
150–500	Upwelling areas; deep estuaries	Pastures; rain forests; moist croplands; lakes
500–1,250	Shallow estuaries; coral reefs	Swamplands; intensively developed agricultural areas

Data in the form of analyzed sea surface temperature and chlorophyll charts are provided daily to the fisheries and shipping industries, whereas information on the location of the north wall of the Gulf Stream and the centre of each eddy is broadcast daily over the Marine Radio Network. Because certain species of commercial and game fish are indigenous to waters of a specific temperature, fishermen can save much money in fuel costs and time by being able to locate areas of higher potential (Cracknell & Hayes, 2007).

A list of key US and European satellites with sensors used to measure ocean colour is shown in Table 10.2. A range of typical ocean colour products derived from SeaWiFS, MERIS, MODIS and similar sensor systems are summarized in Table 10.4 (Ackleson, 2001; 2003; Arnone & Parsons, 2004; Bissette *et al.*, 2004). As shown in the table, ocean colour products define the optical properties of the water for a wide range of applications such as bathymetry, euphotic depth estimation, water quality (turbidity), diver visibility, etc. A map of global ocean chlorophyll concentrations obtained with the SeaWiFS system is shown in Figure 10.1 (Jensen, 2007).

Satellite remote sensors measure the spectral radiances at the top of the atmosphere from which, after atmospheric and other corrections, the spectral radiances emerging from the ocean surface are extracted (Philpot, 2007). The surface radiances are converted to reflectances, providing us with the spectral signatures required for identifying chlorophyll and other water constituents. To produce valid products, such as global ocean chlorophyll concentrations for estimating primary productivity (Figure 10.1), a meticulous calibration and validation approach must be used.

Instrumented ships, buoys and ocean gliders are used to calibrate and validate chlorophyll-a and total suspended sediment maps obtained with ocean colour sensors. Some typical ship or buoy measurements are shown in Table 10.5. In coastal and estuarine waters, this data must frequently be obtained very close to the satellite overpass time and must be statistically representative of prevailing conditions. The water samples are usually taken from the upper half metre of the water column. Sites for calibrating remotely sensed data, such as chlorophyll concentrations in coastal waters, must be located at well-known points representing the entire range of variables to be measured.

Table 10.4 Ocean colour products. Modified with permission from Arnone & Parsons (2004).

Chlorophyll concentration	Biological processes such as algal (harmful and non-harmful) blooms and decay
Spectral backscattering coefficient	90° to 180° particle scattering linked to concentration, composition, index of refraction of organic (marine) and inorganic (terrigenous) particles, re-suspension
Spectral absorption coefficient	Total absorption, changes in water quality
Spectral absorption coloured dissolved organic matter	Conservative tracer of river plumes, linked with coastal salinity, photo-oxidation processes
Spectral particle absorption coefficient	Particle composition (organic and inorganic particles)
Spectral phytoplankton absorption coefficient	Absorption linked to differences in chlorophyll packaging within phytoplankton cells
Remote sensing reflectance	Spectral absolute water colour and water signature
Diffuse attenuation coefficient	Light penetration depth, light availability at depth
Aerosol concentration	Type and distribution, visibility, atmospheric correction methods
Beam attenuation coefficient	Total light attenuation using a collimated beam
Diver visibility	Horizontal visibility, average target size, target contrast, solar overhead illumination
Laser penetration depth	Underwater performance of lasers (imaging or bathymetry systems)

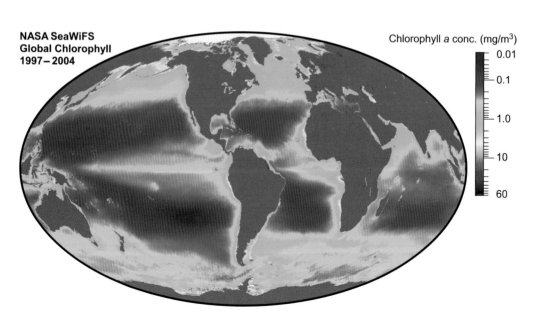

Figure 10.1 Map of ocean chlorophyll concentrations obtained from the SeaWiFS sensor aboard the SeaStar satellite. Credit: NASA/GSFC.

Table 10.5 Key remote sensing related ship measurements.

Direct measurements

- Temperature
- Secchi depth
- Attenuation coefficient
- Spectral reflectance (radiance and irradiance)

- Salinity
- pH

Water sample analysis

- Chlorophyll-a
- Nitrogen

- TSS
- Phosphorus

Ship data acquisition

- Water samples obtained from upper 0.5 m of water column
- Ship data obtained within 20 min of satellite overpass
- GPS used for sample site location

Calibration is frequently defined as the pre-launch characterization followed by the continuing analysis of the onboard sensor calibrators once on-orbit operations commence. Validation is usually thought of as the development of data processing schemes (e.g. atmospheric correction and derived geophysical quantities), plus the verification of product accuracies using ground-truth data.

It is not unusual for these elements to be considered part of the same function. The overlap between calibration and validation occurs because both activities require ground-truth observations. Calibration requires greater accuracy than validation, so applying data from the latter to the former is usually not considered. The Sea-viewing Wide Field-of-view Sensor (SeaWiFS) Project, for example, requires a radiometric accuracy to within 5 per cent absolute and 1 per cent relative, and chlorophyll-a (chl-a) concentrations to within 35 per cent over a range of 0.05–50.0 mg m^{-3} (Hooker & Esaias, 1993).

The difficulty in using validation data for calibration is a consequence of the dynamic range of the two activities. Calibration requires a sampling site wherein the contribution of natural variability – atmospheric and oceanic – is minimized, so that the total uncertainty is properly reduced. This is most simply satisfied at a site with predominantly clear skies and waters, with a simple particle distribution (exclusively marine aerosols for the atmosphere and in-water properties that depend primarily on chl-a). On the other hand, validation requires multiple sites with associated natural variability (Hooker et al., 2007).

Using platforms such as ocean gliders, remotely operated vehicles (ROVs) and autonomous underwater vehicles (AUVs) with advanced optical and acoustic sensors, marine scientists can now perform high-resolution three-dimensional measurements of biological and physical ocean features at various depths. They can view thin layers of high biological productivity at different depths and study the response of planktonic distributions and processes to physical forcing across a wide range of temporal and spatial scales. For example, thin layers require depth resolutions of centimetres, whereas previous measurements were performed at metre intervals, completely missing these biologically active and important layers (Schofield *et al.*, 2004; Cowles & Donghay, 1998).

10.3   Hazardous algal blooms and other pollutants

High concentrations of nutrients exported from agriculture or urban sprawl in coastal watersheds, or produced by coastal upwelling, are causing algal blooms in many estuaries and coastal waters (see for example Figure 7.2, Chapter 7). Algal blooms are harmful in that they cause eutrophic conditions, depleting oxygen levels needed by organic life and limiting aquatic plant growth by reducing water transparency. As shown in Figure 10.2, most algal blooms can be observed with multispectral scanners on satellites because of their distinct colour, location or repetitive seasonal appearance (Ruddick, 2001). Furthermore, hyperspectral sensors with spectral bands fine-tuned for specific pigment analysis allow detection and analysis of algal taxonomy. This can be accomplished because the species-specific algal accessory pigments produce unique spectral signatures, a concept discussed in Chapters 7 (Figure 7.1) and 8 (Figure 8.11).

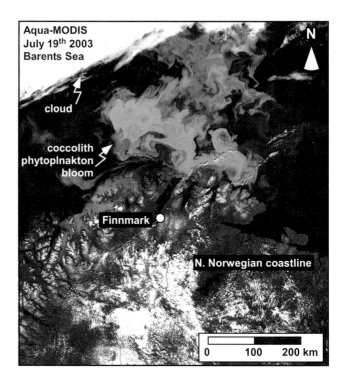

Figure 10.2 Phytoplankton bloom in the Barents Sea (Arctic Ocean) as seen in Aqua-MODIS imagery on July 19, 2003. The bloom is composed of coccolithophores and covers an area of >500 km² off Finnmark, the most northern and eastern county of Norway. Credit: NASA's Earth Observatory.

Remote sensing data can complement the monitoring networks existing in many parts of the world to obtain data on nutrient loading and algal growth. This gives us better insights into overall water quality, distribution of toxin-producing algae and aquatic biogeochemical cycling (Gitelson, 1993).

Some species of algae contain potent toxins that, even in low concentrations, can be extremely detrimental to marine life. This toxicity can propagate through the food chain, posing a serious threat as the contaminated organisms are consumed by predators. Seabirds, marine mammals and even humans are at risk of illness or death if they eat shellfish tainted with algal toxins. Mass occurrences of harmful algae are referred to as 'red tides', for the reason that the bloom reaches sufficient concentration to visually alter the colour of the water. However, this term is misleading, as many non-harmful algae are also capable of altering the appearance of the water, and also the most toxic species can cause severe harm at concentrations far below that needed to induce colouration. For these reasons, a more appropriate term for such phenomena is 'harmful algal bloom' or 'HAB'.

Of concern is the fact that the global frequency and magnitude of toxic blooms seems to be increasing (Hallegraeff, 1993; Fu *et al.*, 2008). Some species recur in the same geographical regions each year, while others are episodic, leading to the unexpected deaths of coral reefs, local fish, shellfish, mammals and birds (Namin *et al.*, 2010). The expansion of harmful algal blooms during the past 20 years is responsible for estimated losses approximating $100 million per year in the United States (Anderson *et al.*, 2000), primarily due to impacts on the aquaculture and fisheries industries.

The triggers for bloom conditions are not fully understood, but nutrient enrichment of waters, especially by nitrogen and/or phosphates, as well as unusually warm conditions, are recognized as precursors. With such diverse causes, prevention of HABs is difficult, so a more efficient means of dealing with this threat is through an effective early warning system.

By virtue of its frequent basin-scale coverage, satellite remote sensing is a well-poised technology for monitoring HABs and is routinely employed for detection and forecasting (Cullen *et al.*, 1997; Tomlinson *et al.*, 2004). However, it is spectrally challenging to identify a bloom to species level under laboratory conditions, let alone through the eyes of a spacecraft. A more promising use of Earth observation is to detect the environmental conditions suitable for bloom development and to track the progress of a bloom as it moves in from offshore. This technique could allow prediction of when and where a coastal region would be affected by a HAB.

One such example is the tracking of SST features, such as fronts, where HAB species are likely to accumulate, using AVHRR data (Tester *et al.*, 1991; Keafer & Anderson, 1993). Alternatively, variations in ocean colour can be used for the detection of anomalously high chlorophyll content that may indicate an impending bloom (e.g. Stumpf, 2001; Miller *et al.*, 2006).

Along these lines, in 1999 the National Oceanic and Atmospheric Administration (NOAA) Coastwatch programme began to acquire SeaWiFS imagery routinely for the purposes of developing an accurate coastal algorithm for chlorophyll. As detailed by Tomlinson *et al.* (2004), this effort was in turn used as an early warning

Table 10.6 Water quality levels.

Water quality vs. chlorophyll-a concentration

- Oligotrophic   < 8 µg / L
- Mesotrophic   8–25 µg / L
- Eutrophic   > 25 µg / L

Water quality vs. total suspended sediments

- Clear   0–10 mg / L
- Moderately turbid   10–50 mg / L
- Highly turbid   > 50 mg / L

Examples

- Delaware Bay is mesotrophic and moderately to highly turbid.
- Chesapeake Bay is mesotrophic to eutrophic and moderately turbid.

system for HAB occurrences off the west Florida coast. Chlorophyll anomalies identified in SeaWiFS imagery were combined with wind vector data in attempts to locate and predict the transport of the toxic dinoflagellate *Karenia brevis*. A bulletin was then developed to warn the state of Florida of potential HAB occurrences (Stumpf *et al.*, 2003).

The advantages of SeaWiFS imagery are that they provide data with a swath-width on the order of 1,000 km and a resolution of 1 km, and repeat coverage of an area occurs approximately every 1–2 days. Beyond the inevitable problem of clouds, the disadvantages of SeaWiFS are that there exist very few species-specific algorithms that can determine concentrations of key harmful algae from water-leaving radiance. As such, the detection of high chlorophyll concentrations does not necessarily indicate the presence of a harmful bloom. Furthermore, high concentrations of algal cells may exist deeper in the water column than can be passively imaged. Even when present at or near the surface, the 1 km spatial resolution of the sensor becomes limiting as the bloom approaches the coast – the point where tracking becomes the most informative.

Concentrations of chl-*a* and total suspended sediments (TSS) can be sensed remotely and used as indicators of the severity of eutrophication and turbidity, respectively. If such general criteria, as shown in Table 10.6, are used to compare estuarine water quality, it is possible to get satisfactory results with sensors that have fewer spectral bands and lower signal-to-noise ratios than the hyperspectral imagers needed for measuring precise concentration levels.

Most riverine and estuarine plumes, and some ocean-dumped waste plumes, can be detected remotely due to their strong surface signatures caused by high turbidity. The drift and dispersion of coastal plumes and ocean-dumped waste have been tracked with Landsat and SeaWiFS imagery (Thomas & Weatherbee, 2006; Dzwonkowski & Yan, 2005; Klemas & Philpot, 1981). To study the dynamics of such plumes, one can use a relatively small number of multispectral bands. However, to detect the composition and concentration of their content is difficult, even with hyperspectral images.

10.4   Sea surface temperature

Accurate large-scale, long-time observations of sea surface temperature (SST) are important to a wide range of global change studies. Sea surface temperatures are necessary, for example, for estimating the source of heat at the air/sea boundary. High-resolution satellite-derived SST measurements are ideal for investigating western boundary currents such as the Gulf Stream and Kuroshio, which exhibit displacements on large temporal and spatial scales. A typical example of thermal infrared mapping of ocean currents is illustrated in Figure 10.3, which shows the Gulf Stream as it meanders northward along the East Coast of the USA.

Another global-scale event which appears linked to elevated SSTs is damage to coral reefs. As described in Chapter 8, long-time-series of accurate, global SSTs are needed to monitor the health of the Earth's coral reefs, which support a large diversity of sea life. Sea surface temperature data has also been used by the fish and wildlife communities to study habitats over many parts of the globe (Smith *et al.*, 1996).

Thermal infrared (TIR) sensors have been deployed for over 40 years on operational meteorological satellites to provide images of cloud top temperatures, and when there are no clouds they observe sea surface temperature (SST) patterns (see Figure 11.1 – Chapter 11). Thermal infrared instruments that have been used for

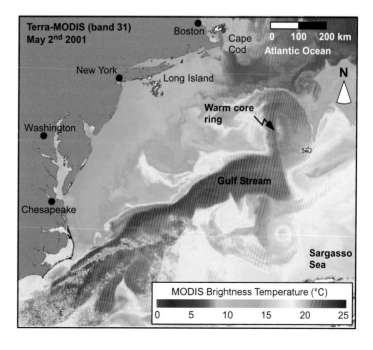

Figure 10.3 Terra-MODIS brightness temperature image for the eastern seaboard of the United States acquired May 2, 2001. The warm waters of the Gulf Stream, with its swirls and gyres, are depicted in reds and yellows. Credit: NASA/GSFC.

deriving SST include the Advanced Very High Resolution Radiometer (AVHRR) on NOAA Polar-orbiting Operational Environmental Satellites (POES), Along-Track Scanning Radiometer (ATSR) aboard the European Remote Sensing Satellite (ERS-2), the Geostationary Operational Environmental Satellite (GOES) imagers, and Moderate Resolution Imaging Spectro-radiometer (MODIS) aboard NASA Earth Observing System (EOS) Terra and Aqua satellites. Thermal infrared was the first method of remote sensing to gain widespread acceptance by the oceano-graphical and meteorological communities.

One reason for the early success of measuring SST is as follows. The thermal infra-red power density emitted by a surface is given by $M = \varepsilon \sigma T^4$, where ε is the emissivity of the target and σ is the Boltzman Constant ($\sigma = 5.67 \times 10^{-8}$ watt/m^2 $^\circ$K^4). Since the TIR radiance depends on both the temperature and emissivity of the target, it is difficult to measure land surface temperatures because the emissivity will vary as the land cover changes. On the other hand, over water the emissivity is known and is nearly constant ($\varepsilon = 0.98$, approaching the behaviour of a perfect blackbody radiator) (Ikeda & Dobson, 1995). Thus the TIR radiance measured over the oceans will vary primarily with the sea surface temperature (SST), allowing the SST to be determined accurately (± 0.5 $^\circ$C) with certain atmospheric corrections (Martin, 2004; Barton, 1995).

Since most SST retrieval algorithms remove the atmospheric radiances by tropo-spheric water vapour but not by aerosols, remotely sensed SSTs must be continu-ously calibrated by surface SST observations. For instance, the AVHRR SST combines satellite-derived surface temperatures with near-surface temperatures measured by moored and drifting buoys or ship observations (Martin, 2004; Elachi & van Ziel, 2006).

Another important application of sea surface temperature sensing is in studies of coastal upwelling, where rising cold water brings nutrients to the surface, inducing phytoplankton and zooplankton to grow and attract large concentrations of fish. The condition of upwelling areas, such as the one off Peru's coast, can be observed by satellites with thermal infrared imagers such as AVHRR (Table 3.6) or ocean colour sensors such as SeaWiFS (Yan *et al.*, 1993; Schofield *et al.*, 2004; Martin, 2004). When wind patterns over the Pacific Ocean change, warm waters from the Western Pacific shift to the Eastern Pacific and the upwelling of nutrient-rich cold water off the Peruvian coast is suppressed, resulting in well-recognized 'El Niño' conditions. A case study of upwelling is presented at the end of this section.

Beginning from the launch of AVHRR/2 on NOAA-7 in 1981, there now exist nearly three decades' worth of infrared satellite SST observations. These contribute to multiyear global climate studies and to regional support of fisheries, ship routing, physical oceanography research and weather forecasting. Examples of long-term studies include the changes in SST patterns associated with such inter-annual climate variation as the La Niña and El Niño cycles in the equatorial Pacific and Atlantic.

Casey & Cornillon (1999), also working from examination of surface SST obser-vations over the past five decades, have shown that if satellite SSTs were accurate to about ± 0.2 K, climate-induced temperature changes would be observable within a two-decade period. Given that the current operational AVHRR accuracy is about ± 0.5 K (Walton *et al.*, 1998) and the MODIS night time SST accuracy is about ± 0.3 K, observation of these changes may soon be achievable. Examples of

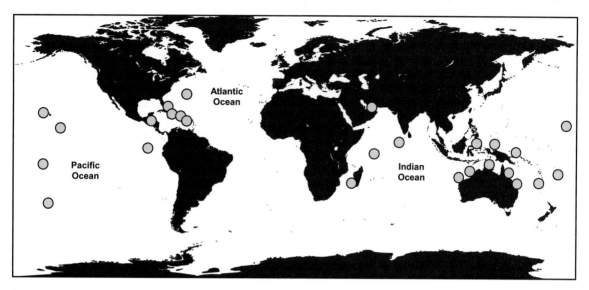

Figure 10.4 Locations of the 24 operational NOAA Virtual Stations, as of 2007. NOAA will expand the network in the coming years to further facilitate the degree to which the threat of coral reef bleaching can be monitored using AVHRR-derived SST data. Credit: NOAA.

short-term SST applications include delineation of ocean fronts, upwelling regions, equatorial jets and ocean eddies. Identification and tracking of such features require accuracies of about ±0.5 K and frequent revisits (Walton *et al.*, 1998). The accuracy requirement is presently met but, because of cloud cover, the revisit requirement is not (Martin, 2004; Smith *et al.*, 1996).

A particularly pertinent use of near-real-time SST data products for monitoring changing climate is the NOAA 'Coral Bleaching Virtual Station Program' (Liu *et al.*, 2005). Here, both a web portal and *Google Earth* are used to freely disseminate information as to the likelihood that selected coral reef sites around the world will be damaged by warmer than usual sea temperatures. The products are entirely derived from the AVHRR satellite sensors, requiring no *in situ* validation (hence use of the term 'virtual'). This aspect of the programme is important, as many of the sites covered are extremely remote to a point that routine ship or buoy monitoring is unfeasible. At the time of writing, 24 virtual monitoring stations were operational, and more information can be found on the programme's Coral Reef Watch website: www.coralreefwatch.noaa.gov/satellite (Figure 10.4).

AVHRR data can be obtained in several ways. As the satellite orbits the Earth, data are broadcast continuously in real time to local ground stations and also recorded onboard for later broadcast to a US ground station. The simplest way to obtain these data is from the Automatic Picture Transmission (APT) mode, which broadcasts the local visible and infrared imagery in an analogue format, with a 4-km pixel size, to any receiving station. The inexpensive APT receivers require only an omnidirectional antenna, and they produce fax-like images. The other source of direct broadcast data is the High Resolution Picture Transmission (HRPT) mode, which broadcasts digital data with 1-km pixels. The HRPT station and its tracking antenna is about an order of magnitude more expensive than an APT station. For regions inaccessible to direct broadcast, there are two options (Martin, 2004):

- First, to obtain 1-km data, the user must request *Local Area Coverage* (LAC) data from NOAA, where the HRPT coverage of the region is recorded for later transmission to a US ground station. The satellite can store about ten minutes of LAC data per orbit.
- Second, the user can obtain lower resolution *Global Area Coverage* (GAC) data. GAC is a reduced data set that is recorded during only one out of every three mirror rotations. During the data-gathering rotation, the data are averaged into blocks of four adjacent samples, then the fifth sample is skipped, then the next four are averaged, etc., so that the data volume is reduced by an order of magnitude. GAC data are recorded continuously around the globe, have a nominal 4-km pixel size, and are downloaded and archived by NOAA. GAC provides global coverage every 1–2 days.

Recognizing the need for improved, long time-series datasets for global change research, NOAA and NASA in 1990 initiated the Pathfinder Program through the Earth Observing System (EOS) Program Office. The datasets are designed to be long time-series of data processed with stable calibration and consensus algorithms to better assist the research community. The Pathfinder Program is also generating important information on the processing of large consistent datasets that can be implemented in the EOS era. This programme is covered in detail in Chapter 5 (Section 5.5.1).

There are multiple processing sites for each Pathfinder. For example, the AVHRR Land Pathfinder is processing the AVHRR data to produce multichannel terrestrial datasets, while the AVHRR SST Pathfinder is processing the data to produce global sea surface temperature. The NOAA/NASA AVHRR Oceans Pathfinder SST dataset is available in a variety of spatial and temporal resolutions and formats to accommodate researchers with varying processing capabilities and scientific interests. The project is charged with reprocessing several decades' worth of archived Global Area Coverage (GAC) data from NOAA polar orbiting satellites.

Measurements of sea surface temperature (SST) can also be made by satellite microwave radiometry in all weather conditions except rain. Passive microwave instruments that have been used for deriving SST include the Scanning Multichannel Microwave Radiometer (SMMR) carried on Nimbus-7 and SeaSat satellites, the Tropical Rainfall Measuring Mission (TRMM) microwave imager (TMI), and data from the Advanced Microwave Scanning Radiometer (AMSR) instrument on the NASA EOS Aqua satellite and on the Japanese Advanced Earth Observing Satellite (ADEOS II). Microwave SSTs have a significant coverage advantage over IR SSTs because microwave SSTs can be retrieved in cloud-covered regions while IR SSTs cannot. However, microwave SSTs are at a much lower spatial resolution than the IR SSTs and typically less accurate.

CASE STUDY: Upwelling and El Niño

The open ocean is biologically quite unproductive when compared to the shallow waters of the continental shelves and coastal upwelling areas (Table 10.3). On the continental shelves, nutrients are supplied by rivers and by wave mixing of surface and bottom water. Upwelling regions owe their high productivity to the slow but persistent upward flow of deep water, which continually charges the photic zone with nutrients. Therefore, some of the world's largest fisheries are located in upwelling areas, such as the main ones located on the west coasts of North and South America, the west coast of Africa and off the coast of Somalia.

One of the best-known upwelling regions is along the coast of Peru in western South America. Here, the northward-flowing Peru Current transports cold water from the south, making the sea surface temperature (SST) unusually cold. SST is lowered even more by Ekman transport. The prevailing southerly winds, blowing parallel to Peru's shore, cause surface water to flow offshore in Ekman transport because the Coriolis deflection is to the left in the southern hemisphere. As a consequence, cold, nutrient-rich water upwells from below, within a relatively narrow coastal zone of 10 to 20 kilometres. Water wells up slowly, but continually, at a rate of about 8 metres per day. This upwelling process continually fertilizes the photic zone and assures a large crop of phytoplankton. The plants, in turn, support very large populations of anchoveta, a small fish that is harvested, dried, pressed into fishmeal and then sold on the world market as feed for livestock and poultry (Pinet, 2009).

In a normal year, the westward-blowing trade winds push warm surface water against the western boundary of the Pacific Ocean near Australia and Indonesia, while nutrient-rich cold water wells up along the west coast of South America, helping fish to thrive. Satellites with radar altimeters have tracked this build-up of warm water in the western Pacific Ocean, which can be as much as 1 metre higher than that in the eastern Pacific (as, for example, shown in Figures 10.8 and 10.9). The actual upwelling areas and their condition can be observed by satellites with thermal infrared imagers such as AVHRR (Table 3.6), or ocean colour sensors such as SeaWiFS (Yan *et al.*, 1993; Schofield *et al.*, 2004; Martin, 2004).

When the trade winds over the equator weaken – and even reverse direction, blowing from west to east – they cause conditions known as El Niño to develop along the Peruvian coast (Yan *et al.*, 1993). This allows the warm 'pile' of water normally held against the western shore of the Pacific to move eastward along the equator and raise the water level off the coast of Ecuador and Peru.

When this bulge of warm water reaches South America, it moves north and south along the coast for hundreds of kilometres, suppressing the normal upwelling of cool nutrient-rich water. As a result, there is little for fish to eat and, in turn, few fish for people to eat. The ocean also affects the atmosphere. With the warm ocean, there is an increase in evaporation and subsequent precipitation over the mountains in that area. The effects of the El Niño reach considerably further than the area surrounding the tropical Pacific: jet streams are altered all over the world, and many places have weather that is very different from normal.

Fishermen along the coasts of Peru and Ecuador have known for centuries about the El Niño. Every 3–7 years, during the months of December and January, fish in the coastal waters off of these countries nearly disappear, causing fishing to come to a standstill. South American fishermen have given this phenomenon the name El Niño, which is Spanish for 'the boy child', because it comes during Christmas (i.e. the time of the celebration of the birth of the Christ child).

During an El Niño, the physical relationships between wind, ocean currents, oceanic and atmospheric temperature break down and form patterns that are as important as the seasons in their impact on weather conditions around the world. For instance, during El Niño, there usually occur increases of rainfall in South America, droughts in Australia, severe winter storms in California, heat-waves in Canada and strong cyclones across the Pacific Ocean (Pinet, 2009). The 1997–1998 El Niño is estimated to have had a total economic impact on the US of about $25 billion, by causing droughts and flooding (Pinet, 2009; Chagnon, 2000; Robinson, 2004).

The coupled atmosphere-ocean phenomenon known as El Niño is usually followed by a period of normal conditions in the equatorial Pacific Ocean. Sometimes, however, but not always, El Niño conditions give way to the other extreme of the El Niño-Southern Oscillation (ENSO) cycle. This cold counterpart of El Niño is known as La Niña, Spanish for 'the girl child'. During La Niña years, the trade winds are unusually strong due to an enhanced pressure gradient between the eastern and western Pacific. As a result, upwelling is enhanced along the coast of South America, contributing to colder than normal surface waters over the eastern tropical Pacific and warmer than normal surface waters in the western tropical Pacific, as shown in the bottom image of Figure 10.5. La Niña also influences global weather, reducing rainfall in the eastern equatorial Pacific, yet increasing it in the west.

The upper image in Figure 10.5 shows thermal infrared images of the westward movement of warm water masses in the Pacific Ocean during the development of an El Niño episode. The temperatures are represented by colours, with the warmest waters in red and coldest in blue. By some measures, this 1997–1998 (super) El Niño event was the strongest on record, with major climatic impacts felt around the world. Waters exceeding 29 °C stretch across the equatorial Pacific basin from the west to the east. By contrast, the lower pane of Figure 10.5 shows that a strong La Niña (cold) event developed by the close of 1998 and a tongue of cool water extends across the Pacific. The La Niña typically (but not always) follows an El Niño.

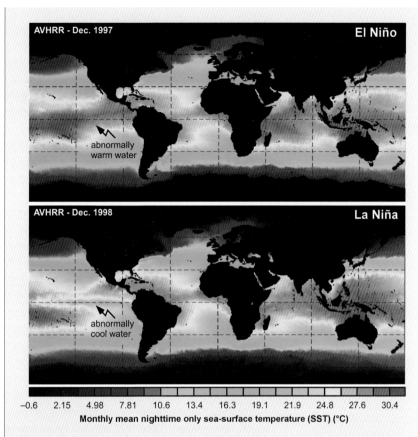

AVHRR - Dec. 1997 — El Niño

abnormally warm water

AVHRR - Dec. 1998 — La Niña

abnormally cool water

−0.6 2.15 4.98 7.81 10.6 13.4 16.3 19.1 21.9 24.8 27.6 30.4

Monthly mean nighttime only sea-surface temperature (SST) (°C)

Figure 10.5 NESDIS AVHRR satellite data. Depicted are monthly mean night-time sea surface temperature (SST) data with a 36 km² resolution. The top pane is from December 1997, the height of the 1997–98 (super) El Niño, while the bottom pane is from December 1998, which was a strong La Niña (cold) event. Credit: NOAA-NESDIS.

The presence of El Niño or La Niña can influence how many hurricanes form in a given year. If a La Niña (cooling of the Pacific) occurs during a northern hemisphere winter and spring, stronger than normal eastern Pacific trade winds pull heat from the ocean, lessening the potential energy for storms there. However, this phenomenon splits the Pacific jet stream and holds the Atlantic portion to the north, allowing hurricanes that form off Africa to advance west toward the Caribbean. That tendency is part of what led to a record year there in 2005 and 2008 (see Figure 11.9, Chapter 11). If an El Niño (warming of the Pacific) grows during spring and summer, the jet stream dips south over North America, creating greater wind shear that will tear apart growing storms that try to organize in the Atlantic. These conditions led to fewer hurricanes there in 2006.

10.5 Ocean salinity

Sea surface salinity (SSS) is critical for determining the global water balance, for understanding ocean currents and for estimating evaporation rates. Sea surface salinity can be measured with microwave radiometers. Airborne microwave radiometers have also been used to determine the structure and influence of river plumes in the Great Barrier Reef, since the input of freshwater plumes from rivers is a critical consideration in the study and management of coral and seagrass ecosystems. Low salinity water can transport natural and man made river-borne contaminants into the sea, and this can directly stress marine ecosystems that are adapted to higher salinity levels.

In microwave radiometry, the power received by the radiometer antennae is proportional to the microwave emissivity and temperature of the ocean surface. Salt dissolves in water, creating charged cations (Na^+) and anions (Cl^-). These charged particles increase the reflectivity and decrease the emissivity of the water. Thus, if the water temperature can be obtained by other means such as thermal infrared radiometers, the salinity can be deduced from the received power.

Salinity is measured in units of parts per thousand (ppt), with average seawater having a salinity of about 35 ppt. This means that the dissolved salt occurs in a concentration of 35 parts per thousand, or 3.5 per cent, with the remaining 96.5 per cent being water molecules. Another set of units used to measure salinity, which are related to the conductivity of the water, are Practical Salinity Units (PSU). Their numerical values are identical to ppt units. An example of a sea surface salinity map obtained with an airborne microwave radiometer is shown in Figure 10.6, which indicates that the surface water salinity across the mouth of Chesapeake Bay ranges from about 12 ppt to 30 ppt.

Airborne scanning low-frequency microwave radiometers (SLFMR) have been used to map the salinities of various bays and estuaries, including their oceanic plumes (Miller *et al.*, 1998; Burrage *et al.*, 2003; 2008; Perez *et al.*, 2006). The characteristics of plumes discharging from major rivers into the coastal oceans reflect river basin properties and processes and significantly impact continental shelf circulation and associated marine ecosystems.

An international project, La Plata, was undertaken within the South American Climate Change Consortium framework to study the seasonal behaviour of the Plata River and Patod Lagoon plumes by obtaining comprehensive measurements of physical, chemical and biological features in summer and winter. Oceanographical and remote sensing datasets were acquired to provide a quasi-synoptic description of the region. The instrument used to map sea surface temperature (SST) and sea surface salinity (SSS) was a Salinity, Temperature, and Roughness Remote Scanner (STARRS).

STARRS comprises L- and C-band microwave radiometers, an infrared (IR) radiometer, an integrated global positioning system (GPS) receiver and a fibre-optic gyro. Measurements from these instruments are combined to retrieve SSS, temperature (SST) and roughness. The L-band radiometer – used primarily to

L-band airborne STARRS salinity

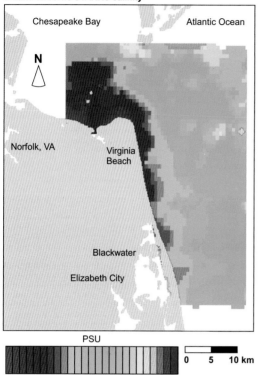

Figure 10.6 Sea surface salinity map of lower Chesapeake Bay produced from the STARRS airborne salinity imager. The instrument is an L-band (1.4 GHz) microwave radiometer. A pronounced gradient in salinity, delivered in Practical Salinity Units (PSU), is evident across the mouth of the bay. Modified with permission from Miller *et al.* (2004).

retrieve salinity – is a multi-beam system sensing natural microwave emission from the sea surface at a wavelength of 21 cm. The six antenna beams point downward and to either side of the aircraft at incidence angles of around 7°, 22°, and 38°. The C-band radiometer, which senses surface roughness, has a single 20°-wide beam pointing directly beneath the aircraft (nadir viewing), which senses natural emission within six microwave channels. A nadir-viewing IR radiometer senses SST from thermal emission in the 8,000 nm to 14,000 nm and 9,600 nm to 11,500 nm bands. Clouds do not block the microwave observations, but they do cause gaps in the IR observations, which are filled by interpolation. Satellite and ship observations are usually obtained to complement surveys conducted using the airborne salinity mapper (Perez *et al.*, 2006).

Sea surface salinity has been the most important oceanic variable which until recently has not been measured from satellites. Surface salinity must be measured by passive microwave radiometers with a 20 cm wavelength, so measuring the SSS for a 50 km resolution at an altitude of 900 km would require an antenna at least 4 m long, which most satellites in the past could not accommodate. However, problems with antenna size have been overcome with new interferometric technology.

There are now instruments designed to provide SSS from satellite orbit. For instance, the European Soil Moisture and Ocean Salinity (SMOS) satellite uses a fixed two-dimensional interferometric antenna. The satellite, launched in 2009, retrieves salinity with an accuracy of 0.1–0.2 PSU. Because both instruments operate over a range of incidence angles, they differ from the old passive microwave imagers. To obtain a SSS accuracy of 0.2 PSU requires a brightness temperature accuracy of 0.1 K, which in turn requires an accurate low-noise radiometer.

SMOS employs a three-armed Y-shaped antenna, where each arm measures 4.5 m in length, contains 24 radiometers and folds for launch. The arms do not rotate but are fixed to the satellite; the polarized signals collected by the radiometers are cross-correlated to construct maps of brightness temperatures over a 100-km swath-width. The swath is divided into pixels with a spatial resolution of 35 km at nadir. The satellite operates in a sun-synchronous orbit at an altitude of 760 km and with a three-day repeat cycle.

Another important satellite microwave instrument is the Advanced Microwave Scanning Radiometer (AMSR-E) on the Aqua satellite (the instrument is further considered in Chapter 7). Launched in 2002, Aqua is a major satellite mission of the Earth Observing System (EOS), an international programme for satellite observations of the Earth, centred at NASA. Aqua carries onboard six distinct Earth-observing instruments, the data from which are being used to examine dozens of Earth system variables and their interactions (Parkinson, 2003):

- the Atmospheric Infrared Sounder (AIRS);
- the Advanced Microwave Sounding Unit (AMSU)
- the Humidity Sounder for Brazil (HSB)
- the Advanced Microwave Scanning Radiometer for EOS (AMSR-E)
- the Moderate Resolution Imaging Spectroradiometer (MODIS)
- Clouds and the Earth's Radiant Energy System (CERES).

Aqua is essentially a sister satellite to Terra, the first of the large EOS observatories, launched in 1999 to monitor the 'health of the planet', with Terra emphasizing land and Aqua emphasizing water. Both satellites, however, measure many variables in the atmosphere, on the land and in the oceans. Aqua data are providing information on water in its many forms (Parkinson, 2003):

- water vapour in the atmosphere;
- liquid water in the atmosphere in the form of rainfall and water droplets in clouds;
- liquid water on land in the form of soil moisture;
- solid water on land in the form of snow cover and glacial ice;
- solid water in the oceans in the form of sea-ice floating in the polar seas.

Aqua and SMOS data are also providing information on soil moisture and vegetation conditions, heavily dependent on water, and on many other aspects of the Earth's climate system. The data obtained from Aqua's microwave radiometer AMSR-E are shown in Table 10.7. Note that Aqua measures soil moisture in addition to various water cycle related properties. Soil moisture plays an important

Table 10.7 Key AMSR-E data products (Parkinson, 2003).

- Water vapour (total atmospheric column)
- Cloud water (total atmospheric column)
- Rainfall
- Sea surface wind speed
- Sea surface temperature
- Sea surface concentration
- Sea-ice temperature
- Snow depth on sea-ice
- Snow-water equivalent on land
- Surface soil moisture

role in studies of global climate change, including the exchange of water and heat between land and the atmosphere. It is also important for agriculture and various ecosystem studies. Using long wavelengths (30 cm) means that the soil data are less obscured by the presence of vegetation.

10.6 Physical ocean features

Conventional remote sensing of the surface of the Earth from aircraft or spacecraft involves using either cameras or scanners that produce images in a direct manner. These instruments are passive instruments; they receive the radiation that happens to fall upon them and select the particular range of wavelengths that have been chosen for the instrument. When these instruments operate at visible or infrared wavelengths, they are capable of good spatial resolution. However, at visible and infrared wavelengths, these instruments are not able to see through clouds. In other words, if the atmosphere is cloudy, they produce images of the top of the clouds and not the surface of the Earth.

By moving into the microwave part of the electromagnetic spectrum, we are able to see through clouds and hence to obtain images of the surface of the Earth or ocean even when the weather is overcast. Furthermore, active microwave sensors, such as radar, enable us to measure many land and ocean properties which we cannot detect with passive visible and infrared sensors.

Three active microwave devices – radar altimeters, scatterometers and synthetic aperture radar (SAR) imagers – are of particular importance to physical oceanography, because they provide accurate global and regional information on ocean elevation, currents, winds and waves. The application of these radars depends on the character of the pulse emitted, which may be long or short, of uniform frequency or of swept frequency. It also depends on what properties of the reflected pulse are measured. For a nadir-pointing radar, the timing of the returned pulse after reflection from the ocean surface, knowing the speed of light (EM waves),

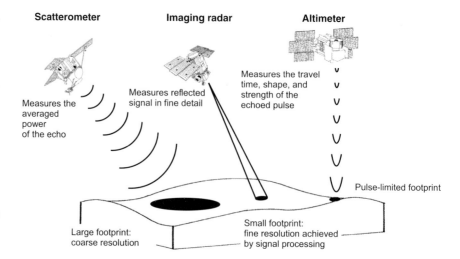

Figure 10.7 Different types of radar for Earth observation.

allows measurement of the distance between the radar and the sea surface. This is the basic principle of radar altimeters.

As shown in Figure 10.7, oblique-viewing instruments which measure the back-scatter from the sea surface can be divided into two types. Those that measure average backscatter from a wide field-of-view are called scatterometers, and are used primarily to measure wind characteristics, which create the surface rough-ness. Radars that have a much finer spatial resolution are called imaging radars. These provide maps of sea surface roughness capable of defining a variety of small and meso-scale ocean characteristics, such as wave fields (Robinson, 2004; Elachi & van Ziel, 2006). A list of satellites with SAR systems is shown in Table 10.8, and the basis of operation of this technology is covered in Chapter 6 (Section 6.2.4).

10.6.1 Sea surface elevation and ocean currents

The variation in the height of the sea surface from place to place reveals important information about weather, climate and sea level rise. Ocean surface topography (OST) data has been used by scientists to calculate the energy imbalance in the world's oceans caused by global warming, to monitor natural climate variations such as El Niño and La Niña, map ocean currents, and to improve predictions of hurricane intensity.

There are several phenomena which can cause water masses to have differing heights from those around them (Martin, 2004; Canton-Garbin, 2008):

- The largest and most consistent deflections of ocean surface come about from the gravitational attraction of sea floor mountains (or canyons). The undulations of the mass of the Earth, and the local differences in gravitation which correspond to these, are referred to as the Earth's geoid; the more massive mountains attract more water above them.

- Another reason can be intense windstorms across the surface of a regional body of water, which can push large quantities of water away from one area and pile it up onto another. The most pronounced example of this is the storm surge that precedes the landfall of a hurricane or cyclone.
- A third cause of differing sea surface height occurs in coastal areas, where the daily excursions of the tides can produce substantial changes in the absolute elevation of the water.

Satellite altimetry produces unique global measurements of instantaneous sea surface heights relative to a reference surface, and it is one of the essential tools for monitoring ocean surface conditions. Scientific results from altimetric data have significantly improved our knowledge of global ocean tides and meso-scale circulation variability in the deep ocean.

Knowledge of coastal and oceanic circulation provides information on the movement of water masses. On a coastal scale, the motion of a patch of water containing a harmful algal bloom or industrial waste is vital for planning appropriate control measures. On the ocean basin scale, knowledge of oceanic circulation is a significant component of planetary heat budget calculations for global climate programmes. One obvious example of the ocean circulation's influence on climate is the Gulf Stream, which transports enormous amounts of heat from the Equator and the tropics to northern Europe and southern Greenland. This causes the average temperature in northern Europe to be about 6 °C to 9 °C higher than at the same latitudes in North America.

There are various physical reasons for the movement of water, including wind stress, tides and water density, but the major ocean surface currents are generated primarily by winds. Atmospheric circulation, on the other hand, including winds, is produced by convection due to variation of temperature with latitude, along with the Coriolis Effect caused by Earth's rotation. Radar altimeters can map the lateral variations in the dynamic height of the ocean's surface which portray pressure gradients and the resulting distribution of geostrophic ocean currents.

As shown in Figure 10.8, a radar altimeter aboard a satellite is a nadir-looking active microwave sensor. Its signal pulse, transmitted vertically downward, reflects from the ocean surface back to an altimeter antenna. The round-trip time and the propagation speed of the electromagnetic waves are used to compute the range between the antenna and the ocean surface. From the altimeter-measured range, the instantaneous sea surfaces relative to a reference surface, such as an ellipsoid, can be determined, provided the satellite orbit relative to the reference surface is known. With knowledge of the oceanic geoid, the sea surface topography relative to the geoid due to ocean dynamic circulation, including the temporal averages, can be mapped. Repeated observations can provide a measurement of the temporal variability of the sea surface height, since the geoid can be treated as time-invariant for oceanographical applications (Ikeda & Dobson, 1995; Robinson, 2004; Elachi & van Ziel, 2006).

Because almost all radar altimeter missions use a repeating orbit, repeat track analysis becomes a conventional approach for application of altimeter data in

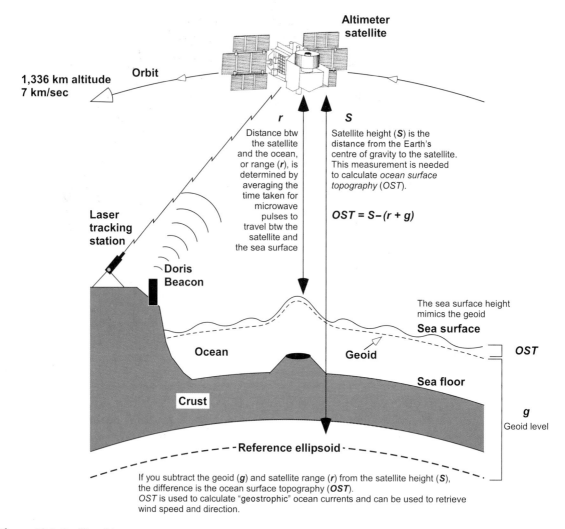

Altimeter satellite

1,336 km altitude
7 km/sec

Orbit

Laser tracking station

r
Distance btw the satellite and the ocean, or range (*r*), is determined by averaging the time taken for microwave pulses to travel btw the satellite and the sea surface

S
Satellite height (*S*) is the distance from the Earth's centre of gravity to the satellite. This measurement is needed to calculate *ocean surface topography* (*OST*).

$$OST = S-(r + g)$$

Doris Beacon

The sea surface height mimics the geoid

Sea surface

Ocean

Geoid

OST

Sea floor

Crust

g
Geoid level

- - -**Reference ellipsoid**- - - - - - -

If you subtract the geoid (*g*) and satellite range (*r*) from the satellite height (*S*), the difference is the ocean surface topography (*OST*).
OST is used to calculate "geostrophic" ocean currents and can be used to retrieve wind speed and direction.

Figure 10.8 Satellite altimeter measurement system.

physical oceanography. A mean at a location is calculated from data available, and then the mean is removed from the data to produce sea surface height anomalies relative to the mean. The repeat track analysis thus eliminates the geoid and its errors, the mean sea surface and a portion of the orbit error (Han, 2005; Martin, 2004; Cracknell & Hayes, 2007).

Oceanographical applications of satellite altimetry require an accuracy of a few centimetres. This requirement constrains not only the altimeter sensor but also pertinent atmospheric and oceanographical corrections that have to be made, the satellite orbit and, in some applications, the reference geoid. However, the accuracy of altimetric sensors has steadily improved to a few centimetres. For instance, the TOPEX/Poseidon mission altimeters have been providing 3 cm vertical resolution with an interval spacing of 25 km.

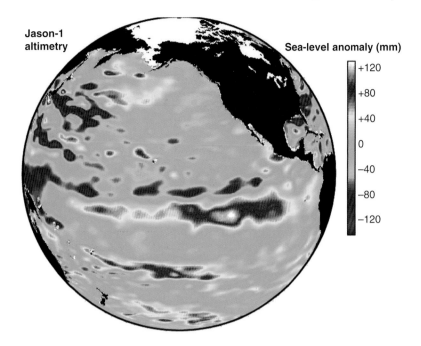

Jason-1 altimetry

Sea-level anomaly (mm)

+120
+80
+40
0
−40
−80
−120

Figure 10.9 Jason-1 satellite altimetry sea surface height anomaly for the Pacific (November 2006). Since warm water expands, it takes up a fraction more space than cool water does, resulting in higher than normal sea surface height. These height anomalies can hence be used as a proxy for sea surface temperature. Credit: NASA's Earth Observatory.

One practical application of radar altimetry has been the observation of Pacific Ocean sea levels in order to predict the onset of and monitor the effects of El Niño. Altimeters on TOPEX/Poseidon and other satellites have shown that, as the warm surface waters move east across the Pacific, the sea level along the coast of Peru can rise by tens of centimetres (Figure 10.9). This associated rise of tropical air and formation of rain clouds contributed to the severe weather and heavy rains in California during the winter of 1995.

In 2008, the Jason-2 satellite was launched, carrying an altimeter capable of providing topographical maps of 95 per cent of the worlds ice-free oceans with an accuracy of 4 per cent. Its data will track not only sea level rise but also reveal how great water masses are moving around the globe. This information will be important to weather and climate agencies for making better forecasts.

Jason-2 is a continuation of the programme started in 1992 with the TOPEX/Poseidon mission and is preceded by the Jason-1 satellite, launched in 2001. The project provides the global reference data for satellite-measured ocean height and forms the basis for the observation that the sea level is rising about 3 mm per year (Figure 10.10). Ocean height, wave height and wind speed can be extracted from the altimeter data. In addition to weather, storm and climate prediction, the data are also used to guide ships in order to avoid storms and conserve fuel, guide undersea drilling and cable-laying operations, and also predict the drift of wreckages and pollutants.

Closer to the coast, shore-based high-frequency (HF) and microwave Doppler radar systems can be used to map currents and determine swell-wave parameters over large areas with considerable accuracy (Graber *et al.*, 1997; Paduan & Graber,

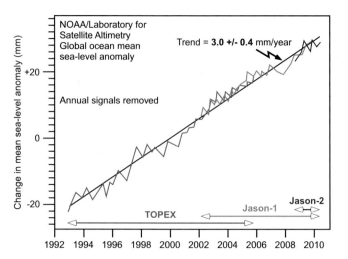

Figure 10.10 Change in mean sea level anomaly, as analyzed by the NOAA Laboratory for Satellite Altimetry using data from radar altimeters on TOPEX/Poseidon, Jason-1 and Jason-2 satellites. Credit: NOAA Laboratory for Satellite Altimetry.

1997; Bathgate *et al.*, 2006). HF radars can determine coastal currents and wave conditions over a range of up to 200 km (Cracknell & Hayes, 2007). While HF radars provide accurate maps of surface currents and wave information for large coastal areas, their spatial resolution, which is about 1 km, is more suitable for measuring meso-scale features than small scale currents. A more detailed description of HF radars is provided in Chapter 9.

Estimates of currents over large ocean areas, such as the continental shelves, can also be obtained by tracking the movement of Lagrangian drifters or natural surface features which differ detectably in colour or temperature from the background waters. Examples of such tracked natural features include chlorophyll plumes, patches of different water temperature and surface slicks. Ocean drifters are specifically designed to track the movement of water (currents) at various depths.

A typical design of global Lagrangian drifters includes a float or surface buoy connected by a cable to a current drogue. The surface float provides buoyancy, contains the electronics and transmits data to satellites. The satellites determine the drifter's position and relay the data to ground stations. The drogue, set for a specific depth, acts like an underwater sail as it is pushed by the ocean current. Ocean drifters may also contain various instruments to measure water temperatures and a variety of other parameters. Ships and aircraft can drop these durable drifter buoys into the sea, where they normally have a survival rate of several hundred days (Davis, 1985; Breaker *et al.*, 1994).

On a planetary scale, one must consider not only the steady general ocean surface circulation caused by prevailing winds, but also variations in this circulation caused by gyres and eddies and also the deep thermohaline circulation, caused by variations in temperature and salinity (see Chapter 12, Section 12.4). Deep thermohaline circulation cannot be directly observed by satellites, yet it has long-term repercussions on the climate via the Great Coveyor Belt (GCB) effect (Canton-Garbin, 2008).

The GBC is formed when, at high-latitudes, warm surface waters such as the Gulf Stream yield their heat to the atmosphere, cool, increase in density enough to sink and return southward at depths of thousands of meters (Figure 12.4). The GCB has become destabilized in the distant and recent past and could again become destabilized in the future, with major consequences to the climate (Anderson, 1997; Broecker, 1997; Rühlemann *et al.*, 1999 – see discussion on the Younger Dryas event in Section 12.4). For example, if, due to global temperature rise, precipitation were to increase in northern Europe, or a considerable amount of ice were to melt in Greenland, this increase of fresh water would slow down the arrival of warm water from the Gulf Stream, which could cause the GCB to collapse. This would cause major temperature drops in northern Europe (Canton-Garbin, 2008).

10.6.2 Sea surface winds

Surface wind measurements are important because winds generate surface waves and currents. Wave conditions must be known for a number of activities, including safe routing of ships, the design of offshore platforms and coastal defence. Winds also drive the ocean currents, modulate the air/sea fluxes of heat, moisture and gases such as carbon dioxide, and influence global and regional weather and climate. The distribution of wind speeds determines the height and direction of ocean swells and allows prediction of their effects on ships and coastal structures (Martin, 2004).

Winds transfer energy to the surface layer of the sea, causing ripples, which can develop into wavelets and waves in proportion to the magnitude and direction of the winds. The most common remote sensor for mapping sea surface winds is the radar scatterometer. Satellites equipped with radar scatterometers use the sea-state to estimate the near-surface wind speed and direction (Figure 10.11). A scatterometer is a sensor that measures the return reflection or scattering of microwave (radar) pulses sent at an oblique angle to the ocean surface from a satellite (Figure 10.12). A rough ocean surface reflects back (backscatters) to the antenna on the satellite a stronger signal than does a smooth ocean surface, because energy from the radar signal is reflected back more effectively when steeper waves are present than when the ocean surface is relatively smooth. Computers estimate wave height, and then the wind speed, from the power differences in the returned signals (Robinson, 2004; Elachi & van Ziel, 2006).

NASA and ESA have launched various satellites that have measured near-surface winds over the ocean. These missions include the initial NSCAT (NASA Scatterometer), launched in 1996, and the QuikSCAT (Quick Scatterometer) launched in 1999 (Robinson, 2004). Figure 10.12 shows the geometry of a scatterometer design. For a steady wind, each scatterometer retrieves the backscatter from the same FOV at two or four different times, azimuth angles and polarizations. For the three beams shown in the figure, the antenna look angles relative to the satellite trajectory are 45° ahead, at right angles to the trajectory and at 45° behind. To obtain wind speed and direction, Bragg scattering must dominate, so the antenna incidence angle must be greater than 20°. Bragg scattering takes place

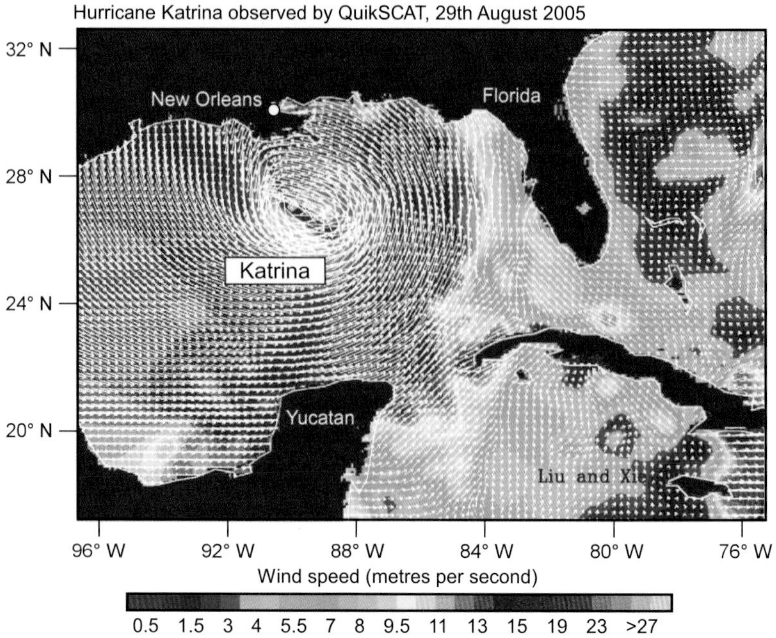

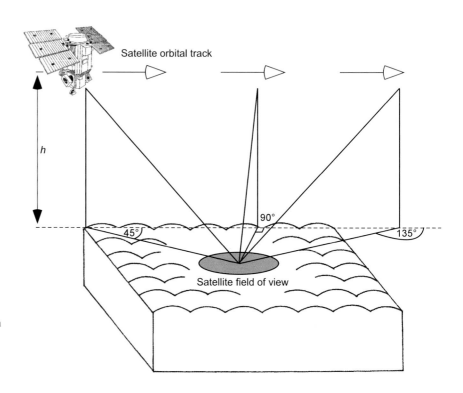

Hurricane Katrina observed by QuikSCAT, 29th August 2005

Figure 10.11 Hurricane Katrina generated surface winds observed by the QuickSCAT scatterometer. The arrows indicate the direction of the surface wind while the colours show the speed. Courtesy W. Timothy Liu & Xiaosu Xie, NASA/JPL.

Wind speed (metres per second)

0.5   1.5   3   4   5.5   7   8   9.5   11   13   15   19   23   >27

Figure 10.12 Example of several looks by a scatterometer at the same field-of-view. Modified from Martin (2004), with kind permission of Cambridge University Press.

for specific incidence angles and frequencies, or when the wave spacing equals half the projection of the incident radar wavelength on the sea surface. If the surface wave spectrum contains a wavelength component with this relation to the incident radiation, Bragg resonance occurs and a strong return signal is obtained.

If the satellite is at an altitude of 800 km, with a surface velocity of 7 km/sec, it will provide three measurements over a period of two minutes. Assuming steady winds, the vector wind speeds can be derived from these observations. The sea winds instruments aboard the QuickSCAT satellite have been measuring winds in this manner over approximately 90 per cent of the ice-free ocean on a daily basis since 1999, with an accuracy of 2 m/s and 20° in azimuth (Martin, 2004).

Altimeters can also measure wind speed. Wind speed observations are based on the measurement of the target cross-section as seen by the zenith-looking radar altimeter. The higher the intensity of the wind, the more the radar pulses are scattered by the waves and the less energy is received, indicating a smaller radar cross-section of the target or ocean surface. Altimeters are discussed in section 10.6.1.

10.6.3 Ocean waves

Wave height is dependent on the velocity of the wind, the distance over which the wind blows (fetch) and the length of time that it blows. Wave direction, average wave height and wave spectrum data are very useful, both as an input to predictive weather forecast models and for real-time information about sea state conditions.

Wave forecasts are crucial not only for marine interests, where high waves could spell disaster for ships on the open sea, but also for residents of coastal communities, where waves and swells could produce high surf, coastal flooding, beach erosion or destruction of shore structures. In recent years, improvements have been made in methods for forecasting wave heights.

In order to predict the sea state and, ultimately the wave height, near-surface wind speeds at frequent time intervals have to be determined. These wind forecasts are obtained by using one of the current operational numerical weather prediction models. The near-surface wind and ocean wave data needed to produce these forecasts come from a variety of sources, including ships, moored automated buoys and orbiting satellites (Martin, 2004; Robinson, 2004).

Spurred by the need for better weather and climate forecasts, the meteorological and oceanographical communities have expanded their monitoring of wind and wave conditions over the open ocean to include more observations on greater spatial and temporal scales. Moored buoys, deployed by various nations in their coastal waters, serve as instrumented platforms for making automated weather and oceanographical observations.

The National Data Buoy Center (NDBC), an agency within the National Weather Service (NWS) of NOAA, operates approximately 80 moored buoys in the coastal and offshore waters of the western Atlantic Ocean, the Gulf of Mexico, the Great Lakes and the Pacific Ocean from the Bering Sea to southern California, around the Hawaiian Islands and in the South Pacific. These buoys have accelerometers or inclinometers that measure the heave acceleration or the vertical displacement provided to the buoy by the waves. These measurements are then processed by an onboard computer that uses statistical wave models to generate wind-sea and swell data, which are then transmitted to shore stations. The data include significant wave height, average wave period of all waves and dominant wave period during each 20-minute sampling interval.

Wave heights can also be obtained by satellite altimeters. Significant wave height, defined as the average height of the highest one-third of waves, is found by fitting a model curve to the slope in the rise of the pulse received by the altimeter antenna. Since the short pulses sent to the sea surface are lengthened due to the first reflection by the crests, and afterwards by the valleys between the waves, the higher the wave, the smaller will be the slope of the return pulse. However, the spatial resolution of satellite altimeters is only about 25 km (Elachi & van Ziel, 2006; Canton-Garbin, 2008).

High-resolution (18 m) SAR instruments can reveal detailed patterns of ocean waves, including wavelength (spectrum) and direction. The reason that waves and swells are visible in SAR images is that the capillary waves associated with Bragg scatter form preferentially on and just ahead of the crests, in part because of the curvature and in part because the troughs are sheltered from the winds while the crests are exposed. This variation in capillary wave amplitude creates the observed bright/dark pattern delineating wave fields on SAR images (Martin, 2004). Since these are active microwave systems, they penetrate clouds and function in all weathers. SAR instruments are carried by satellites in low Earth orbits, usually near-polar, at altitudes between 600 km and 800 km. The beam direction is at 90° to the direction of travel and its axis is tilted between 15° and 60° from the local vertical (Figure 10.7).

As shown in Table 10.8, the Canadian SAR instrument on RADARSAT circles the Earth at an altitude of 790 km and images the Earth in the C-band wavelength of 5.6 cm. Each of RADARSAT-1's seven beam modes offer a different image resolution. The modes include Fine, which covers an area of 50 km² with a resolution of 10 metres; Standard, which covers an area of 100 km and has a resolution of 30 metres; and ScanSAR Wide, which covers a 500-square km area with a resolution of 100 metres.

RADARSAT-1 also has the unique ability to direct its beam at different angles. With an orbital period of 100.7 minutes, it circles the Earth 14 times a day. The orbit path repeats every 24 days, meaning that the satellite is in exactly the same location and can take the same image every 24 days. Using different beam positions, a location can also be scanned every few days.

10.6.4 Oil slicks and other surface features

As we will shortly discuss, usually radar imagery is needed to see an oil spill from orbit. Sometimes, however, conditions conspire to make the slick discernable in the visible spectrum. The Aqua-MODIS image of Figure 10.13 shows the oil slick that

Table 10.8 Characteristics of selected Earth-orbiting Synthetic Aperture Radars (SAR). Modified from Jensen, John R., *Remote Sensing of the Environment: An Earth Resource Perspective*, 2nd edition, Copyright 2007. Printed and electronically reproduced by permission of Pearson Education, Inc.

Parameter	SEASAT	SIR-A	SIR-B	SIR-C/X-SAR	ALMAZ-1	ERS-1,2	JERS-1	RADARSAT
Launch date	June 1978	Nov. 1981	Oct. 1984	April 1994 Oct. 1994	March 1978	1991 1995	Feb. 1992	Nov. 1995
Nationality	USA	USA	USA	USA	Soviet Union	Europe	Japan	Canada
Wavelength (cm)	L (23.5)	L (23.5)	L (23.5)	X (30) C (5.8) L (23.5)	S (9.6)	C (5.6)	L (23.5)	C (5.6)
Depression angle (near to far range)	73–67°	43–37°	75–35°	variable	59–40°	67°	51°	70–30°
Incident angle	23°	50°	15–64°	15–55°	30–60°	23°	39°	10–60°
Polarization	HH	HH	HH	HH, HV, VV, VH	HH	VV	HH	HH
Azimuth resolution (m)	25	40	17–58	30	15	30	18	8–100
Range resolution (m)	25	40	25	10–30	15–30	26	18	8–100
Swath-width (km)	100	50	10–60	15–90	20–45	100	75	50–500
Altitude (km)	800	260	225 and 350	225	300	785	568	790
Latitude coverage	10°N–75°N	41°N–36°N	60°N–60°S	57°N–57°S	73°N–73°S	near-polar orbit	near-polar orbit	near-polar orbit
Mission duration	105 days	2.5 days	8 days	10 days	18 months		6.5 years	

Figure 10.13 Aqua-MODIS visible-spectrum image of the Deepwater Horizon oil slick in the Gulf of Mexico, acquired April 25, 2010. Credit: NASA/MODIS Rapid Response Team.

followed the sinking of the Deepwater Horizon drilling platform off the coast of Louisiana. This image was acquired on April 25, 2010, five days after the explosion that sank the rig. The slick continued to grow for several months as oil escaped into the Gulf of Mexico from a ruptured well on the seabed in 1.5 km of water. The Deepwater Horizon incident rapidly evolved into arguably the worst ecological disaster that the United States has witnessed. The slick in this image is obvious because oil has flattened the sea surface and promoted sun glint. However, such clarity is rare since it demands calm sea conditions and for the sun to be fortuitously low to the horizon as the data are captured, such that glint is imaged (see Chapter 8).

Oil or other slicks on the sea surface are well-resolved in SAR data because they create regions of low backscatter. SAR imagers view the ocean surface at incidence angles between approximately 20° and 30° from the local vertical. Capillary waves and short gravity waves cause the radar pulse to be scattered, including some backscattering to the radar transmitter. As short surface gravity waves or capillary ripples propagate through a region where a surface film is present, their energy is absorbed as the surface film strains, causing damping of these short waves. The film-covered area backscatters less energy to the radar receiver, since most of the radar pulse is reflected from the flatter surface, somewhat like light from a mirror in optics, sending the radar energy in the opposite direction and away from the radar antenna. Ocean surface areas covered by oil or other slicks thus show up as dark in radar images.

For this to work, low to moderate winds must exist to create the short surface waves. Since the short waves being dampened are similar in wavelength to waves used by C– and X–band SARs, Bragg reflection can cause a strong radar signature (Robinson, 2004; Brecke & Solberg, 2005).

The SAR image in Figure 10.14 clearly shows the dark oil slicks off the coast of Wales, near St.Anns Head, after the tanker Sea Empress, carrying North Sea crude oil, ran aground on the rocks in the mouth of Milford Haven in 1996. Until the

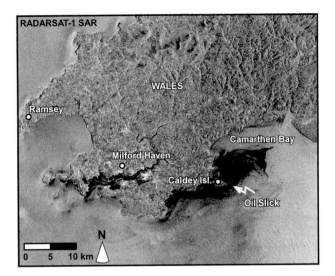

Figure 10.14 RADARSAT-1 SAR image of oil slicks off Milford Haven, Wales, after the oil tanker Sea Empress ran aground on February 15, 1996. Credit: RADARSAT International and Canada Center for Remote Sensing.

2010 sinking of the Deepwater Horizon rig in the Gulf of Mexico, this oil spill was one of the worst such environmental disasters on record, measuring nearly twice as large as the Exxon Valdez spill in 1989. The damaging effects of this spill were felt by the wildlife residing in Milford Haven Estuary and along the southwest coast of Wales, which serve large breeding concentrations of local and migratory seabirds. RADARSAT SAR images were available within hours of data acquisition and were used to assist with clean-up efforts.

When analyzing SAR images to distinguish oil slicks from other surface films, such as organic films generated by natural biological processes or wind-generated slicks, one must consider additional information. This can include the general shape of the slick, its proximity to oil tanker lanes or oil drilling platforms, the local wind speed and other dynamic causes, such as internal waves and ocean fronts.

SAR can also detect other phenomena which modulate the short waves on the sea surface, including ocean fronts and internal waves (Figure 10.15). For instance, large ocean internal waves on continental shelves strongly influence acoustic wave propagation; submarine navigation; mixing of nutrients to the euphotic zone; sediment re-suspension; cross-shore pollutant transport; coastal engineering; and oil exploration.

The water column is frequently not homogeneous but is instead stratified. Internal waves move along pycnoclines, which are surfaces that separate water layers of different densities and temperatures. The larger internal waves can attain heights in excess of 50 m. The period of internal waves approximates the period of the tides, suggesting a cause-and-effect relationship. Internal waves can be detected visually and by radar because they cause local currents which modulate surface wavelets and slicks (Alpers, 1985; Apel, 2003; Zhao *et al.*, 2004).

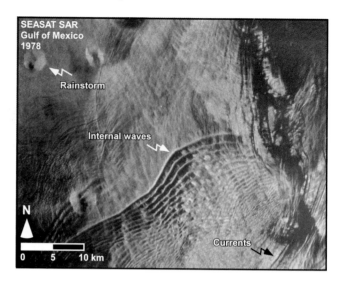

Figure 10.15 SEASAT SAR image of the surface of the Gulf of Mexico. Numerous ocean features are resolved, including rainstorms, internal waves and currents, all of which modify the roughness of the water's surface and hence have a backscatter signature in the microwave spectrum. SEASAT was launched on 26 June, 1978, but failed 105 days later. Credit: NASA.

10.7 Ocean observing systems

In 2004, the US Commission on Ocean Policy and the National Ocean Research Leadership Council identified the Integrated Ocean Observing System (IOOS) as a high priority and emphasized the importance of interagency cooperation for successful implementation. IOOS is part of the US Integrated Earth Observation System, the United States' contribution to the Global Ocean Observing System (GOOS) and a contribution to the Global Earth Observation System of Systems (Piotrowitz, 2006).

IOOS will be a sustained network of sensors on buoys, ships, satellites, underwater vehicles, *in situ* sensors and other platforms that routinely supply data and information needed to detect and predict changes in the nation's coasts and oceans, as well as the Great Lakes. IOOS is intended to draw together the vast network of disparate, federal and non-federal observing systems to produce a cohesive suite of data, information and products at a sufficient geographical and temporal scale to support decision-making by providing scientists with what we need to know about our oceans and coasts in order to fully assess their impact on commerce, transportation, weather, climate and various ecosystems.

Ocean observing data are required to support a wide range of critical decisions, including evacuations, beach and shellfish closures, fisheries catch limits and the identification of safe and efficient shipping routes. The impact of these daily decisions on the US economy is significant, as the coastal states economy is valued at about $10 trillion; the 30 coastal states account for 82 per cent of the total US population and 81 per cent of jobs. In 2005, Hurricanes Wilma, Rita, and Katrina accounted for approximately $157 billion in damages and about 2,000 deaths within coastal communities. Therefore, resource and emergency managers must have ready access to the tools and information needed to support informed and effective coastal and ocean decision-making (Willis & Cohen, 2007).

IOOS is developing as two closely coordinated components – global and coastal – that encompass the broad range of scales required to assess, detect and predict the impacts of global climate change, weather and human activities upon the marine environment. The global component of the IOOS consists of an international partnership to improve forecasts and assessments of weather, climate, ocean state and boundary conditions for regional observing systems in support of GOOS (Hankin *et al.*, 2004). The global component is continuing to make steady progress toward full implementation, with over half of the integrated global array now in place. The global drifting buoy programme has reached 100 per cent of its design goal of 1,250 data buoys in sustained service. The global array of profiling floats (Argo) also achieved over two-thirds of its initial objective of 3,000 floats by 2005.

The coastal component of the IOOS blends national observations in the Exclusive Economic Zone (EEZ) with measurement networks that are managed regionally to improve assessments and predictions of the effects of weather, climate and human activities on the state of the coastal ocean, its ecosystems and living resources and the US economy. The coastal component includes the national aggregate (backbone) of observations, as well as a network of Regional Coastal Ocean Observing Systems (RCOOSs) nested within that backbone. The backbone consists of *in situ* observations and remote sensing, data management and modelling needed to achieve the goals and missions of federal and regional agencies. It also establishes standards and protocols concerning measurements, data management and modelling. The federal agencies are responsible for the system design, operation and improvement, while RCOOSs customize the backbone by increasing the density of observations and the number of variables measured, based on the priorities of their respective regions.

The regional groups are organized into 11 regional associations; Alaska, Pacific Northwest, North/Central California, Southern California, Hawaii/Pacific Islands, Great Lakes, Gulf of Mexico, the Northeast, the Mid-Atlantic, Southeast and the US territories in the Caribbean. These associations represent regional interests of federal and state agencies, the private sector, non-governmental organizations and academia (Pitrowitz, 2006).

For example, the Mid-Atlantic Coastal Ocean Observing System (MARCOOS) covers the US east coast from Massachusetts to North Carolina. Based on the Coastal Ocean Observation Lab (COOL) at Rutgers University, it operates key technologies that allow for fine-scale characterization of the coastal ocean over wide areas, including data streams from US and foreign satellites in space, a network of high-frequency (HF) radars along the shore and a fleet of remotely-controlled robotic gliders moving beneath the ocean surface. The resulting data are delivered in near-real time, analyzed and posted on the World Wide Web (Schofield *et al.*, 2008).

10.8 Marine GIS

Just as for terrestrial applications, the strength of using GIS in the marine realm is that it can be used to organize and store data that describe the spatial arrangement of information. For instance, when mapping the coastal zone, a GIS map

Table 10.9 Public domain coral reef GIS portals.

Organization & guardian	Content	Portal
United States National Oceanic & Atmospheric Administration (NOAA); National Centers for Coastal Ocean Science (NCCOS); Center for Coastal Monitoring and Assessment (CCMA)	Benthic habitat maps of the majority of US territories mapped with a 1 acre minimum mapping unit	http://ccmaserver.nos.noaa.gov/ecosystems/coralreef/welcome.html
Khaled bin Sultan Living Oceans Foundation	Benthic habitat maps of St Thomas, St John, Seychelles and the Southern Saudi Arabian Red Sea	www.livingoceansfoundation.org/
USGS Center for LiDAR Information Coordination and Knowledge (CLICK)	An evolving archive of LiDAR datasets, including data collected over US coral resources of the Florida Reef Tract and Trinidad & Tobago	http://lidar.cr.usgs.gov/
Google Ocean	3-D seabed bathymetry supporting over 20 content layers	http://earth.google.com/ocean/

can be generated to show the structural, zonal and cover characteristics of the seabed, seamlessly integrated with data on the bathymetry, temperature, and hydrodynamic regime. Each dataset occupies a thematic layer in the GIS 'stack' and can be mined for information in the same way that a terrestrial GIS is manipulated. The merging of these layers then results in the three-dimensional characterization of an area that also retains its geospatial qualities. Using the GIS also eliminates the need to define simplistic 'codes' that are shorthand descriptors of map polygon characteristics. Verbose descriptors are easily managed using the GIS.

Although akin to terrestrial examples in many aspects, ocean and coastal GIS is more problematic because the phenomena considered are dynamic and non-discrete. Aquatic features change rapidly with different time constants and are more difficult to define in a meaningful way. For example, coastal plumes and currents which are influenced by tides, river flow, winds, etc. do not lend themselves easily to a map layer. For these reasons, marine GIS typically demands a higher temporal resolution than its land-based counterparts, or is confined to less dynamic features such as seabed character.

Nonetheless, the technology of hosting marine GIS data on the Internet is now advanced, and it offers an efficient means of disseminating spatial information to the public and scientific community. As shown in Table 10.9, there are numerous web-based resources for marine information such as coral reefs. Concerning the Internet, the most significant event at the time of writing is the release of *Google Ocean*, a component of *Google Earth* that includes views of the ocean and portions of the seabed and provides detailed environmental data that will enhance information about the effect of climate change on the world's seas and oceans.

10.9   Key concepts

1 The vastness of Earth's oceans makes them logistically difficult and costly to monitor. Space-based observations are well poised for this task and are used extensively to measure key parameters that influence climate stability.

2 Ocean food webs driven by primary production from phytoplankton lie at the base of the majority of marine ecosystems. As on land, the unique spectral signature of photosynthetic pigments can also be used to monitor the prevalence of chlorophyll in the ocean. This field of remote sensing is termed 'ocean colour' and there exists a long lineage of satellite instruments dedicated to its measurement. This was led by the pioneering CZCS mission, launched in 1978, and followed in subsequent decades by SeaWiFS, MODIS and MERIS. Such missions have utility for measuring oceanic productivity as well as for the detection and tracking of harmful algal blooms in the coastal zone.

3 The temperature of the ocean has a strong bearing on global-scale stability of the atmosphere. Important cyclic phenomena such as El Niño have a pronounced temperature signature. Remote sensing of thermal infrared emissions allows the mapping of patterns in sea surface temperature (SST). Key missions that have delivered SST datasets are the AVHRR, ATSR and MODIS. Measurement of SST can also be achieved using microwave radiometry, a technology employed by spaceborne sensors such as the SMMR, TRMM and AMSR.

4 Oceanic circulation is driven by the spatial distribution of winds, tides, temperature, and salinity. Sea surface salinity can be measured remotely using low-frequency passive microwave radiometers such as the aircraft-mounted SLFMR. The technology relies on the fact that increasing salinity decreases the emissivity of seawater. Recent advances in interferometric technology have allowed salinity to be measured from satellites.

5 Lateral variations in the height of the ocean's surface portray the distribution of geostrophic ocean currents and, under certain circumstances, temperature also. These subtle variations in height are routinely monitored using spaceborne radar altimeters – active remote sensors that utilize the microwave spectrum. Global datasets depicting ocean surface topography are delivered by the TOPEX/Poseidon satellite mission.

6 As wind blows across the ocean surface, it produces short capillary waves. These propagate in the direction of the wind and their height is related to wind velocity. As a radar scatterometer transmits and receives signals of microwave radiation to/from the sea surface, the strength and shape of the return signal can be analyzed to retrieve wind speed and direction by virtue of the impact that these have on short wave geometry. The QuickSCAT satellite employs this technique.

7 Synthetic Aperture Radar (SAR) can provide detailed images of the ocean surface from satellite altitudes. These images contain information on ocean

surface waves, ocean internal waves, oil slicks, ocean fronts and general ocean dynamics. Most of these phenomena modulate the short wavelets on the ocean surface, which interact with the radar pulses.

8 Because ocean observing data are required to support a wide range of critical decisions, the US has prioritized the development of a comprehensive Integrated Ocean Observing System (IOOS). The purpose of the IOOS is to sustain a network of sensors on buoys, ships, satellites, underwater vehicles and other platforms that routinely supply data and information needed to detect and predict changes in the oceans.

11 Monitoring Earth's atmosphere

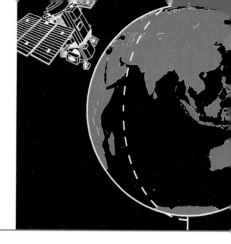

11.1 The status of Earth's atmosphere

The atmosphere protects the biosphere by absorbing ultraviolet solar radiation and reducing temperature extremes between day and night. The organisms that inhabit the Earth employ it as a reservoir for the chemical compounds used in living systems; indeed, the composition of the atmosphere is dictated by biological reactions persisting across geologic history.

At present, the gaseous breakdown of the atmosphere falls to roughly 78 per cent nitrogen, 20.95 per cent oxygen, 0.93 per cent argon, 0.038 per cent carbon dioxide, trace amounts of other gases, and a variable amount (average around 1 per cent) of water vapour. We commonly term this mixture of gases as air. The atmosphere has no outer boundary, just gradually fading into space. Gravity pulls the majority of this air mass earthwards and the densest part (97 per cent by mass) lies within 30 km of the ground. Our atmosphere is thus almost infinitesimally thin – just 1/1,000th of the thickness of the planet – it is but critical for life on Earth.

Earth's climate is largely controlled by a set of fairly simple physical constraints but, as we will see through the course of this chapter, the combination of simple things can have results of almost unpredictable complexity.

Natural Earth processes alter the composition of the atmosphere. Volcanic eruptions inject vast quantities of greenhouse gases and aerosols high into the stratosphere to a point where short-term weather patterns are affected and, in rare cases, long-term climate change is triggered (e.g. Kirchner *et al.*, 1999). These natural effects can be considered polluting. However, the more normal phenomena associated with any mention of atmospheric pollution are the all-too-familiar man-made effects associated with processes such as energy production, waste incineration, transport, deforestation and agriculture.

Remote Sensing and Global Environmental Change, First Edition. Samuel Purkis and Victor Klemas.
© 2011 Samuel Purkis and Victor Klemas. Published 2011 by Blackwell Publishing Ltd.

During the last 200 years, mankind has begun to alter the composition of the atmosphere significantly through pollution. Although air is still made up mostly of oxygen and nitrogen, through this pollution, humankind has increased the levels of many trace gases and, in some cases, released completely new gases to the atmosphere. Many of these trace gases, when present in elevated concentrations, are detrimental to the environment and directly harmful to humans. Conversely, there are some gases, such as stratospheric ozone, which are removed by anthropogenic activity, with equally disastrous consequences.

The effect of aerosols and greenhouse gases on the climate is usually described as 'radiative forcing'. Simply put, this is the perturbation to the net rate of radiative energy flow in the atmosphere caused by atmospheric constituents such as clouds, aerosols particles or gases. Aerosol particles, for example, scatter solar radiation in all directions, so more radiation should be reflected back into space when aerosols are present. Aerosol particles in the 100 to 1,000 nm diameter range are best at disturbing incoming solar radiation due to the similarity of particle size and sunlight wavelength. Such particles have a net cooling effect, and their radiative forcing is negative. Greenhouse gases, by comparison, have a pronounced positive forcing.

Quantification of radiative forcing is relevant, because it is against this factor that trends in atmospheric temperature can be predicted. The atmosphere does not respond to carbon dioxide, for example – it responds to the forcing imparted by this gas. The present-day radiative forcing of the atmosphere is 1.4 to 2.4 Watts per square metre (Wm^{-2}) (Knutti *et al.*, 2002). Climate models predict that the magnitude of this forcing will rise to 5–10 Wm^{-2} by the end of the century, driven by the increased concentrations of atmospheric CO_2 to levels perhaps as high as 1,000 ppm (Hofmann *et al.*, 2009). Because of the long residence time of this gas in the atmosphere, this elevated radiative forcing will persist for millennia, even if the rate at which we consume fossil fuels is rapidly curtailed (Caldeira *et al.*, 2003; Kheshgi, 2005).

The last time the Earth experienced this level of forcing was about 55.8 million years ago, in the Early Paleogene, and also deeper in geologic time such as the Middle Cretaceous (about 100 million years ago) and the Late Permian (251 million years ago) (Knoll *et al.*, 1996; Breecker *et al.*, 2010). As shown in Figure 8.2, several of these time periods correspond to major episodes of extinction for coral reefs.

Of these three episodes, the most recent one where the radiative forcing is presumed to be analogous to that predicted to occur by the end of this century is during an event termed the Paleocene-Eocene Thermal Maximum (PETM). This event occurred about 55.5 million years ago and lasted around 170,000 years. The geological record suggests that at this time there was a large excursion in global temperature, likely related to a massive injection of greenhouse gases into the atmosphere which was possibly released from methane that came out of storage below the seabed (Katz *et al.*, 2001) or massive volcanism (Storey *et al.*, 2007). At that time, global mean temperature is presumed to have been on the order of 35 °C (Weijers *et al.*, 2007), as opposed to around 15 °C today.

In other words, this shows us that a radiative forcing of $5-10\,Wm^{-2}$, which is a realistic value in the near future, is capable of raising global mean temperatures by 20 °C!

Importantly, the PETM was not only a period of higher mean temperatures, but also of particularly strong positive temperature departures at the high latitudes, perhaps as high as 25 °C at 75° North, as opposed to 0 °C at that latitude today (Weijers *et al.*, 2007). The implication of this is that if we take the PETM episode as an analogue to our presently altering climate, polar ice will become particularly vulnerable to slight increases in atmospheric radiative forcing induced by elevated greenhouse gas concentrations over the next century.

Ironically, one of the most powerful agents for negative radiative forcing are sulphates, which are produced from sulphur dioxide released during fossil fuel combustion (Haywood & Ramaswamy, 1998; Haywood & Boucher, 2000; Kravitz *et al.*, 2009). Though considerable effort is made to scrub these emissions from coal-fired power stations, it is very likely that present-day aerosol cooling is suppressing a major portion of current greenhouse warming (Stott *et al.*, 2008). The point is relevant, considering the massive increase in coal use in developing countries such as China. It should also be noted, however, that while cooling our climate, sulphates promote acid rain. Also, during combustion, they are released in association with massive amounts of carbon dioxide – which, as we shall shortly see, has a terribly detrimental effect on the chemistry of our oceans. Predictably, burning coal is not a solution for our global warming woes, although its dampening effect on warming should not be underestimated!

Nobel laureate F. Sherwood Rowland, who, with a colleague discovered that chlorofluorocarbons destroyed ozone molecules in the Earth's atmosphere faster than they could be replenished, eloquently noted: '*Almost all the problems mankind has produced in the atmosphere have a connection with humanity's strong drive for a better standard of living*'. As industrialization continues unabated into the 21st century, we must take stock of the status of our environment and take care of the air that we breathe.

11.2 Atmospheric remote sensing

Atmospheric gases, just like types of land cover, can be differentiated on the basis of their spectral signatures. Gases do not absorb uniformly over the electromagnetic spectrum. Termed absorption bands, there exist some wavelengths where a gas may absorb a great deal of radiation, while the same gas is highly transparent at other wavelengths (e.g. see Figure 2.5 in Chapter 2). This differential pattern of spectral absorption with wavelength is used as a signature to identify both the presence of a gas and its concentration.

Satellite remote sensing proceeds by identifying these narrow absorption bands within light that has twice passed through the atmosphere – once downwards on its path from the Sun to Earth's surface, and again on its outward journey following reflection. This technique forms the basis for satellite monitoring of gases such as ozone, carbon dioxide and water vapour, all critically important for regulating Earth's climate. Even though they may exist in only minute concentrations, even very subtle changes in their prevalence can have huge consequences for the quantity and quality of electromagnetic energy that reaches Earth's surface.

In contrast, atmospheric processes such as rain, snow, turbulence and wind cannot be detected spectrally. Instead, these parameters are derived by observing the Doppler-induced alteration of a radar signal transmitted through the atmosphere and reflected back to an orbiting spacecraft. In the course of this chapter, we will review the suite of space-based missions used to monitor Earth's atmosphere.

11.3   The 'A-Train' satellite constellation

The 'A-Train' is the name given to a constellation of satellite missions which each carry sensors designed to monitor global change. Upon completion, the satellite formation will consist of six low polar orbiting satellites flying in close proximity (Aqua, CloudSat, CALIPSO, PARASOL, Aura and OCO – see Table 11.1). At the time of writing, all but OCO have been successfully lofted and are operational. Unfortunately, the launch of OCO in February 2009 failed to reach orbit and the instrument was destroyed.

Only eight minutes (or a distance of approximately 3,100 km on the Earth) separate the first and last of the five operational instruments. Furthermore, since this constellation is composed of missions with equator crossings in the early afternoon (and also in the middle of the night, at about 1:30 am), it is referred to as the 'Afternoon Constellation' to distinguish it from the 'Morning Constellation' which comprises Terra, Landsat-7, SAC-C and the New Millennium Program's Earth Observing-1 (EO-1), all currently flying.

This fleet of multiple satellites will greatly enhance understanding of the spatial and temporal variation of atmospheric parameters and their ability to enact global change. As things stand at present, the roles of clouds and aerosols in climate change are particularly poorly understood. Taking aerosol sensing as an example, the A-Train features improved versions of instruments with a long heritage (e.g. MODIS, POLDER, TOMS), as well as a LiDAR instrument for vertical aerosol profiling (CALIPSO). The real power of the A-Train mission comes with the synergies among these sensors. By 'synergy' we mean that more information about the condition of the Earth is obtained from the combined observations than would be possible from the sum of the observations taken independently.

The programme involves both overlapping and complementary capabilities in terms of retrieved quantities of data and their sensitivity, resolution, and coverage. The constellation thus offers the first near-simultaneous co-located measurement of atmospheric gases, aerosols, clouds, temperature, relative humidity and radiative

Table 11.1 The NASA A-Train satellite constellation. Modified from NASA-FS-2003-1-053-GSFC.

Spacecraft	Position in A-Train/ formation requirements	Summary of mission	Instruments carried	†Equator crossing time (local) & launch dates
Aqua	Lead spacecraft in the formation until the (re-) launch of OCO	Synergistic instrument package studies global climate with an emphasis on water in the Earth/atmosphere system, including its solid, liquid and gaseous forms	AIRS/ AMSU-A/HSB AMSR-E CERES MODIS	01:38:00 Launched: May 4, 2002
CloudSat	Lags Aqua by between 30 seconds and 2 minutes. Must retain extremely precise positioning relative to both Aqua and CALIPSO to permit synergistic measurements with these partner satellites	Cloud Profiling Radar allows for most detailed study of clouds to date and should better characterize the role clouds could play in regulating Earth's climate	CPR	01:31:00 Launched: April 28, 2006 (lofted with CALIPSO)
CALIPSO	Lags CloudSat by no more than 15 seconds. Must maintain position relative to Aqua to permit synergistic measurements	Observations from spaceborne LiDAR, combined with passive imagery, will lead to improved understanding of the role of aerosols and clouds play in regulating Earth's climate, in particular, how the two interact with one another	CALIOP IIR WFC	01:31:15 Launched: April 28, 2006 (lofted with CloudSat)
PARASOL	Lags CALIPSO by about 1 minute	Polarized light measurements allow better characterization of clouds and aerosols in the Earth's atmosphere, in particular, distinguishing natural and manmade aerosols	POLDER	01:33:00 Launched: December 18, 2004
Aura	Lags Aqua by about 15 minutes but crosses equator 8 minutes behind Aqua due to different orbital track to allow for synergy with Aqua	Synergistic payload studies of atmospheric chemistry, focusing on the horizontal and vertical distribution of key atmospheric pollutants and greenhouse gases and how these distributions evolve and change with time	HIRDLS MLS OMI TES	01:30:00 Launched: July 15, 2004
OCO	Will precede Aqua by 15 minutes when it is launched	Space-based observations of the column integrated concentration of CO_2, a critical greenhouse gas	Three grating spectrometers	01:15:00 Launch failed; OCO-2 in planning

†Note that though Aura crosses the equator 8 minutes behind Aqua, in terms of local time, because it is along a different orbital track, it actually lags Aqua by 15 minutes. Note also that CALIPSO trails CloudSat by only 15 seconds to allow synergy between Aqua, CloudSat and CALIPSO.

fluxes. Utilizing these unique capabilities, the A-Train formation will help to answer five important questions:

- What are the aerosol types and how do observations match global emission and transport models?
- How do aerosols contribute to the Earth radiation budget/climate forcing?

- How does cloud layering affect the Earth's radiation budget?
- What is the vertical distribution of cloud water/ice in cloud systems?
- What is the role of polar stratospheric clouds in ozone loss and de-nitrification of the Arctic vortex?

11.3.1 Dancing on the A-Train

The first problem with integrating synergistic measurements from multiple platforms is devising a data management system that can simultaneously cope with the different spatial resolutions that each instrument delivers. Different instruments in the A-Train have very different resolutions, both vertically and horizontally. Some satellites scan a very large area or swath, while for other instruments the swath is much smaller and only a very small area is viewed. Some instruments have a very large footprint (e.g. they have low spatial resolution), while others have smaller footprints and higher spatial resolution. Also, in terms of vertical information, some instruments (or instrument channels) are designed to penetrate only the top portion of the atmosphere while others are designed to focus on the lower portion of the atmosphere. These differences mean that it is not a trivial matter to 'match' measurements from different satellites, or even from two different instruments on the same satellite.

If the synergy of instruments is to be viable, the orbital characteristics of each satellite must be in tune with the rest of the constellation. This goes beyond placing the payloads in an appropriate orbit and forgetting about them. Orbital paths must be adjusted and tweaked to maintain appropriate satellite spacing. Polar-orbiting satellites, in particular, are influenced by the gravity of both the Sun and the Moon, which can pull them off their intended course into an east-west-leaning orbit. In April 2003, for example, Aqua's orbit underwent a controlled readjustment using the satellite's onboard thrusters to keep its orbit exactly sun-synchronous and in tune with its partner platforms on the A-Train. It is anticipated that the adjustment will have to be repeated at two-year intervals.

Another reason for tweaking the orbit of a satellite is when there exists the possibility of a collision with another instrument. CloudSat, for example, executed a manoeuvre on 4 July, 2007 to avoid a close approach with an Iranian satellite, SINAH-1. Post-manoeuvre, it was confirmed that the miss distance was over 4 kilometres. CloudSat had to conduct another manoeuvre on 7 July to reverse the drift caused by the avoidance manoeuvre and maintain formation with CALIPSO. A more routine manoeuvre of CloudSat is required on a weekly timescale to preserve its position no more than 15 seconds ahead of CALIPSO.

11.4 Remote sensing atmospheric temperature

The first atmospheric temperatures profiles made from space were obtained form observations in the microwave spectrum using the Tiros Operational Vertical Sounder (TOVS). This instrument, which has been flown aboard NOAA satellites since 1978, monitors wavelengths where the atmosphere is (almost) opaque, such that the upwelling radiation measured by the microwave sounder is representative of atmospheric temperature. The AVHRR sensor, an instrument considered several times in this book, is also carried on these satellites. The NOAA mission is termed POES (Polar-orbiting Operational Environmental Satellite) and offers the advantage of daily global coverage. Currently in orbit are NOAA-12 and NOAA-14, which provide global coverage four times daily and carry both AVHRR and TOVS.

Part of the POES mission, the Advanced Microwave Temperature Sounder (AMSU) aboard the NASA Aqua spacecraft currently provides our best knowledge of atmospheric temperature and its global trends. AMSU is a 15-channel microwave sounder designed primarily to obtain temperature profiles in the upper atmosphere (especially the stratosphere) and to provide a cloud-filtering capability for tropospheric temperature observations. Aqua also carries the Atmospheric InfraRed Sounder (AIRS) instrument, which, as the name suggests, monitors infrared wavelengths to deliver atmospheric temperature profiles (Figure 11.1). This technology is fairly new, and AIRS is the first hyperspectral infrared sounder to be provided to users for operational applications.

In the next decade, budget allowing, NOAA plan to have a hyperspectral infrared sounder (GOES-R) in geostationary orbit to provide additional capability. In addition to temperature and moisture profiles, hyperspectral infrared

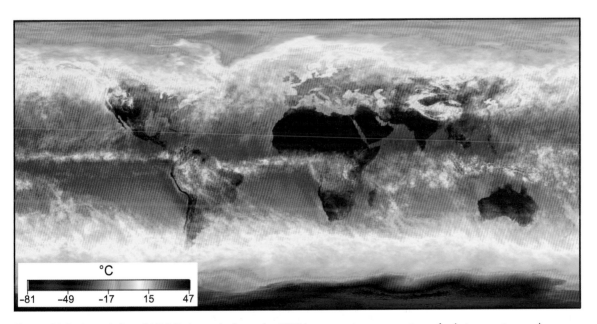

Figure 11.1 Housed aboard NASA's Aqua platform, the AIRS instrument measures atmospheric temperatures using infrared wavelengths. This image shows the temperatures of the Earth's surface or clouds covering it for the month of April 2003. The atmosphere over Antarctica is clearly the coldest on Earth. Since the tropics tend to be freer of clouds, the shapes of the continents can be discerned; this is not the case at higher latitudes. The Intertropical Convergence Zone, an equatorial region of persistent thunderstorms and high, cold clouds, is depicted in yellow. Credit: NASA's Earth Observatory.

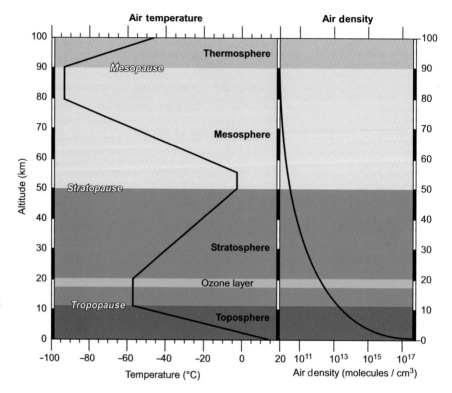

Figure 11.2 The left pane shows how a vertical temperature profile through the atmosphere provides a convenient means of defining its layers by altitude. The right pane shows that, due to gravity, the density of the atmosphere decreases with increasing altitude.

measurements provide information on ozone and other greenhouse gases such as carbon dioxide, carbon monoxide and methane, clouds, aerosols, and surface characteristics such as temperature and emissivity. Unlike microwave instruments, however, infrared measurements are confounded by cloud cover, and their lack of all-weather monitoring capability is a hindrance.

The atmosphere exhibits significant variations in temperature and pressure with altitude (Figure 11.2):

- The lowest 10–15 km of the atmosphere, where temperatures predominantly decrease with height, is termed the **troposphere**. This is where more than 80 per cent of the atmosphere's mass resides and, from a remote sensing perspective, it is the layer with the strongest spectral response. The troposphere is also where the majority of weather occurs.
- Moving upwards, the 'tropopause' demarks the lower boundary of the **stratosphere**. This layer extends from ≈11–50 km altitude and contains 20 per cent of the atmosphere by mass. Little weather occurs at this altitude, although, as depicted in Figure 11.1, the top portions of thunderstorms, such as occur at the Intertropical Convergence Zone, may extend into the lower stratosphere.
- The **mesosphere** lies next in altitude and extends to a height of ≈90 km, at which the atmosphere is at its coldest temperature (−90°C).

- Higher still, the 'mesopause' marks the transition to the **thermosphere**, an atmospheric region that is of high temperature because of the absorption of high-energy solar radiation by oxygen molecules in this region.

Just as for the distribution of heat in the oceans discussed in Chapter 10, the distribution of temperature is a key factor forcing the global circulation of the atmosphere. Furthermore, since temperature dictates the point at which water vapour in the air begins to condense, it is a primary factor controlling the formation of clouds. As we will see in Section 11.10, when we consider ways to forecast the climate using Atmosphere General Circulation Models (AGCMs), clouds are a troublesome factor and limit the accuracy of long-term forecasts. Remote sensing observations of atmospheric temperature are therefore a critical input to AGCMs.

Both tropospheric and stratospheric temperatures are central to the problem of climate warming. The link to the troposphere is obvious, with this being the region that controls the temperature down at the Earth's surface where we live. Predictions made using long-range models are clear that we should expect a significant warming of the troposphere with a rise in greenhouse gas concentrations (Rahmstorf & Ganopolski, 1999). By contrast, the stratosphere is likely to cool as a consequence of the increased trapping of terrestrial radiation down below in the troposphere (Ramaswamy *et al.*, 2006). Accurate knowledge of global-scale fluctuations in stratospheric temperature are necessary to understand the recovery of the ozone layer – a subject we will tackle next.

11.5 Atmospheric remote sensing of ozone

Ozone is an allotrope of oxygen whose molecules contain three oxygen atoms (O_3) rather than the more common two (O_2). Comparatively unstable in comparison to the more 'normal' form of oxygen, ozone is found in much lower concentrations in our atmosphere. Ozone represents a dichotomy in that its presence in the upper atmosphere is hugely beneficial, while in the lower atmosphere it is a dangerous pollutant. Industrialization has served to decrease the prevalence of O_3 in the upper atmosphere, where we need it most, while boosting its concentration closer to the Earth's surface.

Stratospheric (high-altitude) ozone protects the biosphere by attenuating biologically harmful wavelengths of ultraviolet (UV) radiation emanating from the Sun. Here, at an altitude of 15–35 km, relatively high concentrations of O_3 exist in a region commonly termed the 'ozone layer'. Such is the effectiveness of UV absorption by O_3 for radiation with a wavelength of 290 nm that the intensity at Earth's surface is hundreds of orders of magnitude weaker than at the top of the atmosphere. Ozone conducts this service while existing at a concentration of only a few parts per million. This is a tiny fraction compared to the main components of the atmosphere.

Prior to 1979, measurement of atmospheric ozone was conducted from the ground. Indeed, a viable archive of such data exists back to the mid-1950s.

This lengthy time-series is a particularly valuable asset against which to assess modern changes in atmospheric composition. From 1979 onwards, American, Russian and Japanese satellites have been equipped with sensors that collect data on ozone, as well as other atmospheric constituents that affect the amounts of ozone present. The satellites that have flown the Total Ozone Mapping Spectrometer (TOMS) have included Meteor-3 (1991–1994), Nimbus-7 (1978–1993), ADEOS (1996–1997) and Earth Probe (1996–2005). Meteor-3 is worthy of note as being the first (and the last) American-built instrument to fly on a Soviet spacecraft.

TOMS measures the total solar radiance incident on the satellite. By comparing this to the UV radiation scattered back from the atmosphere, it is able to compute total ozone amounts. Ozone measurements given by TOMS are in Dobson units and give the total ozone in a column extending from ground level to the top of the atmosphere.

Because it depends upon scattered solar radiation, TOMS does not work at night. As a result, between the first day of autumn and the first day of spring, there will be parts of the high latitudes that will have no data. This area begins at the pole on the first day of autumn, reaches its maximum on the first day of winter, then shrinks back to zero on the first day of spring. This problem is particularly prevalent as these satellites are polar orbiters and the most drastic declines in ozone have historically occurred above the poles (Randel & Wu 1999).

In 2001, NASA unsuccessfully attempted to launch QuikTOMS (the Quick Total Ozone Mapping Spectrometer). Aboard the same doomed rocket was ORBIMAGE'S OrbView-4 high-resolution imaging satellite.

Beyond TOMS, Europe's polar orbiting environmental satellite, ESA, has been returning ozone data since its launch in 2002 via its GOMOS instrument. Also taking ozone measurement into the 21st century is the Ozone Monitoring Instrument (OMI) aboard the Aura platform (Figure 11.3). OMI will continue the TOMS record for total ozone and other atmospheric parameters related to ozone chemistry and climate. OMI measurements will be highly synergistic with the other instruments carried on the Aura spacecraft.

The MODIS instrument aboard the Aqua satellite delivers a thrice-daily estimate of the quantity of ozone through the troposphere and stratosphere combined. This measurement is critical for deriving a viable atmospheric correction for the other parameters measured by the spacecraft.

It has been well publicized over the last three decades that the concentration of stratospheric ozone has been dropping (Solomon, 1999; Drew *et al.*, 1998). The reported main cause of ozone depletion is the presence of chlorofluorocarbons, or CFCs, in the stratosphere (Andrews *et al.*, 1987; Holton *et al.*, 1995). CFCs are compounds of chlorine, fluorine and carbon and, because these gases are stable, inexpensive, non-toxic, non-flammable and non-corrosive, they have been historically used as propellants, refrigerants, solvents, etc. However, it is this stability that causes CFCs to persist within the environment. These molecules eventually find their way to the stratosphere, where they undergo a series of chain reactions which ultimately damage the ozone layer.

Figure 11.3 The concentration of ozone over Antarctica in 2006, 2007 and 2008, as measured by the Ozone Monitoring Instrument (OMI) aboard NASA's Aura satellite. The three panes were acquired in September, the time of year when ozone levels are typically at their lowest. The blues and purples that cover most of Antarctica illustrate where ozone levels were low (the 'ozone hole'), while greens, yellows and red point to higher concentrations. In 2006, the ozone hole covered an area of 20 million square kilometres, more severe than any year since 2000. Credit: NASA's Earth Observatory.

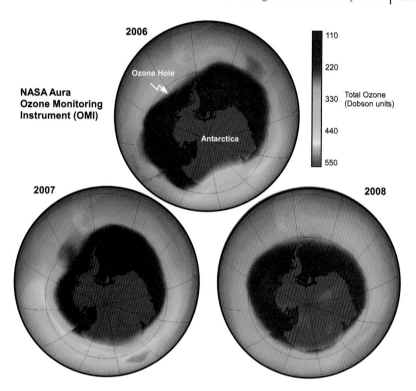

This mechanism of destruction was identified in several seminal papers that appeared throughout the mid-1970s (e.g. Molina & Rowland, 1974). Following the discovery of a giant 'ozone hole' over the Antarctic in 1985 (Farman *et al.*, 1985), this theoretical concern was pushed into the international limelight. Worldwide efforts to reduce the use of CFCs were focused by the 1989 Montreal Protocol (http://ozone.unep.org/) which, during its lifetime, has been amended several times to impose stricter regulation in an effort to suppress the runaway loss of stratospheric ozone over the poles. It is acknowledged that these efforts were largely thwarted by the contraband manufacture of CFCs by developing states that caused a steady rise in stratospheric CFC concentration up until at least 2000 (Montzka *et al.*, 1999).

At the time of writing, the Montreal Protocol is celebrating its 20th birthday. Public opinion holds that the CFC problem has been largely dealt with but, in reality, this is far from true and the ozone hole that sparked the rush to find alternatives to CFCs may not heal for another 50 years, if at all (Weatherhead & Andersen, 2006). The ozone-friendly chemicals that replaced CFCs came with their own suite of drawbacks – the manufacture of HCFC-22 (chlorodifluoromethane, a refrigerant used in air-conditioning units) has inadvertently caused the production and release of large quantities of HFC-23 (fluoroform, CHF_3) into the atmosphere. This gas does not deplete ozone, but has a greenhouse potential a staggering ten thousand times more powerful than CO_2 (Velders *et al.*, 2009).

Scientific opinion strongly points to the fact that the Montreal Protocol is in desperate need of yet another re-write. In September 2007, progress was made in this regard when the Protocol was further amended to influence developing countries to phase out HCFC production and use by 2030, 10 years ahead of the previous target date. The United Nations Environment Programme (UNEP) further designated 2007 the International Year of the Ozone Layer.

Down below in the troposphere, ozone is a naturally occurring greenhouse gas formed as a product of photochemical reactions with such gases as NO_x, CH_4, CO and volatile organic carbons. However, the past decades have seen a rapid rise in tropospheric ozone concentration through anthropogenic activity. Indeed, Sitch *et al.* (2007) note that emissions associated with fossil fuel and biomass burning have acted to approximately double tropospheric ozone concentration, and further increases are to be expected through the remainder of the century. This trend is worrying, because of the damaging effect that ozone has on the living tissues of plants and animals (Wang & Mauzerall, 2004; Ashmore, 2005) and the strong negative effects it imparts on human health (World Health Organization (WHO), 2003).

Tropospheric ozone is mainly produced during the daytime in polluted regions such as urban areas. Significant government efforts are under way to regulate the gases and emissions that lead to this harmful pollution, and smog alerts are regular occurrences in polluted urban areas. Recent work by Hocking *et al.* (2007) using radar has shown that, under certain conditions, stratospheric ozone may intrude down into the troposphere. Thus, the concentration of low-level ozone can no longer be considered as decoupled from the concentration of the gas in the upper atmosphere.

Although *in situ* tropospheric ozone measurements are rather fragmentary, as are observations of tropospheric ozone precursors, there has been substantial progress in obtaining relevant information due to the development of satellite remote sensing (Kondratyev & Varotsos, 2002). Most commonly used in this regard is the aforementioned TOMS instrument (Fishman *et al.*, 1996; Fishman & Brackett, 1997; Hudson & Thomson, 1998), as well as the application of airborne differential absorption LiDAR (Browell *et al.*, 1988). The previously discussed Aura-OMI sensor is capable of returning relative tropospheric concentrations of compounds such as oxides of nitrogen or volatile organic compounds such as formaldehyde that, through photochemical reactions, lead to ozone production. As such, the OMI can be used to indirectly assess the potential occurrence of ozone in the troposphere (Ziemke *et al.*, 2006), which is of great value when assessing the atmospheric impact of phenomena such as regional-scale biomass burning and low-altitude ozone emanating from urban centres.

11.6 Atmospheric remote sensing of carbon dioxide

There is currently no dedicated Earth Observation mission to map Earth's atmospheric CO_2 concentrations. There were plans to close this gap with the scheduled 2009 lofting of the US$ 300 million Orbiting Carbon Observatory (OCO); however, this failed during launch when the Taurus rocket's clamshell fairing protecting the satellite in flight failed to separate at the designated time. The vehicle and OCO were therefore too heavy and travelling too slowly to make orbit, and minutes later they fell into the ocean near Antarctica.

This was to be a new satellite mission and one of several sponsored by NASA's Earth System Science Pathfinder (ESSP) programme (covered in Chapter 5, Section 5.5.1). The spacecraft was designed to collect precise global atmospheric measurements of CO_2 in partnership with Japan's Aeroscience Exploration Agency's GOSAT (Greenhouse gases Observation Satellite), which was successfully lofted in early 2009 and is now starting to deliver data to Earth. This satellite measures methane, water and ozone as well as CO_2. Unlike OCO, GOSAT utilizes energy further into the infrared and is capable of operating in the absence of sunlight.

The power of the synergy between OCO and GOSAT was to arise from the fact that they were to have complementary sampling strategies. OCO was to measure atmospheric CO_2 over a small area of approximately $10\,km^2$ along-track with a repeat cycle of 16 days. The instrument, which was predicted to provide precise data about the origin and fate of anthropogenic carbon emissions, would have been capable of gathering as many as 74,000 soundings on the sunlit side of any orbit. This is sufficiently dense to obtain high-quality measurements even in those regions where clouds, aerosols and topographical variations are present. By contrast, GOSAT now monitors a wider area of atmosphere (about $50\,km^2$) and provides a series of snapshots, spaced by 100 km and repeated every 3 days (Chevallier et al., 2009).

The mode of operation of these instruments relies on rays of sunlight entering a spectrometer. This light will have passed through the atmosphere twice – once as it travels from the Sun to the Earth, then again as it travels from the Earth's surface back to the orbiting instrument. As sunlight shines down and is reflected back, various molecules absorb some of it at distinctive wavelengths. By comparing the different bands associated with CO_2 and with other gases (which serve as calibrations), the instruments come up with an estimate of the number of CO_2 molecules in a column of air.

The OCO spectrometer was designed to deliver retrievals of the column-abundance (Figure 11.4) of atmospheric CO_2 with a precision as high as 1–2 ppm against a background of 380 ppm. To achieve this accuracy, the instrument's detectors were to be shielded from energy sources that would generate measurement error. This was to be achieved by chilling the spectrometers to a frigid $-150\,°C$ using an onboard cryo-cooler. To maximize the amount of sunlight entering the spectrometers (and thus the signal to noise ratio), OCO was set to cross the equator at approximately 1:18 pm local time. Since column-averaged CO_2 measures tend to be near their daily average value at this time of day, the collected data were to be highly representative of the region where they were acquired.

Even if the launch had been successful, OCO would have faced two major problems. First, its spectrometers rely on sunlight and thus can not operate at night or during polar winters. Second, since OCO senses absorption features in the infrared (4,260 nm), measurements are confounded by cloud cover. The instrument

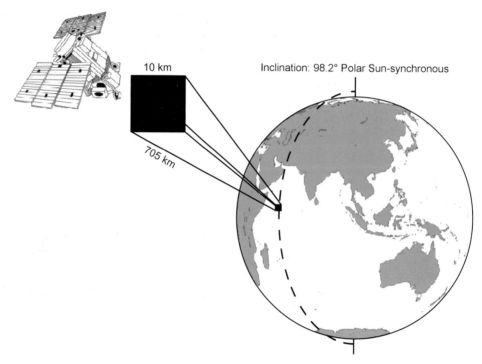

Figure 11.4 Had the mission reached orbit, the Orbiting Carbon Observatory (OCO) spacecraft was to measure the 'column abundance' of atmospheric CO_2, i.e. the fraction of CO_2 in the atmosphere in a column projected over a fixed area on the Earth. The column extends from ground level to the top of the atmosphere with a cross-section of 10 km.

was to have a small viewing window to counter this problem. A thin column of air is less likely to be beset by clouds than a broader swath (such as adopted by GOSAT). However, cloudy areas such as the Amazon basin would still have proven troublesome to monitor. Once launched, OCO was to orbit in the previously discussed A-Train constellation.

1957 marked the year that Sputnik-1 instigated the legacy of artificial satellites. Though not part of the national consciousness, in the same year, an equally historic scientific experiment was embarked upon. Far removed from the celebrations on the streets of Moscow, Charles David Keeling collected his first flask from the frigid air of the South Pole. Fifty years later, Keeling's painstaking measurement programme has yielded the longest continuous recording of atmospheric carbon dioxide (Figure 11.5). Long-term and precise data collection is a precarious activity. Due to funding cuts in the mid-1960s, Keeling was forced to abandon continuous monitoring efforts at the South Pole, but he scraped together enough money to maintain operations at Mauna Loa, Hawaii, which have continued to the present day. With the benefit of hindsight, the resulting 'Keeling curve' of CO_2 concentrations ranks very high indeed among the achievements of

Figure 11.5 The Keeling curve depicting the historical concentration of atmospheric carbon dioxide as directly measured at Mauna Loa, Hawaii. Data are reported as a dry mole fraction, defined as the number of molecules of carbon dioxide divided by the number of molecules of dry air, multiplied by one million (ppm). Inset shows the seasonal fluctuation of CO_2 as caused by the breathing of the biosphere, dominated by springtime CO_2 uptake and autumn release in the northern hemisphere. Keeling measured the isotopic abundance of carbon-13 to show that seasonal changes were caused by land plants.

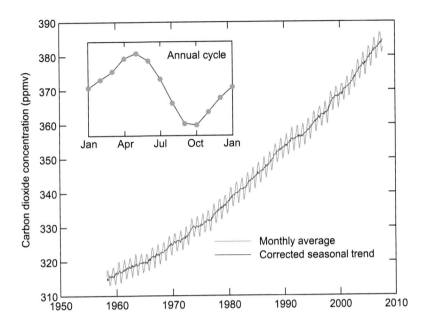

twentieth-century science. The failed OCO mission was to have been the venue that would have reunited the descendants of Sputnik and Keeling's flask on the scientific podium.

As early as 1970, Keeling's record was sufficiently long to reveal two key facts. Firstly, the atmospheric concentration of CO_2 fluctuates seasonally. Further work on isotopes revealed this to arise through seasonal variations in carbon dioxide uptake by land plants. Since many more forests are concentrated in the northern hemisphere, more CO_2 is removed from the atmosphere during northern hemisphere summer than southern hemisphere summer.

The second pattern revealed in the curve had lasting and dire ramifications. Atmospheric carbon dioxide concentrations were rising. The connection between rising CO_2 and fossil-fuel burning had been established. It was clear that a substantial fraction of CO_2 added by humans remains in the atmosphere and is not removed by the biosphere. The gravity of this finding lies in the fact that CO_2 has an influence on our climate far out of proportion with its atmospheric concentration.

Carbon dioxide (CO_2) is considered a trace gas. Today its globally averaged concentration of is only ≈ 385 ppm by volume in the Earth's atmosphere. However, due to its ability to absorb many infrared wavelengths of the Sun's light, it is an important greenhouse gas. Subtle changes to the amount of CO_2 in the atmosphere have major repercussions on Earth's temperature. Indeed, high atmospheric concentrations of CO_2 caused by emissions from fossil fuel burning are now recognized to be the major cause of global warming (Crowley & Berner, 2001; Kump, 2002; Hofmann *et al.*, 2009). In recent decades, this has become a politically charged subject. However, as scientists we must understand that just because a subject can be used politically, it does not prevent us from objectively considering the scientific evidence in order, through application of the scientific method, to arrive at sensible, robust and defensible conclusions.

In addition to having detrimental effects in the atmosphere, emissions from fossil fuel burning also acidify our oceans (IPCC, 2007; Brierley & Kingsford, 2009). The oceans are a massive reservoir of CO_2, and there is a constant flux of the gas across the interface between the ocean surface and atmosphere (Goyet *et al.*, 2009). The oceans are also a sink for anthropogenic CO_2. They have already taken up nearly half of the atmospheric CO_2 that humans have produced over the last 200 years and will continue to do so as emissions increase (Caldeira & Wickett, 2003; Sabine *et al.*, 2004). However, while ocean CO_2 uptake is essentially buffering climate change, ocean pH is falling as dissolving CO_2 produces carbonic acid (H_2CO_3). If carbon dioxide emissions continue to rise at current rates, upper-ocean pH will decrease to levels lower than have existed for tens of millions of years and, critically, at a rate of change 100 times greater than at any time over this period (Caldeira & Wickett, 2003).

Recent evidence suggests that this acidification is, in fact, exacerbated by depletion of stratospheric ozone (previous section) through changing weather patterns that serves to accelerate the drawdown of CO_2 into the oceans (Lenton *et al.*, 2009). Along with the drop in surface ocean pH, a substantial change to carbon chemistry will occur and, most importantly, a decline in carbonate ion accretion (Kleypas *et al.*, 1999, 2001; Feely *et al.*, 2004; Royal Society, 2005; Turley *et al.*, 2007). This change has a profound affect on the ability of calcifying marine organisms to manufacture their shells, liths and skeletons from calcium carbonate crystals (Cohen *et al.*, 2009; Ellis *et al.*, 2009).

As such, rising atmospheric CO_2 may have a detrimental influence on marine food webs (Hare *et al.*, 2007; Tortell *et al.*, 2008) and, in particular, the status of the world's coral reefs (Hoegh-Guldberg *et al.*, 2007; Manzello *et al.*, 2008). This subject is contentious, however, as recent research shows that some calcareous marine organisms in fact display increased rates of growth under elevated CO_2 (Gooding *et al.*, 2009; Ries *et al.*, 2009). Compelling arguments do, nevertheless, exist to suggest that changes in ocean chemistry related to CO_2 have been a factor in most, if not all, major extinctions through Earth's history (Veron, 2008 presents a convincing argument to this effect).

Work to model the rate at which the ocean absorbs carbon dioxide requires knowledge of the total ocean surface, the saturation state of the waters beneath the surface and the concentration of atmospheric CO_2. Although these parameters are relatively easy to quantify, based on our CO_2 emissions, the atmospheric concentration is rising at a rate slower than climate models predict – the oceans appear to be absorbing more gas than they should (Orr, 1993; Orr *et al.*, 2001; Steinacher *et al.*, 2009).

This disjoint between our predictions and reality is relevant to the all-important evolution of Earth's climate over the 21st century. This depends on the rate at which anthropogenic carbon dioxide emissions are removed from the atmosphere by the ocean and land carbon cycles. Coupled climate-carbon cycle models suggest that global warming will act to limit the land-carbon sink (Friedlingstein *et al.*, 2006),

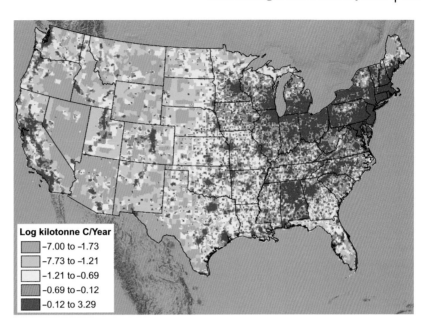

Figure 11.6 The map shows annual CO_2 emissions in 2002 (units: $\log_{10} GgC\ 100km^{-2}\ year^{-1}$) on a 10 km × 10 km grid for North America. The data were assembled by the Vulcan Project. Modified with permission from Gurney *et al.* (2009). *High resolution fossil fuel combustion CO_2 emission fluxes for the United States: Environmental Science and Technology* [**43**], 5535–5541. Copyright 2009. American Chemical Society.

Log kilotonne C/Year

	−7.00 to −1.73
	−7.73 to −1.21
	−1.21 to −0.69
	−0.69 to −0.12
	−0.12 to 3.29

pushing emphasis towards absorption of CO_2 by the oceans (Sitch *et al.*, 2007). However, these 'first generation' models neglect the impacts of changing atmospheric chemistry and, in particular, the rising concentrations of tropospheric ozone associated with biomass burning and fossil fuel emissions, which may serve to alter the balance of marine versus terrestrial sequestration.

Bearing in mind the profound implication of rising atmospheric CO_2, it is paramount that we should be able to make Earth-scale measurements of the gas with a spatial and temporal resolution fine enough to characterize sources and sinks on regional scales. Today, a network of ground-based stations strung across the globe measure CO_2 and other greenhouse gases at Earth's surface with high precision but patchy coverage. Typically, the weekly air samples collected have to be shipped to central laboratories for analysis. These stations are costly to maintain, require skilled staff for their operation and, since they only exist in adequate density in Europe and North America, the seasonal, interannual, and longer-term variability of carbon fluxes have not been quantified satisfactorily in several key locations (Hirofumi *et al.*, 2009). Large expanses of the Earth, including Africa, India, Siberia and much of South America, have very few, if any, monitoring stations.

Satellites, by comparison, can provide high spatial coverage and a global overview of atmospheric CO_2 concentration. The state of the art for real-time regional-scale carbon dioxide monitoring is the NASA/DOE-funded Vulcan Project for North America. The project measures fossil fuel CO_2 emissions on a 10 km × 10 km grid for the country and relies on a medley of remote sensing and *in situ* data. Vulcan has achieved the quantification of the 2002 US fossil fuel CO_2 emissions at the scale of individual factories, power stations, roadways and neighbourhoods on an hourly basis (Figure 11.6) (Gurney *et al.*, 2009). This spatial and temporal resolution is sufficient to pinpoint sources and sinks of CO_2 within the bounds of the monitored area. The Vulcan data are publically available through *Google Earth* from the project's website at Purdue University (www.purdue.edu/eas/carbon/vulcan/index.php).

11.7   Remote sensing atmospheric dust

Besides gases (air), the atmosphere also includes small amounts of material in a solid state. Unlike the majority of gases, which are distributed relatively evenly around the Earth, these solid particles are relatively sparse over the oceans, dense over land, and found in very high concentrations over urban centres. Termed airborne 'dust' or 'particulates', these constituents arise from wind-lofted soils, sand, spores and pollen, ash from fires and volcanoes, salt from sea spray and soot from combustion. By virtue of their mass, these particles are typically confined to the lower few kilometres of the atmosphere, though explosive volcanic eruptions may loft large quantities of solids as high as the stratosphere, which may serve to induce rapid and prolonged cooling of the global climate (Lamb, 1969; Bradley & England, 1978; Briffa *et al.*, 1998). With the terms 'dust' and 'particulates' used synonymously, this section will focus on wind-blown 'dust' in the lower atmosphere (the troposphere – Figure 11.2), with combustive particulates being covered in Chapter 6 and volcanic ejecta discussed in a previous section.

Though it is a minor atmospheric constituent, dust is an important modulator of global climate. Termed a 'dust aerosol', plumes of dust are a radiative force in the atmosphere (Yoshioka *et al.*, 2007). Intuitively, one would think that the forcing would be negative because of the scattering of solar radiation. However, the situation is complex, first because the mineral particles both scatter and absorb in the visible and infrared spectrum, and second because their scattering properties are tied to the distribution of particle sizes (Dentener *et al.*, 1996; Sokolik *et al.*, 1998).

Dust aerosols have a limited lifetime in the atmosphere, which is typically on the order of a week or so (Gobbi *et al.*, 2000), with the heaviest particles settling out first. This ensures that the mineral and size distribution of particles in the plume is in constant flux. Coupled with the fact that the dust sources and sinks are unevenly distributed, the dust burden is spatially and temporally complex and may be a force for both radiative warming and cooling during its residence in the troposphere. In addition, without dust particles, water vapour cannot condense or freeze to form fogs, clouds, and precipitation from clouds (Went, 1964; DeMott *et al.*, 2003). Dust, in this respect, is an important source of 'condensation' or 'Aitken' nuclei.

Cohesive forces in damp soils prevent their blowing into the atmosphere but, once desiccated, fine particles are winnowed by the wind, made airborne and transported large distances before eventually settling back to Earth through turbulent deposition or precipitation processes. Drought-affected and poorly vegetated lands are hence a point source of airborne dust. It was prolonged drought that in the 1930s turned the once-productive fields of the American mid-west into the infamous 'dust bowls'.

CASE STUDY: Spaceborne monitoring of African dust events

The African Sahel region, which has suffered varying degrees of drought since 1970 and is covered in Chapter 5, is presently a major dust source. In periods of drought, many millions of tonnes of dust blow out of the Sahel, across the Atlantic and into the Caribbean (Swap *et al.*, 1996; Prospero *et al.*, 2002). A feedback loop has been proposed, whereby the airborne dust actually suppresses rainfall, promoting drought and accelerated desertification (Rosenfeld *et al.*, 2001).

The magnitude of dust transported out of North Africa has varied by a factor of 4 between the wet 1960s and the drought-ridden 1980s (Prospero & Nees, 1986; Prospero & Lamb, 2003). This increase in dust export from the Sahel has been attributed to widespread land degradation and desertification, driven by a combination of climatic desiccation and human impacts: overgrazing, deforestation, population pressure and inappropriate land use practices (Tegen & Fung, 1995; Tegen *et al.*, 1996). It has even been suggested that the Sahel has even become a more significant source of dust than the Sahara (N'Tchayi *et al.*, 1997).

Just as essential nutrients in Hawaiian rainforests owe their origin to Asian dust, the iron- and clay-rich soils found on many Caribbean islands originate from Africa. This flux has a pronounced effect on the Atlantic's biogeochemistry, which, being iron-limited, witnesses a boost in primary production from the iron-rich African dust that falls by gravity into the ocean (Fung *et al.*, 2000; Sarthou *et al.*, 2003; Jickells *et al.*, 2005; Mahowald *et al.*, 2005). By virtue of the solar photo-reduction of iron, the longer the dust plume spends airborne, the greater the iron's solubility when it eventually reaches water, and the greater the effect on ocean biogeochemistry (Zhu *et al.*, 1997; Hand *et al.*, 2004).

Every dust particle also has a certain resident microbial community on it that is imparted to the ocean as the particle falls in. This has been implicated in the increased incidence of disease in marine organisms, including the unprecedented degradation of coral reefs in the Caribbean (Harvell *et al.*, 1999; Shinn *et al.*, 2000).

By virtue of their vast scale, African dust events are not easy to monitor from the ground but are well resolved by satellites. Early remote sensing efforts concentrated on tracking the airborne dust plume. The AVHRR sensor proved useful, though not particularly accurate, providing the basis for the first estimates of the mass of dust that was emanating from Africa (Prospero & Carlson, 1972;

Prospero & Nees, 1977; Carlson, 1979). It was later discovered that these values were likely biased by calibration problems with AVHRR (Holben *et al.*, 1990), which, in the instrument's defence, was due to the fact that it was not originally designed for aerosol measurements.

Since the 1999 launch of Terra and 2002 of Aqua, MODIS has come to the forefront of this research (Gao *et al.*, 2001; Kaufman *et al.*, 2005) (Figure 11.7). The Terra spacecraft also carries the Multi-angle Imaging SpectroRadiometer (MISR) instrument that returns global data products on atmospheric aerosol loadings, including airborne particulates (Diner *et al.*, 2001) (Figure 11.8). The satellite observes Earth at nine widely spaced viewing angles through three visible bands and one infrared band. Spatial resolution varies between 275 m and 1,100 m, and each measurement takes 7 minutes for MISR's nine cameras to image one location.

There are two challenges that face the remote sensing of atmospheric dust. First, a strategy must be found to distinguish accurately dust from other aerosol types (Jones & Christopher, 2007). Second, any quantitative retrieval of aerosols – dust or otherwise – is difficult over land because of its non-Lambertian surface (Kaufman *et al.*, 1997; Diner *et al.*, 2001; Ramon & Santer, 2001). For the case of African dust, the second challenge is most easily avoided by monitoring the plume while it is above the Atlantic (Tanré *et al.*, 1997).

While a dust plume is well-defined from scattering in the visible spectrum (Figure 11.7), six MODIS bands positioned between 550 nm and 2,100 nm are

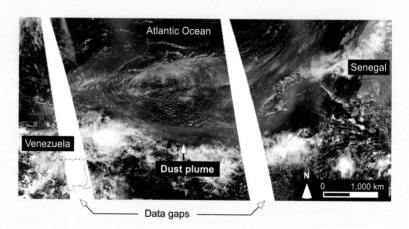

Figure 11.7 An Aqua-MODIS image acquired June 24, 2009 shows a vast plume of dust blowing out of the African Sahel and departing across the Atlantic from the border region of Senegal and Mauritania. The dust cloud is near-continuous all the way from the west coast of Africa to the Caribbean coast of Venezuela and the Lesser Antilles. Credit: NASA.

required to retrieve quantitative aerosol information. Products of 1×1 km are delivered to Earth daily for the oceans globally and a portion of the land. From these, the normalized difference dust index (NDDI) (Qu *et al.*, 2006) can be derived from MODIS reflectance measurements.

As well as atmospheric ozone, as covered in a previous section of this chapter, the TOMS sensor which has flown on a variety of satellites from 1978 to 2005 has also been used to monitor dust (Herman *et al.*, 1997; Prospero *et al.*, 2002). This instrument capitalizes on the absorption of ultraviolet wavelengths by dust. This property is particularly useful for measurements over the ocean, where airborne sea salt contributes to the total aerosol column but does not absorb in the UV and hence can be differentiated from dust absorption. Smoke, however, does absorb in this region and can confound the retrieval of dust abundance (Hsu *et al.*, 1996).

Despite the application of multiple satellite sensors, it remains difficult to constrain the relative magnitude of dust in a plume even over the ocean, and ground calibration is still important. The cutting edge of this technology involves ground-based upward-looking LiDAR (Gobbi *et al.*, 2002).

More recent work has utilized satellite data to pinpoint the sources of African dust, and the Bodélé Depression, once the lakebed of a much larger Lake Chad, has been identified as one of the most active. Close examination of the dust events in the Bodele show that the dust cloud is comprised of a multitude of discrete plumes that blow off of dry lake beds (playas). MODIS and Landsat show that the plumes are long and narrow, with very little dispersion over a distance of 100 km or more, and that they tend to originate from the eastern (windward) edge of the playa (Mahowald *et al.*, 2005).

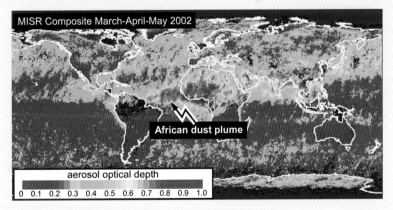

Figure 11.8 An African dust event extending across the breadth of the Atlantic is clearly resolved in this MISR image of aerosol optical depth. The algorithm retrieves aerosol loading, along with information about particle properties, from the variation in scene brightness over MISR's nine widely spaced view angles and four wavelengths (446, 558, 672 and 866 nm). Credit: NASA.

11.8   Clouds

Although the role of clouds in the Earth's energy budget is still poorly constrained, it is clear that the path to unravelling how clouds will mitigate or amplify climate change is through intensive observation. Remote sensing is well poised for this task, and in 2006 alone, NASA launched two satellites to focus on cloud distribution and dynamics. Lofted simultaneously, CloudSat and the cloud-aerosol LiDAR and infrared pathfinder satellite (CALIPSO) use state-of-the-art instruments to reveal detailed information about clouds and their effect on climate. CloudSat is a joint venture by NASA and the Canadian Space Agency; CALIPSO is a joint effort between NASA and CNES, the French space agency. Both fly in formation with four other satellites in NASA's A-Train constellation for coincident observations.

Clouds do far more than signal the day's weather – they also regulate Earth's climate. In particular, a cloud's temperature, depth, size and shape play a significant role in how much of the Sun's radiation reaches Earth's surface and what fraction of heat energy reflects back into the atmosphere. Energy reflected by clouds is dissipated back into space so, when cloud patterns alter, Earth's energy balance is changed and temperatures on the Earth's surface are affected. Of particular importance is the fact that changes in cloud properties can accelerate climate change (Arking, 1991; Yao & Del Genio, 1999).

Intuitively, a warming climate will cause greater evaporation from our oceans and a cloudier atmosphere. The newfound abundance of cloud will reflect more of the Sun's energy back into space, thus promoting cooling. In this simple scenario, Earth's atmosphere will 'reorganize' itself to mitigate the original input of energy that initiated the climate change. Some scientists, however, caution that we cannot rely on these thick and dense clouds to come to our aid:

- First, a warmer atmosphere needs more water vapour molecules to become saturated and to condense into clouds, so it is hard to anticipate exactly how clouds respond to human-induced climate perturbations (Ramanathan *et al.*, 1989; Carslaw *et al.*, 2002).
- Second, although summer is warmer and more humid than winter, the sky is not noticeably cloudier on average in summer (Tselioudis *et al.*, 1998).
- Third, until recently it had been assumed that the most efficient reflectors of energy would be dense clouds in the warmest latitudes. More water vapour can be held aloft in warm air, leading to thicker and denser cloud decks. This assumption is, however, refuted by work carried out by Tselioudis *et al.* (1998), who found that over more than half the Earth, and especially in the warmer places (both in the tropics and in mid-latitude summertime), stratus and stratocumulus clouds were actually brighter when the air was cold.

In short, by virtue of the fact that clouds are mobile and temporary, it is difficult to predict their future locations and type of formation, or whether they will be an overall force for warming or cooling.

11.9   Forecasting Earth's atmosphere

Arising largely from the elegant work of Edward Lorenz in the 1960s and early 1970s, 'chaos theory' represented a paradigm shift in our understanding of non-linear dynamical systems such as weather (Lorenz, 1969). By today's standards, working on an unsophisticated computer running a simple weather simulation, Lorenz discovered that minute changes (differences in the fourth or higher decimal places of input data) in initial model conditions produced large changes in the outcome in periods as short as a few weeks.

The implication was of great consequence. With such chaotic behaviour in even a simple model, would the real-world weather, with all its infinite complexity, ever be predictable? Later to become entitled by Lorenz as the 'butterfly effect', the experiment indicated that tiny fluctuations in initial conditions of a (weather) system can cause a chain of events leading to large-scale phenomena. At the time of its discovery, a poignant analogy was used. Could the flap of a butterfly's wings in Brazil set off a tornado in Texas? Had the butterfly not flapped its wings, the trajectory of the system might have been vastly different.

Lorenz's finding harked of an unpleasant truth. While the specifics of a weather forecast may be reliable for a couple of days, even when armed with today's Earth-observing and computer modelling technologies, predicting weather beyond a week in advance is little more than speculation.

11.10   Atmospheric models and reality

Atmospheric computer models attempt to mimic the complexity of Earth's climate system by breaking the hydrosphere and atmosphere down into layers and the geography of Earth into manageable pixels. These massive programs are computationally intensive and demand supercomputers to run. The need for vast computer power arises from the complexity of the models and the fact that useful forecasts of climate change can only be attained by producing climate projections that extend many centuries into the future.

Provided that the computer model truly reflects the way that energy balances between land and sea, and the way that the oceans and the atmosphere exchange heat and moisture, the computer predictions should be in precise concert with reality. Needless to say, however, this rarely occurs. The models are based on the integration of a variety of fluid dynamical, chemical, and sometimes biological equations. These equations govern atmospheric motion within the model, which in turn may be further parameterized for effects such as turbulent diffusion, cloud formation and the effect of surface terrain. Most atmospheric models are numerical, i.e. they discretize equations of motion and they are able, depending on the type of the model, to predict microscale phenomena such as tornadoes and boundary layer eddies as well as synoptic and global flows.

Table 11.2 Key atmospheric parameters useful to climate research models and the present-day spacecraft used for their measurement.

Atmospheric parameter	Satellite
Surface wind speed and direction	ERS-2 (Europe), QuikSCAT (US)
Upper-air temperature	Aqua (US), Metop (Europe)
Water vapour	GOES series (US), Metop (Europe)
Cloud properties	CloudSat (US), CALIPSO (US), Metop (Europe)
Precipitation	TRMM (US/Japan), Aqua (US)
Earth radiation budget	Aqua (US), Meteosat (Europe), SOURCE (US)
Ozone	Aura (US), Terra (US), ERS-2 (Europe), Envisat (Europe)
Aerosols	Parasol (France), Envisat (Europe), Terra (US), Aqua (US)
CO_2, methane and other greenhouse gases	Terra (US), Aura (US), Metop (Europe), Envisat (Europe)
Upper-air wind	Meteosat (Europe), GOES series (US)

Atmospheric models relevant to climate prediction can be split into three families:

1 Atmosphere general circulation model (AGCMs);
2 Ocean general circulation models (OGCMs);
3 Coupled atmosphere-ocean general circulation models (AOGCMs).

As their names imply, the first two pertain to the atmosphere and ocean respectively. These three-dimensional models require as input details of sea surface temperatures and sea-ice coverage, which would typically be obtained through satellite observation (Table 11.2). Run at their finest spatial resolution, the models can be used for short-range weather forecasting. To produce projections for decades or centuries, rather than days, they are used at a coarser level of detail in order to reduce computational overhead. This coarsening is required because every twofold increase in spatial resolution requires a tenfold increase in the number of teraflops required to run the model (Wehner *et al.*, 2008). The computational bottleneck may, however, be short-lived, with machines offering petaflop computing predicted before the close of the present decade. These platforms will likely run at peak speeds exceeding 3,000 teraflops and will offer the possibility of century-long simulations of the climate at resolutions less than 10 km (Wehner *et al.*, 2008).

Today, AOGCMs represent the state of the art and consist of an AGCM coupled to an OGCM. They are used to explore the variability and physical processes of the coupled climate system. At the time of writing, global climate models typically only have a resolution of a few hundred kilometres.

Climate projections from the Hadley Centre (UK) make use of the HadCM2 AOGCM model, developed in 1994, and its successor HadCM3 AOGCM, developed in 1998. Greenhouse gas experiments with AOGCMs have usually been driven by specifying atmospheric concentrations of the gases but, if a carbon cycle model is included, the AOGCM can predict changes in carbon dioxide concentration, given the emissions of carbon dioxide into the atmosphere. At the Hadley Centre, this was first done in 1999.

Similarly, an AOGCM coupled to an atmospheric chemistry model is able to predict the changes in concentration of other atmospheric constituents in response to climate change and to the changing emissions of various gases. From there, first-cut predictions on future climatologies such as hurricane frequency and intensity can be made (Emanuel *et al.*, 2008; Saunders & Lea, 2008).

Although cited as weak evidence for global warming by climate-sceptics (e.g. Singer, 1999), it is coupled AOGCM models that provide the best insight into future global warming. Looking several decades into the future and the advent of truly global models with resolutions approaching 1 km, predictions of regional climate change are likely to improve greatly with the modelling of fine-scale cloud processes, a key component missing from current coupled AOGCM models (Randall *et al.*, 2007; Wild, 2009).

11.11 Hurricanes

Nowhere does the use of atmospheric models get more press time than for the prediction of hurricane tracks. Few things in nature can compare to the destructive force of a hurricane. Capable of annihilating coastal areas with sustained winds of 250 kilometres per hour or higher, hurricanes deliver intense areas of rainfall and storm surge. In fact, during its life cycle, a hurricane can expend as much energy as 10,000 nuclear bombs!

The birth of a hurricane, known as 'cyclogenesis', occurs over tropical waters (between 8° and 20° latitude) in areas of high humidity, light winds and warm sea surface temperatures (typically >26.5°). These conditions usually prevail in the summer and early autumn months of the tropical North Atlantic and North Pacific Oceans. For this reason, the hurricane 'season' in the northern hemisphere runs from June through November. The tropical North Atlantic Ocean, the Caribbean Sea and the Gulf of Mexico are regions where the hurricanes that threaten the US Eastern seaboard build and intensify (Thorncroft & Hodges, 2001). Here, the ocean dynamics are highly variable in space and time and are characterized by the presence of warm currents, meanders and eddy formation, often with very high hurricane heat potential values during the summer and early autumn months. This part of the ocean is thus vigilantly monitored for signs of cyclogenesis (Figure 11.9).

There exist three portions of the electromagnetic spectrum that are useful for monitoring and tracking storm systems:

1 **Radar**: since radar is reflected by raindrops, it works well as a means to locating and tracking rainstorms, tornadoes and hurricanes. The SeaWinds instrument on NASA's Quick Scatterometer spacecraft, also known as

Figure 11.9 Multiple 'named' tropical depressions and hurricanes in various states of genesis linger across the Tropical Atlantic and Eastern Pacific in this GOES satellite image acquired September 3, 2008. The GOES data has been overlaid on the NASA 'Blue Marble' dataset for realism. Credit: NASA's Earth Observatory.

QuikSCAT, is a specialized microwave radar that measures both the speed and direction of winds near the ocean surface (see Chapter 10). Doppler systems can additionally be used to measure changes in frequency of the signal sent to a target, from which wind characteristics can be derived.

2 **Infrared**: clouds, temperature and humidity all have a signal in the infrared, and these wavelengths are used to detect a variety of pertinent storm parameters. Housed aboard the recently lofted NASA Aqua satellite is the AIRS (Atmospheric Infrared Sounder) sensor, which is the central part of the AIRS/AMSU (Advanced Microwave Sounding Unit)/HSB (Humidity Sounder for Brazil) instrument group set to obtain global temperatures and humidity records throughout the atmosphere. It is hoped that these data will lead to improved weather forecasts and improved determination of cyclone intensity, location and tracks, and the severe weather associated with storms.

3 **Visible light**: sensors utilizing the visible spectrum offer useful information as to the physical structure and size of a storm. The Medium Resolution Imaging Spectrometer (MERIS) is an imaging spectrometer aboard the European Envisat platform that measures the solar radiation reflected by the Earth at a ground spatial resolution of 300 m, with 15 spectral bands in the visible and near-infrared and programmable in width and position. MERIS is particularly suited for storm tracking since it offers global coverage of the Earth every three days. Alternative platforms such as Landsat, SPOT, etc. may be spectrally well poised, but poor temporal resolution renders them as sensors of opportunity and therefore they are not reliable platforms for storm observation.

Remotely sensed data, including sea surface temperatures and ocean heights (topography) are used routinely as part of a suite of tools to predict cyclogenesis. Sea surface temperature (SST) is critical to storm formation as warmer water exists in partnership with a layer of overlying warm air (measurement of this parameter is covered in Chapter 10, Section 10.4). The existence of the latter means higher dew points, consequently giving the growing hurricane more energy to be released in the heat of condensation during cloud and precipitation development. SSTs are easily measured on a clear day using infrared satellite remote sensing techniques. Many Earth observation satellites carry suitable infrared instrumentation offering adequate spatial and temporal resolution for storm forecasting.

With a birds-eye view of the Atlantic, the meteorology satellites GOES-East and METEOSAT-9 are particularly well poised for the task. The former is part of the Geostationary Operational Environmental Satellite (GOES) programme operated by the United States' National Weather Service (NWS), and is currently stationed at 75°W, at an altitude of 35,790 km above the Amazon Basin. Also geostationary, but positioned above the African subcontinent, METEOSAT-9 is part of a European meteorological programme.

Both spacecraft have swaths wide enough to image the entire Earth and provide the capability to continuously observe and measure meteorological phenomena in real time. In addition to short-term weather forecasting, these enhanced operational services also improve support for atmospheric science research and numerical weather prediction models, while providing estimates of the location, size and intensity of severe storms with their visible and thermal infrared imagers (Figures 11.9, 11.10). These instruments deliver data at very frequent intervals (e.g. every 20 minutes) and their images can be supplemented by once-daily passes of sensors with higher spatial resolution, such as the AVHRR.

CASE STUDY: Hurricane Katrina

During its entire passage across the Atlantic and the Gulf of Mexico, Hurricane Katrina was tracked by the GOES-12 geostationary satellite, NOAA polar orbiters using AVHRR sensors, other satellites and many aircraft flights by NOAA and USAF pilots. For instance, NOAA WP-3D Hurricane Hunter aircraft obtained airborne Doppler radar-derived wind speed cross-sections as Katrina crossed the Gulf and made landfall on the Louisiana coast on August 29, 2005 (Figure 11.10). Other data gathered included the location of the hurricane, its size, movement speed and direction, the size of the eye, wind velocity and minimum central pressure.

Figure 11.10 shows a satellite image of Katrina as it passes over the Gulf of Mexico. The image was enhanced using data from several sensors, including the Tropical Rainfall Measurement Mission (TRMM) satellite and its microwave imager (TMI). By measuring the microwave energy emitted by the Earth and its atmosphere, TMI is able to quantify the water vapour, the cloud water and the rainfall intensity in the atmosphere.

The image in Figure 11.10 was taken on August 28, 2005, when Katrina had sustained winds of 185 km/h and was about to become a Category 4 hurricane in the central Gulf of Mexico. The image reveals the horizontal distribution of rain intensity within the hurricane, as obtained from TRMM sensors.

Figure 11.10 Enhanced satellite image of Hurricane Katrina over the Gulf of Mexico, obtained by the Tropical Rainfall Measuring Mission (TRMM) satellite. The satellite's microwave imager measured the rain rates, while its visible infrared scanner shows the extent of the hurricane. Credit: NASA's Earth Observatory.

Rain rates in the central portion of the swath are from the TRMM precipitation radar (PR). The PR is able to provide fine-resolution rainfall data and details on the storm's vertical structure. Rain rates in the outer swath are from the TRMM microwave imager (TMI). The rain rates were overlaid on infrared data from the TRMM visible infrared scanner (VIRS). TRMM reveals that Katrina had a closed eye surrounded by concentric rings of heavy rain (red areas) that were associated with outer rain bands. The intense rain near the core of the storm indicates where heat, known as latent heat, is being released into the storm (Pierce and Lang, 2005).

The NWS National Hurricane Centre was able to feed the satellite data into several different numerical computer and statistical models to attempt to forecast the path, speed and strength of the hurricane. The centre also used computer storm surge models such as the SLOSH model (Chapter 9) to provide guidance on storm surge height and extent of predicted flooding.

Days before Katrina made landfall, NOAA's storm surge model predicted that Katrina's storm surge could reach 5.5 m to 6.7 m above normal tide levels, and in some locations as high as 8.5 m. The model predictions were sent in near real time to various disaster management agencies at the local, state and federal levels. Emergency managers used the data from SLOSH to determine which areas must be evacuated (NOAA, 2005; 2006; 2008; NOAA/NHC, 2008). The devastating storm surge that accompanied Katrina and the means for remotely monitoring the resulting flooding are covered in Chapter 9.

Table 11.3 Key parameters useful to storm detection and present-day spacecraft used for their measurement.

Storm parameter	Satellite
Tropical rainfall intensity	TRMM (US/Japan), Aqua (US)
Wind strength and direction	QuikSCAT (US), ERS-2 (Europe)
Ocean and air temperatures and humidity	Aqua (US)
Atmospheric temperature and water vapour profiles	COSMIC/FORMOSAT-3 (US/Taiwan)
Sea surface temperature	Aqua (US), Envisat (Europe), GOES series (US), METEOSAT (Europe)
Sea surface roughness	ASAR (Europe)
High atmosphere cloud structure and pressure	Envisat (Europe)
Damage assessment	QuickBird (commercial), IKONOS (commercial)

With the hurricane fuelled by the presence of warm water, the most important area of ocean to monitor lies directly below the storm. Here, shrouded in dense cloud, the infrared eyes of the meteorological satellites are blind. Beneath the storm, radar altimetry may be used to indirectly estimate SST. As a rough estimate, a 1 cm anomaly in the height of the ocean surface is deemed to represent a heating of 1°C (Zumpichiati *et al.*, 2005). Combined with prior temperature data obtained by more conventional means, an estimate of upper layer ocean heat content may be retrieved.

Altimetry products delivered from TOPEX/POSEIDON were suitable for such forecasting until the instrument's demise in 2006 (Mainelli-Huber, 2000; Shay *et al.*, 2001). More recently, the successor to this programme, the Jason-1 altimeter, provides the majority of meteorological ocean height data. Linked by more than their classical names, data collected by the fleet of 3,000-odd autonomous oceanic floats comprising the ARGO array are used in conjunction with Jason-1. Both TOPEX/POSEIDON and Jason-1 are, however, hindered by their rather large

(4 km) footprint of the radar pulse, since data closer to the coast cannot be used due to land contamination. It is in this realm, just prior to the hurricane making landfall, that SST estimates are particularly valuable. With a footprint of 70 m, the recently launched ICESat laser altimeter is well poised to retrieve SST data much closer to the coast, and will perhaps shortly fill a niche far from its intended mission of polar monitoring (Leben *et al.*, 2003).

Beyond prediction and monitoring, remote sensing is the primary tool used to track the legacy of hurricanes upon landfall – namely the impact of storm surge, massive winds, and torrential rain. In this realm, spacecraft with metre-scale spatial capability have found particular utility, with IKONOS and QuickBird used most often. Typically, the image data are rapidly melded into a GIS that is used to guide rescue and clean-up efforts.

11.12 Key concepts

1 The atmosphere protects the biosphere by absorbing ultraviolet solar radiation and reducing temperature extremes between day and night. At present, the gaseous breakdown fall to roughly 78 per cent nitrogen, 20.95 per cent oxygen, 0.93 per cent argon, 0.038 per cent carbon dioxide, trace amounts of other gases and a small but variable amount of water vapour. The dominant gases of the atmosphere are essentially climatically inert; only the minor components, the greenhouse gases, play a role in regulating the temperature of Earth's surface.

2 Our reliance on the burning of fossil fuels for energy production has released vast quantities of carbon dioxide into the atmosphere. Though still present at a very low concentration compared with air's main components of nitrogen and oxygen, CO_2 has an influence on the climate far out of proportion to its prevalence in the atmosphere.

3 Carbon dioxide absorbs many infrared wavelengths of the Sun's light and is therefore termed a greenhouse gas. Other important greenhouse gases are methane, nitrous oxide and, in the lower atmosphere (troposphere), ozone. Greenhouse gases are essential to maintaining the temperature of the Earth; without them, the planet would be so cold as to be uninhabitable. However, an excess of greenhouse gases raises the temperature of the atmosphere, as we are currently witnessing.

4 In spite of its importance to the greenhouse effect, there is presently only one dedicated satellite mission to monitor CO_2 concentrations in the atmosphere. NASA's Orbiting Carbon Observatory (OCO) was planned to fill this shortfall, but failed to reach orbit. This leaves a single mission, the Japanese GOSAT, which targets the infrared absorbance bands of carbon dioxide.

5 Stratospheric ozone protects the biosphere by attenuating biologically harmful wavelengths of ultraviolet (UV) radiation emanating from the Sun.

This region is termed the ozone layer. The concentration of ozone in the high-atmosphere is depleted by the presence of CFCs, which are non-volatile compounds used as propellants.

6   Remote sensing of ozone is conducted by measuring the total solar radiance incident on the satellite as compared to the total UV radiation scattered back from the atmosphere. The concentration of ozone is measured in Dobson units and the principle instrument used is the Total Ozone Mapping Spectrometer (TOMS). The TOMS sensor has been flown on multiple satellites and, in particular, Meteor-3 (1991–1994), Nimbus-7 (1978–1993), ADEOS (1996–1997) and Earth Probe (1996–2005). Measurements into the future will be made with the Ozone Monitoring Instrument (OMI), aboard the Aura platform.

7   Besides gases (air), the atmosphere also includes small amounts of material in a solid state. Termed a 'dust aerosol', these solids are typically confined to the lower troposphere and are an important modulator of global climate. The radiative forcing of dust is complex and dependent on the distribution of mineralogy and particle size in the plume. A major source of atmospheric dust is the Bodélé Depression in Chad, where strong winds loft vast clouds that extend across the breadth of the Atlantic, dropping sufficient dust in the Caribbean to alter soil composition.

8   By reflecting the Sun's energy away from the Earth, the quantity and type of clouds present in the atmosphere have a strong impact on its energy budget. It is presently unclear whether clouds will have a net positive or negative radiative forcing on the Earth's climate, and this unknown factor seriously confounds the accuracy of atmospheric models. Two satellites are currently in orbit for the monitoring of clouds, namely CloudSat and CALIPSO. Both are operated by NASA.

12 Observing the cryosphere

12.1   Introduction

A significant acceleration in ice shelf degradation had been likely initiated at least by the early 1980s (Vaughan & Doake, 1996; Comiso and Parkinson, 2004), but was not widely recognized as being part of a systematic trend until years later (Walsh *et al.*, 1996; Serreze *et al.*, 2000; Krabill *et al.*, 2004). Though recognized by scientists for nearly two decades, acceptance of the degradation was not embraced by the general public until recently, when it finally percolated into the mainstream consciousness following the release of several key reports from the IPCC (Intergovernmental Panel on Climate Change). The slowness of societal acceptance of climate change can be attributed to the subtleties of the science, whereby balanced conclusions have to be made on the basis of an incomplete understanding of the rates and seriousness of shifting patterns. Also, it is undeniable that climate change has become ever more politically unpalatable as the evidence mounts for modern human society as its primary cause.

The vastness of the polar ice caps and their inhospitable conditions, with seasonal darkness, does not lend itself well to field study. As early as the 1960s, satellite remote sensing emerged as the only technology capable of making repeated observation of these harsh environments. Our understanding of the dynamics of the high latitudes in response to changing climate therefore owes much to remote sensing.

12.2   The history and status of the polar ice sheets

Earth's poles receive less solar energy than the equatorial regions and therefore are colder. Although they have not been a permanent feature through Earth's history, we find ourselves in a time where water ice is abundant at high latitudes but is

Figure 12.1 Composite images of the Arctic Circle complied from Special Sensor Microwave Imager (SSM/I) flown onboard the United States Air Force Defence Meteorological Satellite Program. A clear reduction in ice extent is visible between 1979 and 2003. The SSM/I will be considered several times throughout this chapter in the context of the monitoring of glaciers, sea-ice and permafrost.
Credit: NASA's Earth Observatory.

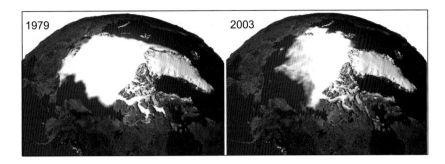

evidently diminishing. On geological timescales, ice caps may grow or shrink due to climate variation. A glacial period is signalled by the building up of polar ice. As the ice mass grows, Earth's supply of water becomes increasingly trapped in glaciers and sea level falls, land area increases, rainfall diminishes and the Earth grows colder. Finally, the process is reversed and the polar ice starts to melt, the seas rise and temperatures again become moderate.

Through the 1970s, breakthroughs in ocean coring capability, coupled with advances in radiocarbon dating, revealed repetitive occurrences of ice ages through geologic history, with a dominant cycle of 100,000 years (figure 12.2) (Rutherford & D'Hondt, 2000). This periodicity suggests that astronomical forcing promotes variation in the prevalence of ice cover (Muller & MacDonald, 1997), though natural oscillations of the ocean/continent/atmosphere system are also recognized to impart a degree of control (Oerlemans, 1982). The eight glacial cycles that have occurred over the last 600,000 years are clearly correlated with fluctuations in the atmospheric concentration of carbon dioxide (figure 12.2 – upper pane). The rise and the fall of the sea over these time periods can therefore also be related to atmospheric CO_2 concentrations (Rohling et al., 2009; Siddall et al., 2009).

The Earth is in an interglacial period now, with the last ice retreat (termed the Wisconsin glaciation) ending about 10,000 years ago. Presently, global sea-ice coverage averages approximately 25 million km^2, with an additional 15 million km^2 of Earth's land surface covered by ice sheets and glaciers. Substantial reductions in the extent of Arctic sea-ice since 1978 (2.7 ± 0.6 per cent per decade is the annual average, with 7.4 ± 2.4 per cent per decade for summer), increases in permafrost temperatures and fairly rapid reductions in glacial extent globally have also been observed in recent decades. This is particularly true for the Greenland and Antarctic ice sheets (Collins et al., 2007).

The presence of ice is fundamental to the operation of the ocean-atmosphere system and, as geologic history has made clear, changes in ice cover significantly alter both our climate and sea level. The task of remote sensing is to put the ice loss that we are currently witnessing in context and to aid our understanding of whether the melt is part of a natural cycle or related to human-caused global changes.

Figure 12.2 The red line is the concentration of atmospheric CO₂ for the last 600,000 years inferred from the analysis of air bubbles trapped in Antarctic ice cores (e.g. Siegenthaler *et al.*, 2005). The blue line depicts global ice volume for the same period. The 100,000 year glacial-interglacial cycle is clear, as is the correlation between CO₂ and ice volume.

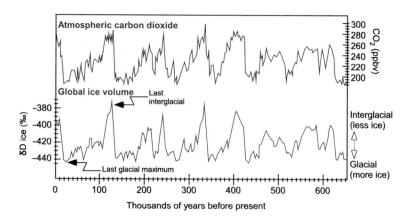

Figure 12.3 Sea level (SL) change during the past 140,000 years. The figure shows that generally the sea has been at, or below, today's level. There has been at least one period – the Oxygen Isotope Substage 5e – where it has been higher. This event is of relevance to the rise in temperature we may be currently facing.

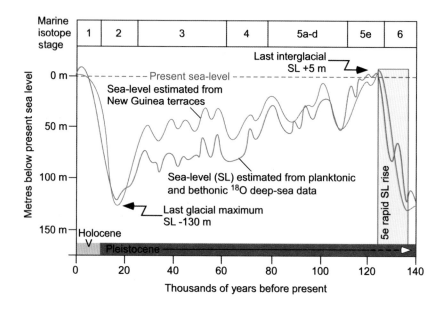

12.3   Ice and sea level

As the extent of Earth's ice sheets have waxed and waned over geologic history, sea level has risen and fallen in accordance. As shown in Figure 12.3, high sea levels correspond to periods when little ice existed at Earth's poles (interglacials), whereas low sea level correspond to times when vast quantities of water were locked up within extensive ice sheets (glacial periods).

As recently as 20,000 years ago, at the height of the last ice age, sea level dropped to ≈130 m below that of today (Lambeck & Chappell, 2001; Hanebuth *et al.*, 2009). During the last interglacial period, ≈130,000 years ago, it was several metres higher than present. This event, termed Marine Isotope Stage 5e (where 'e' signifies the Eemian Interglacial period), was characterized by global mean surface temperatures at least 2 °C warmer than present and represents a sea level overshoot (or series of multiple overshoots) following a rapid rise as the Earth deglaciated from ice age conditions (Sherman et al., 1993). At this time, the sea level likely rose by more than 1.5 m per century (Neumann & Hearty, 1996).

The 5e stage is particularly well studied, as the prevailing global mean temperatures compare well with projections for future climate change under the influence of anthropogenic greenhouse gas emissions (Blanchon et al., 2009). Perhaps the rate of sea level change during the 5e carries a forewarning as to what we will witness in the coming century?

At the present time, the majority of Earth's ice is located at the poles, with significant amounts also held within high altitude glaciers. The Greenland and Antarctic ice sheets average a thickness exceeding 2 km and contain more than 75 per cent of the Earth's fresh water. If both were to melt, 33×10^6 km^3 of fresh water would be released into the oceans, resulting in a sea level rise of approximately 80 m. Such a drastic scenario is fortunately not foreseen as likely to occur in the near future, but even small reductions in ice thickness, by the order of 0.1 per cent (2.4 m), would cause sea level to rise by approximately 10 cm. Such a rise would be sufficient to inundate vast tracts of land which are home to tens of millions of people, and it would seriously damage coastal cities and ecosystems (Nicholls et al., 1999). It is for this reason that such importance is placed on tracking the relatively subtle changes in sea level believed to be occurring presently.

12.4   Ice and climate

Unlike land-ice (Greenland and the Antarctic), melting ice shelves that float on the surface of the sea do not change sea level a great deal. However, because of the density difference between fresh water and salt water, a slight increase in sea level, on the range of several centimetres, will result because fresh water released by melting ice has a slightly greater volume than an equivalent weight of salt water. Thus, when freshwater ice melts in the ocean, it contributes a greater volume of melt-water than it originally displaced, so the melting of the northern polar ice cap, which is composed of floating pack ice, would offer a minor positive contribute to raising sea levels. Furthermore, because they are fresh, the melting of the polar ice cap would cause a significant decrease in the salinity of surface waters (Jacobs et al., 2002).

The presence of sea-ice is therefore relevant to global climate for two reasons:

1 **Albedo.** The high albedo of the ice surface reflects solar energy away from the Earth and back into space, thus promoting cooling and the growth of ice sheets. This is termed the 'ice-albedo feedback' and is one of the many positive feedback loops that coerce the Earth to evolve ice age conditions (Curry et al., 1995). The feedback works both ways, though, and, once a melt

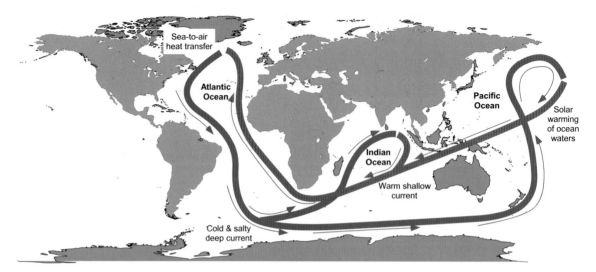

Figure 12.4 The oceans play a major role in the distribution of the planet's heat through deep sea circulation. This simplified illustration shows thermohaline circulation which is driven by differences in heat and salinity. Records of past climate suggest that there is some chance that this circulation could be altered by the changes projected in many climate models, with impacts to climate throughout lands bordering the North Atlantic.

is initiated, it can be sustained through changes in albedo. The albedo of seawater is low (≈ 0.1) compared to sea-ice (≈ 0.6) or fresh snow (≈ 0.9). Reduction of sea-ice therefore greatly decreases the fraction of reflected incoming radiation, causing further warming and melting. The 'albedo' effect is the main reason why the influence of warming is significantly more pronounced at high northern latitudes than in temperate or tropical regions. It is also this contrast in albedo which allows ice to be easily differentiated from open water using remote sensing in the visible spectrum.

2 **Ocean Circulation.** Ice and fresh melt water from the Arctic Ocean have profound effects on ocean circulation patterns in the North Atlantic and therefore affect climate patterns on a global-scale. As ice forms, pure water is preferentially frozen, producing high-salinity brines beneath the ice sheets that resist freezing (high salinity depresses the freezing point of water). The dense saline water sinks down the continental slope, separating from a low-salinity surface layer and forming North Atlantic deep water. The sinking of dense cold water at high latitude, coupled with low latitude heating, drives thermohaline circulation, which in turn has a significant impact on Earth's radiation budget. In theory, the thermohaline circulation could be influenced by large influxes of low density melt-water from the Greenland and Arctic ice sheets, and deep water formation may be disrupted to a point that the thermohaline circulation is retarded (Rahmstorf, 1997). This disruption occurs because the melt-water is not sufficiently dense to sink and in turn push the thermohaline conveyor.

Such an event, the Younger Dryas, is thought to have occurred at the end of the Pleistocene epoch, between approximately 12,800 to 11,500 years ago, when North America deglaciated, causing a sudden influx of fresh water into the high-latitude Atlantic (Anderson, 1997; Broecker, 1997; Rühlemann *et al.*, 1999). Recent research indicates that a rapid and complete shutdown of the thermohaline circulation would require an enormous volume of fresh water to enter the North Atlantic in a short time (Vellinga & Wood, 2002). Such a scenario appears unrealistic, and a more gradual weakening of the thermohaline circulation in response to climate warming seems most plausible (Stocker & Schmittner, 1997; Barreiro *et al.*, 2008). This said, work by Bryden *et al.* (2005) worryingly indicated that a 30 per cent decrease in circulation had occurred during the past five decades. More recent studies, however, suggests that this decrease resulted from a natural lull in thermohaline vigour (Cunningham *et al.*, 2007) and should not be attributed to changing climate.

12.5 Present ice loss in context

Twenty-one thousand years ago at the height of the last ice age, much of North America and Europe were covered by glaciers over 2 km thick. With sea level so low, land bridges existed between the Americas and Asia across the Bering Straight, and the British Isles were connected to mainland Europe. The chill was global; land and sea-ice combined to cover 30 per cent of the Earth's surface, more than at any other time in the last 500 million years. Between 14,000 and 6,000 years ago, the ice was in retreat. The ensuing melt raised sea level by 100 m, inundating great tracts of land and shaping the coasts to the form that we observe today. The march of the sea was rapid at up to 1.25 cm per year.

Today we are, fortunately, dealing with a much less pronounced inundation. Recent predictions peg the rise to be in the order of 2 cm every 10 years (Cabanes *et al.*, 2001; Church *et al.*, 2001). This is more gradual, but clearly at a level perceivable over a human lifetime and highly significant when we project changes over many generations (Figure 1.1, Chapter 1). In context, there have certainly been times when the globe has been devoid of polar ice, and arguably also episodes when the Earth has been entirely ice-covered for long periods (600–700 million years ago – the somewhat controversial Neoproterozoic Snowball Earth hypothesis) (Hoffman *et al.*, 1998). The challenge now upon us is to verify whether the present warming is a part of a long-term and natural trend, or an anomaly forced by human activity. This is a difficult task, as our physical observations of ice status stretch back less than a century, with longer-term records being derived through proxies and all the uncertainty that this entails. We are dealing with the classical problem of inferring long-term trends from very short records.

This is not to say that we are unable to meaningfully reconstruct the history of ice on our planet, nor make predictions as to its future status. It just means that we must be creative in our tackling of the problem; a skill that is fortunately abundant in the scientific community.

Figure 12.5 Landsat images of Mount Kilimanjaro draped over a digital elevation model. The time-series depicts the radical reduction of ice extend between 1993 and 2000. Ice on the summit of the mountain formed more than 11,000 years ago but has dwindled by 82 per cent over the past century (Thompson *et al.*, 2002). Courtesy: NASA's Earth Observatory.

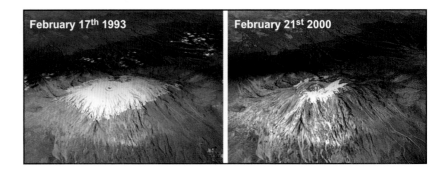

Figure 12.5 depicts the pronounced decrease in ice atop Tanzania's Mount Kilimanjaro. The melting is part of a trend of glacial retreat throughout Africa, India, and South America and has been proposed as compelling evidence of Earth's recent warming (Thompson *et al.*, 2009). Melting is also occurring on Mount Kenya, the Rwenzori Mountains in central Africa, and also on tropical glaciers high in the Andes and Himalayas.

12.6 Remote sensing of the Earth's ice sheets

The advent of remote sensing has accelerated our understanding of Earth's ice sheets perhaps to a greater degree than any other Earth system. The vast tracks of inhospitable terrain render traditional fieldwork impossible, and orbital systems provide the only means of routinely monitoring these areas. Sea-ice has a higher albedo than the surrounding ocean, which makes it easy to detect from visible remote sensing instruments. However, since these sensors measure reflected radiation from the sun, visible data can only be collected during the daytime.

The inability to measure at night is also a significant hindrance for remote sensing the poles, which are shrouded in six months of winter darkness. This dark area begins at the pole on the first day of autumn, reaches its maximum on the first day of winter and then shrinks back to zero on the first day of spring. Because clouds also reflect visible radiation, a cloudy sky prevents satellites from viewing visible light reflected from sea-ice. The ice-covered polar regions tend to be cloudy, with clouds obscuring sea-ice 70 per cent to 90 per cent of the time (Hahn *et al.*, 1995). Against these difficulties, a wide variety of remote sensing solutions have been developed that encompass all of our available satellite technologies.

12.6.1 Passive optical and thermal remote sensing

Despite the limitations posed by seasonally dark and cloudy poles, measurement of ice albedo and temperature are possible under favourable conditions using optical sensors such as MODIS (Riggs *et al.*, 1999; Hall *et al.*, 2008), AVHRR

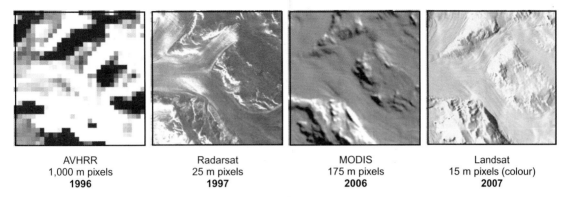

AVHRR	Radarsat	MODIS	Landsat
1,000 m pixels	25 m pixels	175 m pixels	15 m pixels (colour)
1996	**1997**	**2006**	**2007**

Figure 12.6 The enhanced resolution of the LIMA mosaic (Bindschadler *et al.*, 2008) is shown in comparison to three previous Antarctic mosaics constructed from AVHRR, Radarsat and MODIS. The same area is depicted. Each pixel from the new mosaic is multi-band colour and 15 m × 15 m. Credit: NASA.

(Key & Haefliger, 1992; Lindsay & Rothrock, 1993; Comiso, 2003) and Landsat TM (Key *et al.*, 1997), but passive optical technologies are frequently considered in unison with data from active remote sensing instruments in order to construct a more viable time-series of imagery (Bell *et al.*, 2007). Not limited by sunlight, and operating in the visible wavelengths, ICESat is an active LiDAR optical instrument specifically designed for high-latitude work. This instrument will be discussed shortly.

Produced through US-UK collaboration and marking a major contribution to the International Polar Year (www.ipy.org), the first ever true-colour high-resolution satellite view of the Antarctic continent was released in 2007. The composite, termed the Landsat Mosaic of Antarctica (LIMA), is made up of imagery that has a spatial resolution of 15 metres per pixel, the most detailed satellite mosaic of the continent yet created (Bindschadler *et al.*, 2008). LIMA is comprised of more than 1,000 Landsat ETM+ images acquired between 1999 and 2001. Before its release, the highest-resolution satellite mosaic of Antarctica was the MODIS Mosaic of Antarctica (MOA – resolution 175 m) and RADARSAT Antarctic Mapping Project (RAMP – resolution 25 m) created from radar data (Figure 12.6). LIMA covers the entire continent except from the South Pole at 90° to 82.5° south latitude, where Landsat has no coverage because of its near-polar orbit. A combination of LIMA and MOA imagery can display the entire continent, albeit with coarser resolution. The LIMA mosaic can be accessed and downloaded through a web portal at http:\lima.nasa.gov.

12.6.2 Passive microwave remote sensing

The microwave portion of the spectrum covers wavelengths of approximately 1 cm to 1 m. As discussed in Chapter 7 (Section 7.6), all objects emit microwave energy of some magnitude, though the amount is typically very small. Unlike the shorter wavelengths of visible and infrared energy, microwaves are situated in a portion of the electromagnetic spectrum that is not particularly susceptible to atmospheric scattering (see Figure 2.5), so microwaves are therefore of great utility to monitoring portions of the Earth shrouded by dense cloud cover and often perpetual darkness. A passive microwave sensor detects naturally emitted microwave energy, and the crystalline structure of ice typically emits more microwave energy than the liquid water in the ocean, making it easily discernable (Fahnestock *et al.*, 2002). Passive microwave sensors are employed by both radiometers and scanners and use antennae to detect the prevalence of microwaves within a defined field of view.

Passive microwave sensors have the advantage of having been flown on early satellite platforms, so there exists a comprehensive time-series of data with which to study changes in ice distribution since the early 1970s (our knowledge of sea-ice extent prior to this time is very limited). In 1972, observation began with the Electrically Scanning Microwave Radiometer (ESMR) aboard NOAA's Nimbus-5 satellite. From 1972–1976, the instrument provided daily- and monthly-averaged sea-ice concentrations for both the Arctic and Antarctic at 25 km gridded resolution (Crane *et al.*, 1982). Following a two-year hiatus, the Scanning Multichannel Microwave Radiometer (SSMR) provided data from 1978–1987, followed by the SSM/I (Special Sensor Microwave/Imager) from 1987 until present (Smith, 1996; Bjørgo *et al.*, 1997).

The Advanced Microwave Scanning Radiometer for EOS (AMSR-E) is a twelve-channel, six-frequency, total power passive-microwave radiometer system that is housed aboard the Aqua satellite platform. This instrument was lofted in 2002 and provides daily global coverage of sea-ice extent and temperature with a 25-km resolution (Spreen *et al.*, 2008) (Figure 12.7). Reassuringly, during periods of overlapping coverage, the AMSR-E data are shown to provide basically the same spatial and temporal variability in ice coverage as those from the SSM/I (Comiso & Nishio, 2008). When cloud and light conditions are favourable, MODIS (also riding aboard Aqua) provides a further level of validation to the AMSR-E sea-ice product (Heinrichs *et al.*, 2006). These passive microwave instruments are discussed in Chapter 7 in the context of measuring soil moisture.

Despite an all-weather capability, passive microwave radiometry is hindered by the fact that the emitted energy level that has to be detected is very low. To combat this problem, it is necessary that the receiving instrument collects energy over a large field of view, yielding data of a resolution too coarse to discern the detailed structure of ice fields. The extensive time-series of data produced by passive microwave sensors is, however, sufficient to outweigh this limitation. Indeed, satellite passive microwave observations have been used to detect trends in ice dynamics which now form the basis for many predictions of the effect of future warming scenarios (Cavalieri *et al.*, 1999; Parkinson *et al.*, 1999).

It is known that the Arctic ice melts and thins in the summer, reaching its minimum in September. Microwave data assembled by Stroeve *et al.* (2005) show that the minimum extent of sea-ice has, however, decreased since 1979 from a long-term average of more than 7 million km^2 to less than 6 million km^2 in 2002. Merging a SMMR and SSM/I time-series, Comiso (2006) and Comiso & Nishio (2008) showed that the reduction in extent is accompanied by a decrease in

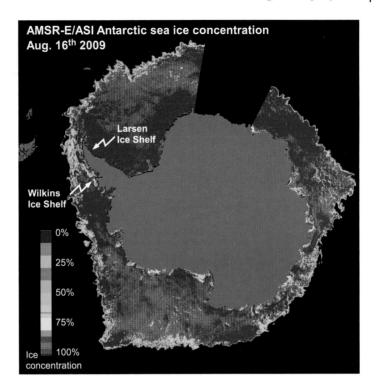

Figure 12.7 AMSR-E/ASI (89 GHz) sea-ice concentration data acquired Aug. 16, 2009 for Antarctica. Credit: Spreen & Kaleschke (2008).

thickness – a trend which now appears to persist in winter. The observed rate of ice decline equates to nearly 9 per cent per decade (e.g. Meier *et al.*, 2007), which, if it continues unabated, would render the Arctic Ocean ice-free by 2060.

The microwave energy emitted by snow and ice can indicate whether the surface is wet or dry. Wet surfaces imply melting. Research on this property confirms previous reports of particularly extensive melting far inland of the ice-sheet in 2005 (Fettweis *et al.*, 2007).

12.6.3 Active microwave remote sensing

In addition to passively sensing Earth's microwave emissions, satellite sensors can actively emit microwaves toward the Earth, which reflect off the surface and return to the sensors. This type of remote sensing is termed 'active microwave' or 'radar' and is covered several times in this book (see Section 3.2.4 for an introduction). Possessing the same virtues as passive microwave sensing (all-weather capability unhindered by cloud and the extended polar night), active sensors are not constrained by a coarse resolution because they are detecting a much stronger (reflected) signal. Sea-ice is typically detected using one of three permutations of active microwave sensing: imaging radar, non-imaging radar or altimetry (i.e. Figure 10.7 in Chapter 10).

Imaging radar

As the name suggests, imaging radar is used to obtain an image of a target using microwaves. Such images are far removed from how the Earth would appear when viewed in the visible spectrum, and thus require quite different processing and interpretation. The character of ice in the microwave spectrum is discernable by virtue of differences in roughness and texture, caused, for example, by the inclusion of air bubbles (Barber & LeDrew, 1991; Collins *et al.*, 1997). Such differences can be used to separate thicker multiyear ice from younger thinner ice, and thus have considerable utility to aid in the interpretation of ice sheet dynamics (Fahnestock *et al.*, 1993).

Having been aloft for nearly 30 years on a variety of platforms, Synthetic Aperture Radar (SAR) is the imaging radar most commonly employed for polar observation (Rignot *et al.*, 2000; Dall *et al.*, 2001). SAR uses the distance that the radar moves between emitting a pulse and receiving back its echoes to add detail to the image produced (see Chapter 6, Section 6.2.4). In contrast to more traditional optical remote sensing, the spatial resolution of a SAR is unrelated to the distance between the radar instrument and the target. For this reason, a SAR in orbit can provide an equally high resolution to an instrument mounted on an aircraft. Also, unlike visible imagers, radar penetrates the snow surface, thus providing glaciologists with a tool capable of sensing internal properties of the ice sheet or glacier (König *et al.*, 2001). SAR is treated in detail in Section 6.2.4 of this book (Figure 6.8).

The United States is conspicuous by its lack of a satellite-based SAR and the RADARSat, managed by the Canadian Space Agency, is the primary SAR mission today used to produce timely and accurate information on sea-ice (Stern & Moritz, 2002; Kwok *et al.*, 2003). The European ERS-1 (1991) and ERS-2 (1995) SAR instruments also provide a decade of historical data and have been particularly useful in the quantification of ice loss in Antarctica and its contribution to sea level rise (Rignot, 2002; Rignot *et al.*, 2008). A limitation of ERS-1 and 2 is that they only work over very flat areas and tend to lose track of the radar-echo over steeper areas around the continent's coast, limiting assessment of ice thickness in this crucial area. Over steep slopes, the altimetric systems are unable to maintain a lock on the returning signal as it becomes deflected away from the orbiting sensor. The result is lost data and poor across-track resolution.

SAR interferometry is the term for the process of extracting information from multiple SAR images of the same area. Despite the fact that the same target is imaged on different orbital passes, the positions of the satellite on each pass is never identical. As a result, in each case there is a difference in distance between the satellite and ground (the 'path'). Termed a phase shift, the path difference also means the phase of the signal in each image is inconsistent. SAR interferometry makes use of this information by subtracting the phase value in one image from that of the other for the same point on the ground. This is, in effect, generating interference between the two phase signals and is the basis of interferometry.

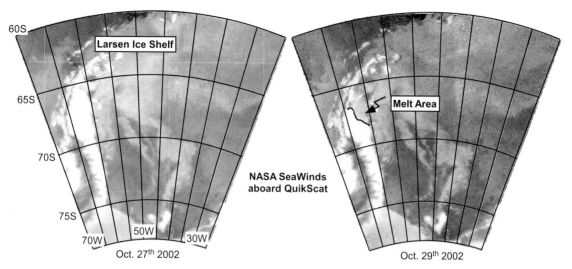

Figure 12.8 The SeaWinds instrument on NASA's Quick Scatterometer (QuikSCAT) spacecraft captured these near real-time backscatter images of melting on the Larsen C ice shelf in Antarctica's Weddell Sea between October 27 (left) and October 29 (right), 2002. The melting extended as far south as 68° south and resulted from a cyclone that delivered warm air to the region. The image on the right also shows a noticeable recession in the sea-ice margin to the west of the Antarctic Peninsula. Credit: NASA/JPL-Caltech.

In recent years satellite-based SAR interferometry has emerged as a technology that not only can provide continuous high-resolution topographical data in polar regions (critical for mass balance calculations) but is also well proven to be able to measure rates of ice flow (Fahnestock *et al.*, 1993; Goldstein *et al.*, 1993; Rignot *et al.*, 2000; Dall *et al.*, 2001; Kenyi & Kaufmann, 2003; Strozzi *et al.*, 2004). Interferometry will be considered later in this chapter in the context of monitoring permafrost, and is also discussed in Chapter 6 (Section 6.3) for quantifying urban subsidence.

Non-imaging radar

Sensors using non-imaging radar are also termed scatterometers. These instruments operate by measuring the amount of reflected energy, or backscatter, from the Earth's surface. Providing less spatial detail than SAR and a level roughly equivalent to that obtained from passive microwave systems, scatterometers provide daily data about sea-ice during day and night and through cloud cover.

The SeaWinds sensor aboard NASA's Quick Scatterometer (QuikSCAT) satellite provides daily global views of ocean winds and sea-ice and has been particularly powerful in documenting a drastic decline in the area of Arctic perennial sea-ice (Nghiem *et al.*, 2006; Kwok *et al.*, 2009) (Figure 12.8). The mission has been returning quality data since its launch in 1999. It should be noted that significant ground-processing is required to extract accurate wind vectors from this instrument.

Altimetry

Akin to an acoustic fathometer on a boat, platforms carrying an altimeter send a pulse of radar energy toward the Earth and measure the time it takes to return to the sensor. Much the same as considered in Chapter 10 in the context of measuring ocean surface topography, the distance between the target and the sensor is determined from the pulse's round trip duration and against a known reference (such as the ellipsoid). This information is used to measure the height of a sea-ice surface above sea level, from which can be derived the total thickness of the sea-ice.

Ice thickness controls the exchange of heat between ocean and atmosphere and is hence a crucial parameter for understanding polar thermal fluxes. Since the 1950s, submarine-mounted upward-looking sonar has provided data on ice draft, the submerged portion (roughly 89 per cent) of the ice sheet. But those data are largely anecdotal, limited to one-dimensional transects under certain regions of Arctic ice; some of the data even remain classified.

Early satellites with radar altimeters were not in orbits that adequately covered the poles and thus did not collect substantial sea-ice data. Nonetheless, a reasonable time-series of soundings does exist from 1978 through to the present of the Greenland and Antarctic Ice sheets (Herzfeld *et al.*, 1997). This deficit was intended to be met with the 2005 launch of CryoSat, a European Space Agency (ESA) satellite specifically designed for monitoring sea-ice. Regrettably, the launch of this important instrument was not successful, but the follow-up mission, CryoSat-2, launched without trouble in April 2010. With the same mission objectives as the failed CryoSat, the replacement instrument monitors the thickness of land-ice and sea-ice with an emphasis on investigating the connection between polar ice-melt and sea level rise. These data are particularly valuable, given that the comparable NASA ICESat mission lacks the capability to monitor sea-ice thickness.

With the present lack of a dedicated satellite mission for polar altimetry, the use of unmanned aerial vehicles (UAVs) are being used to make high-resolution maps of ice surface topography. Two instruments of note are the L-band wavelength Uninhabited Aerial Vehicle Synthetic Aperture Radar (UAVSAR), and the proof-of-concept Ka-band wavelength radar called the Glacier and Land Ice Surface Topography Interferometer (GLISTIN). The latter instrument carries two receiving antennas, separated by about 25 centimetres. This gives it stereoscopic vision and the ability to simultaneously generate both imagery and topographical maps. The topographical maps are accurate to within 10 cm of elevation on scales comparable to the ground footprint of ICESat, a satellite mission covered in the next section.

12.6.4   Active optical remote sensing – ICESat

ICESat (Ice, Cloud, and land Elevation Satellite) is the benchmark Earth Observing System mission for measuring ice sheet mass balance, cloud and aerosol heights, as well as land topography and vegetation characteristics.

Figure 12.9 Preparation of Europe's CryoSat-2 spacecraft shown nearing completion in 2009, ahead of what was to be a successful launch on 8 April, 2010. The precursor to this mission, CryoSat-1, was destroyed on launch in 2005. Its mission to map the world's ice caps was considered so important that space ministers ordered a re-build.

Launched from California in 2003, the mission is designed to detect changes in ice sheet surface elevation as small as 1.5 cm per year over areas of 10,000 km^2 (Schutz *et al.*, 2005). Such precision is provided by a space Light Detection and Ranging (LiDAR) instrument termed GLAS (Geoscience Laser Altimeter System), which combines a surface LiDAR with a dual wavelength cloud and aerosol LiDAR. As discussed elsewhere in this book (laser ranging technology mounted aboard aircraft is detailed in Chapters 3, 6, and 8), LiDAR operates in a similar way to a microwave altimeter by relating pulse travel time to topography, but it uses energy in the visible or infrared portion of the electromagnetic spectrum. GLAS consists of three lasers that emit infrared and visible laser pulses at 1,064 nm and 532 nm, 40 times a second. This operation produces a series of approximately 70 m diameter spots that are separated by nearly 170 m along track.

ICESat is a platform deemed capable to rapidly address the present lack of a baseline with which to quantify future changes in the polar regions. It is predicted that the rate of greenhouse warming will increase. However, hindered by a lack of understanding of the natural variability of the system, we are not yet in a position to make a balanced assessment of changes in ice dynamics. It is imperative that such a baseline is installed rapidly.

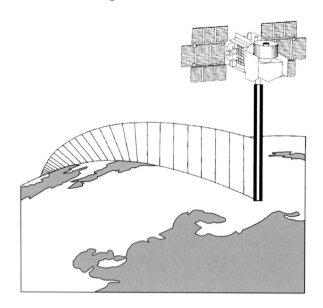

Figure 12.10 ICEsat is a platform carrying the GLAS LiDAR instrument. This laser altimeter measures the time required for a laser pulse of 5 nanoseconds duration to complete the round trip from the instrument to the Earth's surface and back to the instrument. This time interval can be converted into a distance by multiplying by the speed of light and the one-way distance can be obtained as half the round trip distance. The mission measures ice-sheet topography and associated temporal changes as well as cloud and atmospheric properties.

Of particular urgency is ascertaining whether the Greenland and Antarctic ice sheets are growing or shrinking. Present uncertainty is ± 30 per cent of mass input, which equates to ± 5 cm per year average ice thickness. This translates to an uncertainty of ± 2.3 mm per year in sea level change. ICESat is capable of measuring <1 cm per year changes in ice thickness over vast areas (this equates to <5 per cent of mass input and <0.4 mm/yr global sea level change).

Work published by Kwok *et al.* (2009) is likely the most accurate and comprehensive set of maps yet of the entire Arctic basin, based on ten LiDAR surveys taken between 2003 and 2008. The same authors unequivocally show Arctic ice sheets to be both shrinking and thinning. Proof of this capability is vital, as thinning of sea-ice is a precursor to collapse (Scambos *et al.*, 2004), and this will be covered in the next section in the context of the Larsen and Wilkins shelves. Similar mechanics were witnessed in the collapse of portions of the Arctic's Ayles and Ward Hunt shelves in 2005 and 2008, respectively. Elevation changes are also associated with perturbations in ice-stream flow and grounding.

As sea-ice ages, brines drain out of it. Older, multiyear ice, which has not melted for several seasons, is hence less salty than fresh, younger ice. This loss of salinity imparts a greater concentration of air bubbles to the old ice, so consequently it becomes up to one-third more reflective to solar energy than newly frozen ice. By processing ICESat soundings in unison with QuikSCAT microwave scatterometer data, Kwok *et al.* (2009) demonstrated that it is possible to distinguish ice thickness data into two classes; multiyear (old) ice, and first-year (fresh) ice. The results are highly relevant as, because older ice is thicker than

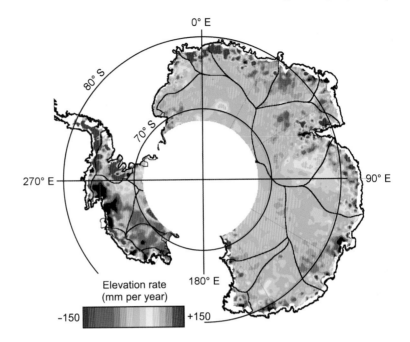

Figure 12.11 Rate of elevation change of the Antarctic ice sheet, 1992–2003, derived from the European altimeters ERS-1 and ERS-2. Modified with permission from Wingham *et al.* (2006). *Mass balance of the Antarctic ice sheet. Philosophical Transactions of the Royal Society A* [**364**], 1627–1635.

younger ice, it acts as an insulating buffer at the end of each melt season. 2008 was the first year on record that the areal coverage of multiyear ice was exceeded by fresh ice.

By virtue of its lower albedo and thinness, the higher the coverage by fresh ice at the start of the melt season, the greater the quantity of open water that will be present by the summer. As previously discussed, open water has a much lower albedo than any sort of ice, and its presence leads to excessive warming of both the atmosphere and ocean.

In addition to topographical measurement, the GLAS LiDAR directly measures cloud heights for energy balance calculations and to obtain unique information on polar clouds, especially during the polar winter.

12.7   Ice sheet mass balance

The melting of land-ice has great implication for projected sea level rise. The mass balances of the two large ice sheets of the Antarctic and Greenland is the difference between the mass gained by snow accumulation and that lost by ice ablation and calving. If the accumulation rate is greater than the ablation rate, then the mass balance will be positive and the ice sheet will grow. If the mass balance is negative, the ice sheets will shrink, and this is certainly the case at the present time in Greenland (Luthcke *et al.*, 2006; Thomas *et al.*, 2009).

The status of Antarctica is less clear and in a state of flux. Though far from being fully understood, it is accepted that the mass balance is likely close to zero (Giovinetto & Zwally, 2000). The determination has proved difficult for glaciologists, since a warming climate does more than simply melt ice. Although it seems counterintuitive, it is now apparent that ice sheets may thicken when warmed (Johannessen *et al.*, 2005). Increasing temperatures will melt ice at the periphery of ice sheets, a process clearly active in Greenland (Abdalati & Steffen, 2001); conversely, however, as the atmosphere warms, it can carry more moisture, leading to increased precipitation and more snow falling – which in turn leads to thickening the ice caps. In a warmer world, storm severity and frequency may increase (Emanuel, 1987; Idso *et al.*, 1990; Knuston *et al.*, 1998; Emanuel, 2005), also serving to add more snow to the poles.

Evidence for this has been provided by satellite surveys conducted using the European Space Agency's radar altimeters ERS-1 and ERS-2. The work showed that, between 1992 and 2003, the East Antarctic gained about 45 billion tonnes of ice – enough to reduce the oceans' rise by 0.12 mm per year (Davis *et al.*, 2005). However, it should be noted that the study showed the smaller West Antarctic shelf thinning by 0.9 centimetres per year, so any thickening of the eastern ice sheet should not be seen as a long-term protection against a rise in sea level. The latter point is strengthened by the fact that Antarctic temperatures now routinely rise above freezing in the summer months, causing a release of armadas of icebergs from the shelf (Streten, 1973; Doake *et al.*, 1998; MacAyeal *et al.*, 1998; Rack & Rott, 2004). Recent work indicates that the fragile marine ecosystems of Antarctica face a much higher level of disturbance as increasing quantities of free-floating ice grind the shallow seabed (Smale *et al.*, 2008).

CASE STUDY: Disintegration of the Larsen and Wilkins ice shelves

In the last decade, there have been two spectacular events that indicate warming temperatures on the Antarctic Peninsula: the catastrophic disintegration of the Larsen-B ice shelf in early 2002 and the collapse of the Wilkins ice shelf in 2008. These collapses are worrisome as there is compelling evidence that both shelves have been intact for at least the past 12,000 years (Domack *et al.*, 2002).

Both events were captured in detail by MODIS aboard NASA's Terra and Aqua satellites, allowing an appraisal of the mechanisms leading to collapse. The time-series presented in Figures 12.12 and 12.13 tell the same story. The first evidence for the breakup is the appearance of melt-water ponds that appear blue in the MODIS scenes (intact ice is bright white). A series of sufficiently warm summers can change permeable snow into impermeable ice. This change in the ice shelf surface allows melt ponds to form during subsequent warm summers, and these ponds may find and fill pre-existing surface cracks. Since water is denser than ice, it deepens pre-existing cracks, eventually slicing completely through the shelf (Doake *et al.*, 1998; Rack & Rott, 2004).

In both cases the collapse of these shelves was extremely rapid (a matter of weeks) and led to a considerable increase in the rate that their inland glaciers started to empty into the ocean (De Angelis & Skvarca, 2003; Scambos *et al.*, 2004). For Larsen B, 2,300 km² of the ice shelf fragmented to icebergs within one week (Rack & Rott, 2004), with 1,800 km² occurring in a similar timeframe for Wilkins during 2008. The collapses both occurred at the end of exceptionally warm summers.

Both the Larsen and Wilkins Shelves consisted of floating ice, sod their disintegration did not contribute significantly to rising sea level. The glaciers

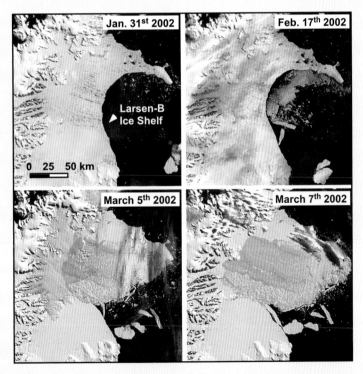

Figure 12.12 A time-series of Terra MODIS scenes shows the disintegration of the Larsen B Ice Shelf over February and March, 2002 (location shown on Figure 12.7). Credit: National Snow and Ice Data Centre, University of Colorado.

that they were holding back, however, were land-based, and the subsequent acceleration of deposition on their seaward flank is a force for perceptible sea level rise.

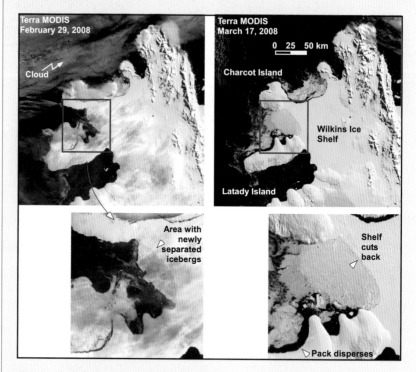

Figure 12.13 The beginning of the collapse of the Wilkins Ice Shelf on the Antarctic Peninsula (location shown on Figure 12.7), as imaged by Terra MODIS. In February 2008, a large bluish area of iceberg formation is clearly visible against the white intact ice shelf. By March 17, the shelf has cut back more than 25 km into the ice bridge connecting Charcot Island to the Wilkins Ice Shelf. This bridge effectively formed a barrier pinning back the northern ice front of the central Wilkins Ice Shelf. By April 2009, the bridge was to collapse entirely, leading to the uncontrolled disintegration of the Wilkinson shelf. Credit: NASA.

12.8 Remote sensing permafrost

Permafrost – ground that has remained frozen for two or more years – forms in any climate where the annual mean temperature is below the freezing point of water. The largest areas are in the polar latitudes, though high-altitude alpine lands may also be sufficiently chilled to remain permanently frozen. Permafrost is

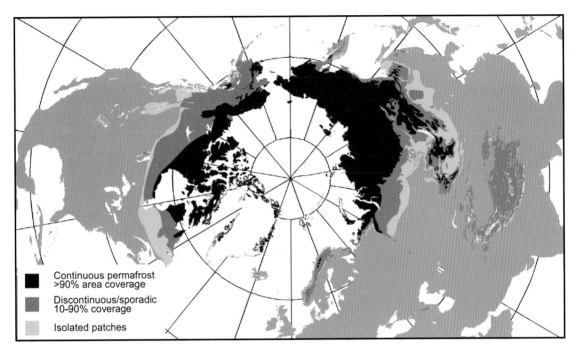

Figure 12.14 Permafrost extent in the northern hemisphere (June 2007). In UNEP/GRID-Arendal Maps and Graphics Library assembled by Hugo Ahlenius. Retrieved August 15, 2009 from http://maps.grida.no/go/graphic/permafrost-extent-in-the-northern-hemisphere (Brown *et al.*, 1997).

relevant for discussions of global environmental change, as it is a vast store of carbon on geological timescales. These frozen ecosystems contain almost twice as much carbon as is present in the atmosphere and cover 24 per cent of Earth's land surface (Brown *et al.*, 1997) (Figure 12.14). If released, this vast reservoir has the potential to create a positive feedback to climate change, exacerbating global warming (Zimov *et al.*, 2006; Schuur *et al.*, 2009).

As atmospheric CO_2 accumulates and the Arctic region warms, the extent of permafrost may retreat and ultimately disappear. In areas of thicker and colder permafrost, there is a high probability that warming will result in an increase in permafrost temperature, a decrease in permafrost thickness and a thickening of the upper active layer that seasonally cycles through freeze and thaw conditions (Smith & Riseborough, 1996; Smith & Burgess, 1999).

To appraise the risk posed by carbon-release from thawing permafrost, it is necessary to have an audit of its global prevalence. Remote sensing typically falls into two categories for monitoring these environments: visible and infrared imagery with high spatial resolution for monitoring small areas; and coarse-resolution active/passive microwave sensing at regional- to global- scale.

Visible-spectrum imagery from high-resolution satellite sensors such as IKONOS and QuickBird, as well as declassified images generated by the CORONA spy satellite, are used in conjunction with older aerial photographs to identify changes that have occurred in permafrost terrain in recent decades (Strozzi *et al.*, 2004; Duguay *et al.*, 2005). Such time-series are, however, challenging to assemble because of persistent cloud cover in the northern latitudes, seasonal darkness and radical shifts in sun angle through the seasons. Time-series data collected in the passive microwave spectrum, by contrast, are not confounded by cloud, darkness or solar geometry. As detailed in Chapter 7, this region of the electromagnetic spectrum can be used to detect soil moisture (Gloersen and Barath, 1977; Van de Griend & Owe, 1994; Njoku & Entekhabi, 1996; Narayan & Lakshmi, 2006; Gruhier *et al.*, 2008). An increase in wetness in a time-separated image series can be attributed to thawing of permafrost (Zhang *et al.*, 2004; Zhang & Armstrong, 2005).

As well as simple ice presence/absence, passive microwave radiometry can be used to determine land temperature on the basis of microwave brightness – a measure of the radiation emitted by an object (McFarland *et al.*, 1990; Aires *et al.*, 2001; Fily *et al.*, 2003). Relevant sensors in this regard have been covered earlier in this chapter (Section 12.6.2). Though their spatial resolution is coarse (tens of kilometres), these passive instruments are weather-independent, have a short revisit time and offer 20-year time-series over vast scales.

These virtues have seen observations from passive microwave satellites form the backbone of climate change studies in the northern latitudes (Chang *et al.*, 1987; Grody & Basist, 1996; Cavalieri *et al.*, 1997; Zhang *et al.*, 1999). Of particular note are the Special Sensor Microwave Imager (SSM/I) and the Advanced Microwave Scanning Radiometer-Earth Observing System (AMSR-E), which detect surface soil freeze or thaw based on microwave brightness. The strategy is effective, since there is a large brightness contrast between water and ice, and therefore also between freezing and thawing ground conditions, and this forms the basis for image classification (England, 1990; England *et al.*, 1992).

Furthermore, since the observed microwave energies originate from slightly beneath the ground, they return information that extends a little way into the subsurface. Long wavelength microwave radiation emanates from a greater soil depth than short, and hence portrays the soil status from deeper below the surface. The penetration of microwaves into the ground, however, does not exceed a few tenths of a wavelength. With useable microwave wavelengths ranging from ≈5 – 30 cm, the depth of penetration is a matter of centimetres only, though permafrost may persist to depths of tens to hundred of metres. Passive microwaves, therefore, only provide a very limited glimpse into the ice dynamics below the surface, although this is still further than that delivered by visible or infrared energy.

Active remote sensing using radar provides its own illumination and, like passive microwave systems, sensors allow observations day and night throughout the year, regardless of cloud cover and solar illumination effects. The ability of radar to observe freezing and thawing of a landscape using backscatter has its origin in the distinct changes of surface dielectric properties that occur as water transitions between solid and liquid phases (Kraszewski, 1996).

Using NSCAT, a NASA microwave radar scatterometer that sat aboard the 1996-launched Advanced Earth Observing Satellite (ADEOS), Kimball *et al.* (2004)

assessed temporal change in the 1997 seasonal thaw between January and June across one million km² of central Canada. A failure in the ADEOS's solar array at the end of June that year forced the termination of NSCAT.

Such work is pertinent as boreal ecosystems, such as this one in Canada, cycle between winter-dormant and summer-active states. With the arrival of spring, the upper few metres of the ground thaws, liquid water becomes available, the growth of green vegetation accelerates and soil decomposition initiates, imparting a net flux of CO_2 to the atmosphere (Black *et al.*, 2000). A thorough understanding of the spatial and temporal aspects of this thaw is valuable as it has important implications for hydrological, energy exchange, and biogeochemical cycling (Kimball *et al.*, 2000; 2004). The NSCAT data were proven capable of detecting frozen versus non-frozen ground, but were unable to deliver information on the degree of wetness or soil moisture content.

Synthetic Aperture Radar, another active remote sensing technique and one that is discussed in almost every chapter of this book, also has application for monitoring permafrost. Changes in topography often accompany the thawing of frozen ground. This may be rather subtle and hard to detect in the case of subsidence, or pronounced and obvious for slope failures and landslides. Both can be detected by satellite using Interferometric SAR (InSAR) (Kenyi & Kaufmann, 2003; Strozzi *et al.*, 2004). This technology is covered in Chapter 6 in the context of quantifying subsidence in urban centres, but the application is essentially the same for permafrost. InSAR capitalizes on the phase difference between two time-separated SAR images from the same scene. The technique generates a two-dimensional displacement map, the interferogram, which is capable of measuring sub-centimetre vertical changes in ground deformation. The technique does not work well in areas that are highly vegetated, but the frozen boreal regions, with their lack of liquid water, are relatively hostile to plant growth. Snow cover is a more common factor that prevents construction of the interferogram. As covered in Chapter 6, paired observations from the European ERS-1 and ERS-2 satellites, which carry identical radar instruments (wavelength = 5.3 cm), are used for regional-scale permafrost monitoring.

A disadvantage of InSAR is that, in order to detect the subtle subsidence that accompanies melt, a high-resolution DEM is required for processing the radar data (Alasset *et al.*, 2008). InSAR is further limited by atmospheric effects. Changing tropospheric water vapour content between the paired acquisitions serves to decorrelate and destroy the interferogram. This is a particular problem in the polar regions, where atmospheric moisture loading is seasonally variable.

From-orbit measurements of permafrost status are, however, not without inaccuracy, and must be ground-truthed. Since the depth to which the ground is frozen is relevant to both the age of the permafrost and its carbon storage capacity, verification is also conducted by remote, though field-based, methods such as ground-penetrating radar and coring (Brandt *et al.*, 2007). These assessments may have to be repeated on a daily basis to capture adequately the dynamics of the frozen ground.

12.9   Key concepts

1 Receiving a fraction of the solar energy compared to the rest of the Earth's surface, the poles are radically colder. Through geologic history, the quantity of water ice stored in the polar regions has waxed and waned. For at least the last 140,000 years, ice has been abundant at high latitudes. In the last several decades, however, strong evidence has emerged that the ice is starting to diminish at a rate that is likely faster than can be explained by natural processes. The presence of polar ice is fundamental to the operation of the ocean-atmosphere system and changes in its cover will alter both our climate and sea level.

2 Being vast, remote and inhospitable, the polar ice caps are especially difficult to monitor from the ground. Satellite remote sensing technologies also face numerous hurdles. First, the inclination of a sun-synchronous, orbit favoured by many spacecraft, does not pass directly over the poles. In addition, these high-latitude areas are shrouded in clouds for a major part of the year and lie in perpetual darkness from the first day of autumn until the beginning of spring.

3 Unhindered by cloudiness and seasonal darkness, both passive and active microwave instruments have particular utility for polar observation. The microwave portion of the spectrum covers wavelengths of approximately 1 cm to 1 m. Passive sensors are used to detect the small quantity of microwave energy naturally emitted by the Earth. These instruments are notable in that they have been launched since the early 1970s and hence provided our earliest assessments of sea-ice extent. Key passive instruments are the ESMR (1972–1976), the SSMR (1978–1987), the SSM/I (1987 until present) and the AMSR-E (2002 until present). In contrast, active microwave sensors emit a signal toward the Earth and measure the reflection off the surface. The three families of active instruments are imaging radar, scatterometers and altimeters.

4 When conditions allow, passive optical sensors have also provided viable time-series of polar imagery. Examples include data from MODIS, AVHRR and Landsat TM. Of particular relevance is the true-colour mosaic of the entire Antarctic continent (LIMA) that was released in 2007. This composite consists of more than 1,000 images acquired between 1999 and 2001 with a 15 m resolution. The data are freely available for download from the LIMA website.

5 Launched in 2003, ICESat is presently the leading mission capable of measuring subtle changes in ice sheet elevation. It can detect changes as small as 1.5 cm per year over areas exceeding 10,000 km^2. ICESat is conspicuous in being the only spaceborne LiDAR instrument, a technology that is more commonly deployed on aircraft. The platform also carries a sensor to measure the properties of both clouds and atmospheric aerosols.

6 Vast areas of the Earth's land regions are covered by permafrost, a term that describes land that remains continuously frozen for two or more years. These areas are vast reservoirs of carbon and thus, if thawed, have the potential to release CO_2 to the atmosphere and accentuate global warming. Aspects of permafrost dynamics can be observed in visible and infrared data, but passive and active microwave radiometry is particularly valuable because it can operate in the polar latitudes, unhindered by seasonal darkness and cloud cover.

13

Effective communication of global change information using remote sensing

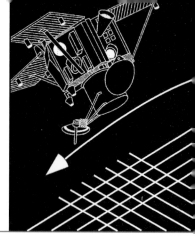

The scientific investigation of global environmental change has a chequered past. Today we witness an often furious, highly politicized public debate as to what the scale and seriousness of global change really is. The boundary between fact and supposition is broad and fiercely contested between so-called 'climate sceptics' and those who believe that human modification of the climate is indisputably linked to extensive consumption of fossil fuels. Meanwhile, present media coverage suggests that attacks on global warming science may be broadening into a generalized anti-science movement.

As eloquently pointed out by Chuvieco (2008), other critical components of global change, such as tropical deforestation, biodiversity loss, water pollution and soil erosion are often less reported on, perhaps because they have less direct impact on the developed economies. However, they are the most evident signals of global change, and are – unlike global warming – clearly beyond scientific dispute as to whether or not they are human-caused.

The media conflation of these factors with the issue of global warming makes it even harder to communicate global change information to policymakers and the public. Communicate we must, though, and the channels must be clear and mature. There is a paramount need for positive interaction between researchers, policymakers and journalists in order to improve communication about science and climate change. Information must be allowed to flow freely, transcending the disciplinary and institutional borders that traditionally have partitioned the Earth sciences. This requires new mechanisms for interaction, such as forums for dialogue among them.

Links to the media are particularly necessary because, as demonstrated by Hwang & Southwell (2009), television news exposure, especially for stories framed as relevant to the everyday lives of individual audience members, radically affect people's beliefs about science. In addition, such exposure strongly crystallizes the beliefs of multiple individuals, even if they have not seen a particular news piece, through interpersonal conversation with those who have seen it.

Remote Sensing and Global Environmental Change, First Edition. Samuel Purkis and Victor Klemas.
© 2011 Samuel Purkis and Victor Klemas. Published 2011 by Blackwell Publishing Ltd.

Recent studies in the social sciences reveal that younger generations tend to assign responsibility for the environment to governments and consumers rather than accepting personal responsibility (Wray-Lake *et al.*, 2010). With governments strapped for funds in time of recession, this demographic leaves the very real possibility that we will be remembered as the generation that saved the banks but squandered the biosphere. It is undeniable that, despite credible forecasts and warnings from the scientific community about climate change for a quarter of century, greenhouse gas emissions have continued to grow, signals of human-induced climate change have clearly emerged, and a preponderance of scientists studying the issue project more adverse consequences to come unless stronger actions are taken.

From the policy-making level down to personal voting and purchasing decisions, public action has simply not been commensurate with the threat as characterized by mainstream science. This disjoint in public reaction in the face of environmental change is set to become more pronounced, considering that the younger generations, as exposed for example by the work of Wray-Lake *et al.* (2010), seem to be even less motivated than their elders to confront the climate change challenge. This signals the urgent need for a renewed focus on young people's views through environmental education.

Several international programmes have been initiated to fill the void that separates the science of global change from the public. The Center for Climate Change Communication (4C) at George Mason University, Virginia, is one such programme. It endeavours to conduct public engagement and behaviour change research that can be used to improve climate change communication and social marketing programmes. More information about 4C can be found through their website (http://climatechangecommunication.org). A similar programme is the Yale F&ES Project in Climate Change (http://environment.yale.edu/climate), which aims to close the gap between science and action.

In the United States, federal organizations such as NOAA, NASA and the National Science Foundation (NSF) all have programmes dedicated to the promotion of education about global change with a focus towards the younger generation. The NOAA education programme (www.education.noaa.gov) offers resources for teachers to promote scientific literacy, including coverage of the importance of the use of satellite observation for understanding and predicting Earth's climate.

13.1 Global environmental change as an interdisciplinary issue

A characteristic of the Earth sciences is that Earth scientists are unable to perform controlled experiments on the planet as a whole and then observe the results. This is an important consideration, because it is precisely such whole-Earth, system-scale experiments, incorporating the full complexity of interacting processes and feedbacks, that might ideally be required to fully verify or falsify climate change hypotheses (Schellnhuber *et al.*, 2004). Nevertheless, countless empirical tests of numerous different hypotheses, carried out across all the disciplines of Earth system science, have built up a massive body of interdisciplinary knowledge. This repeated testing has refined the understanding of numerous aspects of the climate system, from deep oceanic circulation to stratospheric chemistry (Le Treut *et al.*, 2007). Sometimes a combination of observations and models can be used to test planetary-scale hypotheses. For example, the global cooling and drying of the atmosphere observed after the 1991 eruption of Mount Pinatubo in the Philippines provided key tests of particular aspects of global climate models (Hansen *et al.*, 1992).

Global environmental change and the conservation of biodiversity are complex problems which often link local and global issues. Few would disagree that the critical needs and challenges of environmental management require several disciplines working in concert (Robinson, 2006; Morse *et al.*, 2007; Purkis *et al.*, 2008; Margles *et al.*, 2010). Adopting an interdisciplinary approach yields numerous benefits, including the promise of an outcome that resonates better with a wider range of stakeholders (Meine, 1998) and the ability to achieve more insight than from any single discipline (Yaffee, 1998; Purkis *et al.*, 2010), as well as the potential to address intractable and complex problems (Roberts & Bradley, 1991). The insights provided by scientists from different disciplinary backgrounds, as well as policymakers and the public, are hence key to helping us move forward in developing interdisciplinary exchanges on complex global change issues.

The smooth passage of information pertaining to environmental change that has been acquired via remote sensing is hindered by the fact that it must travel across many scientific disciplines. The processing of the raw satellite imagery is typically carried out by a remote sensing scientist with a strong numerical background, using high-end software. The processed imagery will probably then be communicated to a scientist with a broader academic foundation, most likely a specialist in some particular aspect of global change, trained as a biologist, geologist, oceanographer, chemist, climate scientist and so forth. This second-tier individual may not hold a full understanding of the steps that have been conducted by the first tier to convert the raw imagery to a more user-friendly format. Importantly, the second-tier scientist may be unaware of assumptions that may have been made during processing and the artefacts that therefore may lurk within the processed data. With this workflow completed, the findings of the study will be published in a peer-reviewed journal or presented at a symposium. If it is perceived to be of wide interest, at this point a journalist may pick up on the work and distil it further into a popular article that can be readily digested by the general public.

Through each step of this (minimum) three-tier hierarchy, critical information on the accuracy and precision of the underlying data will likely be lost. The consequence is the induction into the public domain of global change 'facts' which, through no fault of any of the scientists involved, do not stand up to scrutiny.

Even without media misrepresentation of facts, an inevitable Catch-22 situation develops that will pitch scientists against the public. In trying to understand the climate system more fully through Earth observation, scientists will reveal greater uncertainty about the range of possible climate outcomes. At the same time, policymakers and the public will demand greater certainty so that they can plan accordingly. The only antidote to this inevitable confrontation is effective communication of the limitations of Earth observation data and enhanced collaboration between scientific disciplines and, through the media, the public. We cannot call into question the premise of global warming simply because we have a cold winter, as was the case in 2009–10.

Perhaps the most structured and interdisciplinary organization dealing with global environmental change is the Intergovernmental Panel on Climate Change (IPCC). This organization was jointly established in 1988 by the United Nations Environment Programme and the World Meteorological Organization to conduct periodic assessments of the state of knowledge concerning global climate change. The assessment reports prepared by the IPCC provide a comprehensive statement of the state of knowledge concerning topics such as scientific information, environmental impacts, response strategies and other issues concerning climate change. The IPCC is structured into three working groups to study various aspects of climate change:

- Working Group I: Climate system and climate change;
- Working Group II: Climate change: impacts, adaptation, and vulnerability;
- Working Group III: Mitigation of climate change.

While Working Group I is primarily focused on the physical mechanisms of climate change, the second and third groups also consider social and economic dimensions. Each of the three is charged with issuing periodic assessments.

To communicate its findings effectively, the IPCC releases, at six-year intervals, a brief report entitled 'Summary for Policymakers' or more simply the 'SPM'. The last was published in 2007 and the next is due out in 2013. The SPM follows the production of a much more comprehensive full 'assessment report' by each of the aforementioned working groups. It differs from the full report in that, although the content is determined by scientists, the form is approved line by line by governments, including non-scientist civil servants.

The purpose of the SPM is to communicate the panel's findings clearly and their message constructively. The adoption of such a two-layered approach is wise, making the findings more accessible to governments and policymakers, who otherwise would have been ill-equipped to distil them from an in-depth and science-loaded full report. The IPCC model is easily transferable, and such an interdisciplinary approach should be encouraged.

13.2 Effective communication through accessibility of data

Owing to its global scope, much of global change science relies on the interpretation of data acquired from remote sensing instruments, which deliver one of the most systematic ways of collecting worldwide data in a fully comparable and repeatable way. As we have repeatedly discussed in this book, however, such data can be inaccurate, suffering from noise and bias that must be accounted for if valid conclusions are to be distilled from Earth observation sensors. It is for this reason that the success of a satellite mission depends as much on systematic long-term calibration of its sensors as placing the instrument in orbit. Indeed, several of the early Landsat satellites ended their careers through the radiometric degradation of a sensor, as opposed to an outright mechanical failure.

While fraught with calibration challenges, the products of satellite remote sensing are typically visually engaging, colourful and interesting, offering themselves easily to effective communication. The beauty of Earth observation data must not be confused with reliability and the public, as for scientists, must be helped to understand the message that the data are carrying. Fortunately, several programmes exist to aid in this task.

The first step of effective communication using remote sensing is the provision of public access to the data. Section 3.6 of this book, entitled 'Existing image archives' introduces the reader to a number of such initiatives. While high-end GIS and image manipulation packages are typically required to manipulate and process Earth observation data, there now exist several no-cost software systems that are available to the public (Table 13.1). The most accessible of these, because of their simplicity of use, are *Google Earth*, *GeoPDF* and *ArcGIS Explorer*. These offer an effective means of communication from scientists to the public to promote training and the transfer of knowledge.

TerraGo Technologies (www.terragotech.com) developed the free downloadable software tool *GeoPDF*, which preserves the geospatial framework and coordinates of GIS layers in Adobe's free *Acrobat Reader* software. Those interested in generating GeoPDFs are required to invest in conversion software wherein GIS layers (raster images and maps along with shapefiles) are exported using the 'ArcMap-to-GeoPDF' tool.

A GeoPDF that is 'GeoMark-enabled' allows an end-user to draw points, lines and polygons while viewing the GIS layers in free *Adobe Reader*. When the drawing is complete, the end-user can export these vectors as shapefiles that can be loaded back into the GIS. The knowledge transfer potential of GeoPDFs is unlimited; for example, students could receive a stack of images and shapefiles as a GeoMark-enabled GeoPDF, view the stack in *Reader*, interpret features of interest and send the interpretation back to the instructor as a shapefile.

Google Earth is a remarkable image and map visualization tool which is covered in detail in Section 3.6 of this book. So powerful is this free software that it is possible to look at high-resolution images of the entire planet through *Google Earth*, to admire the shapes of mountains and valleys and to 'walk' down the streets of unfamiliar cities.

ESRI's *ArcTools*, as well as a host of other geospatial software, can export GIS layers as .kml and .kmz files. For users with the basic free *Google Earth*, receiving GIS layers as .kml and .kmz files is a necessity. However, for users with *Google Earth*'s professional license (available for US$ 400/yr), ESRI shapefiles and GeoTIFFs can be imported directly into the software. GIS layers can be stacked in

Table 13.1 No-cost software for visualizing remote sensing and GIS data.

Software	Developed by	Further info.
GeoPDF	TerraGo Technologies	www.terragotech.com
Google Earth	Keyhole, Inc. & Google	www.google.com/earth
ArcGIS Explorer	ESRI	www.esri.com/software/arcgis/explorer
DIVA-GIS	R. Hijmans	http://diva-gis.org
GMT	P. Wessel & W. Smith	http://gmt.soest.hawaii.edu
GRASS	Open Source Geospatial Foundation	http://grass.fbk.eu
SPRING	Brazil's National Institute for Space Research	www.dpi.inpe.br/spring/english
Bilko	UNESCO	www.noc.soton.ac.uk/bilko

Google Earth and saved as a new .kml or .kmz. Depending on size, the .kmz can be emailed as an attachment, transferred over the Internet via ftp or burned onto a CD or DVD. The .kmz is secure in that only people with access to the file can see it on *Google Earth*; it remains on the local computer while being displayed on the imagery and maps that are streaming over the Internet to the computer.

ESRI created the free downloadable *ArcExplorer*, a package with different capabilities compared with *Google Earth*. Both visualization tools stream a 3-D model of the globe that has imagery and a terrestrial DEM over the Internet. Shapefiles, GeoTIFFs and .kml/.kmz files can be loaded into *ArcExplorer*. When a stack of GIS layers is built, the user can save the data package with a project .nmf file.

ArcExplorer was superseded in 2010 by *ArcGIS Explorer*. The new product continues to have a downloadable software package that can be used for authoring, but there is also a version hosted on ESRI's web server, *ArcGIS Explorer Online*, that can be used by anyone to view GIS maps without having to download any software. GIS layers and mapping projects that are created for *ArcExplorer* or *ArcGIS Explorer* using desktop GIS software can be uploaded to an Internet GIS Service so that anyone with access to the Internet can view the maps. Many tools are made available to the Internet-based user to interrogate and interact with the GIS maps.

With data access facilitated, the next step is to encourage the public to understand, interact, and use the available resource. To this end, Congress put before the US government in 2005 a bill 'to encourage the development and integrated use by the public and private sectors of remote sensing and other geospatial information'. As of 2010, however, opposition still exists for greater public access to remote sensing data that are collected by the US government through satellite and aerial reconnaissance. One side of the argument pitches these data as a public resource that should be more fully and openly exploited in the public interest. The other side calls for restricted public access in the name of combating terrorism and national security.

Certainly a middle ground exists in that, as discussed with relation to *Google Earth* in Chapter 3, a large amount of information derived from remote sensing can be communicated without the public release of the raw data from which this information was derived. Through the availability of software such as *Google Earth*, *ArcExplorer* and *GeoPDF*, at no cost, it is safe to assume that the public's appetite for geospatial information has been whetted, and that barriers to public-access of remote sensing products will be overcome by the private sector, even if state and local governments are not forthcoming. The release of *Google Earth* and *ArcExplorer* stand in testament to this fact. If governments are brought on board through pressure from the public and scientists alike, then access to geospatial information will be all the more rapid.

14 Looking ahead: future developments

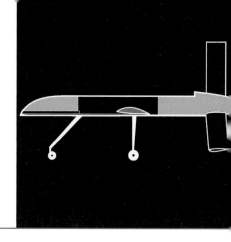

14.1 Emerging technologies

14.1.1 Fusion in remote sensing

The power of satellite constellations arises from the ability to fuse data from multiple platforms to yield information on a common theme. Examples include the simultaneous consideration of optical and radar data, or hyperspectral imagery from the coastal zone with LiDAR bathymetry. Similarly important is the capability to fuse data acquired in the same portion of the EM spectrum but with dissimilar spatial or temporal resolutions. Most common in this realm is the fusing of multispectral coarse-resolution imagery with a more highly resolved panchromatic image. The goal is to obtain a high-resolution multispectral image which combines the spectral characteristic of the low-resolution data with the spatial resolution of the panchromatic image.

Fused products are routinely derived for Landsat ETM+, SPOT, QuickBird, IKONOS and WorldView-2. In the latter two cases, this delivers sub-metre spatial resolution while retaining four or more spectral bands. Future advances will enhance this technique through more rigorous preservation of the spectral information within each pixel for tasks such as classification and un-mixing.

The routine integration of complementary datasets from multiple sensors remains hampered by many factors, but most notably:

- the lack of robust algorithms to automate fully the registration of data captured with several sensors at dissimilar resolutions and using different acquisition modes;
- problems with reconciling information attained in the optical spectrum versus radar – this is pertinent, considering radar's all-weather imaging

Remote Sensing and Global Environmental Change, First Edition. Samuel Purkis and Victor Klemas.
© 2011 Samuel Purkis and Victor Klemas. Published 2011 by Blackwell Publishing Ltd.

capability and the importance to monitor areas such as the poles and tropical forests that are frequently shrouded in clouds;
- the science of object recognition, classification and change detection from data fusion products is still in its infancy.

A realm in which considerable progress has been made is the fusion of hyperspectral airborne imagery and LiDAR for shallow water applications (Tuell *et al.*, 2004; Kopilevich *et al.*, 2005; Tuell *et al.*, 2005). The merging of data here is perhaps less challenging than when considering imagery from multiple satellites, as both the LiDAR and spectral sensor can be mounted in close proximity to one another on the same aircraft, so that the data are acquired in concert and are coincident in time and space.

The market leader of this technology is the CHARTS system (Compact Hydrographic Airborne Rapid Total Survey), which consists of an Optech, Inc. SHOALS-3000 LiDAR instrument integrated with an Itres CASI-1500 hyperspectral imager. CHARTS collects either 20 kHz topographical LiDAR data or 3 kHz bathymetric LiDAR, each concurrent with digital red-green-blue and hyperspectral CASI imagery. The package marks an early step in the evolution of fused airborne sensors.

As always, technological challenges are first solved using aircraft and later evolve into spaceborne instruments. The evolution of the CHARTS system is particularly pertinent to the rapidly evolving science of remotely sensing coral reefs (see Figure 8.22). With the LiDAR evaluating depth and certain water body apparent optical properties, simultaneous to the CASI recording upwelling radiance, all necessary ingredients to obtain optical closure on the light field above the reef are present. With this, CHARTS represents a true breakthrough in remote sensing technology and opens up many possibilities that are not available to single instruments or poorly parameterized radiative transfer models (see Chapter 8, Section 8.4.3).

As proven by CHARTS, fusion products from airborne sensors offer considerable cost savings as compared to flying multiple sorties with different arrays of instruments. The ADS Mk II Airborne System, operated by Tenix LADS, combines a digital camera with a LiDAR instrument capable of collecting topographical and bathymetric LiDAR data soundings as well as LiDAR reflectivity. When used in the marine environment, relative reflectivity data is a measure of the reflectance of the seabed in a single wavelength (green/blue, 532 nm). The numerical values for the relative reflectivity are scaled logarithmically to an 8-bit integer range (0–255). As depicted in Figure 14.1, the system has been trialled by NOAA for the purpose of coral reef mapping in Puerto Rico. Due to the high intensity of the LiDAR laser beam, this fused system has an advantage over a passive airborne scanner in that it can collect data down to water depths of 50 metres.

14.1.2   Hyper-spatial satellites

The availability of metre-scale satellite imagery has revolutionized many disciplines of Earth observation. For coastal and marine applications, there are currently three civilian sensors operational that provide extremely fine spatial resolution (<5 m) in

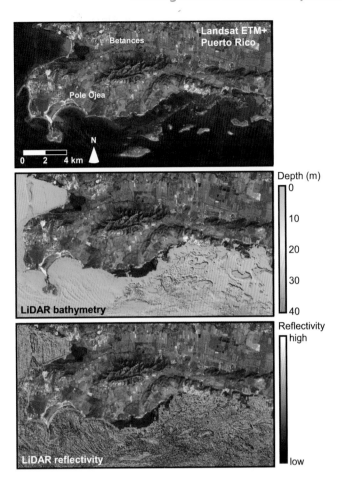

Figure 14.1 Top pane is a Landsat ETM+ image of southwestern Puerto Rico. Middle and lower panes are LiDAR bathymetry and LiDAR sea floor reflectance, respectively, acquired in 2006 using an ADS Mk II Airborne System operated by Tenix LADS. Landsat image: NASA, LiDAR. Credit: NOAA.

the water-penetrating bands (400–600 nm). These are IKONOS, operated by GeoEye, and QuickBird and WorldView-2, operated by DigitalGlobe.

Launched in 2009, WorldView-2 delivers an extra channel in the blue spectrum (8 bands in total, 6 in the visible, 2 near-infrared) and a 1.8 m × 1.8 m size, as compared to QuickBird's 2.8 m² pixels and only four spectral bands (3 visible and 1 near-infrared) (for further details see Table 6.1 (Chapter 6)). The additional visible bands translate to a greatly enhanced ability to derive bathymetry from WorldView-2 scenes, as well as increased benthic mapping capability and analysis of water quality. The advantage comes from the inclusion of a 'coastal' band (400–450 nm), capable of penetrating 25–30 m in clear water, and a 'yellow edge' band (585–625 nm) between the visible green and red channels. The coastal band is disadvantaged by a greater atmospheric influence than the standard blue band (450–510 nm) that is also carried by the instrument.

Relying on a 30° off-nadir look angle, WorldView-2 also offers a target revisit time of less than two days. This temporal coverage may prove sufficient to routinely capture rapid dynamic events such as coral bleaching.

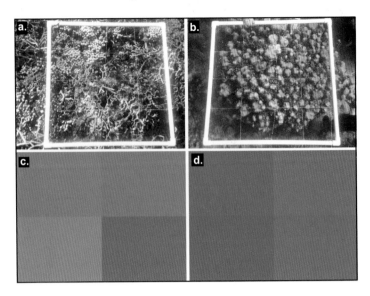

Figure 14.2 1×1 m digital photo quadrats of reef habitat: (a) A mixture of live coral colonies (predominantly *Acropora cervicornis* and *Agaricia tenufolia*) (b) This contains a single massive *Montastrea annularis* colony. (c) and (d) are a and b resampled to a 0.5 m × 0.5 m pixel size. In both cases, the spatial resolution is clearly insufficient to discern the extent of individual coral colonies.

Maintaining a competitive market place, GeoEye are also planning a next-generation system. GeoEye-1 will launch within the next two years and will provide multispectral imagery at 1.65-metre resolution – a two-fold improvement over the now-aging IKONOS.

If the current trend continues, there is a high likelihood that the civilian market will shortly be offered metre-sized pixels by a multispectral imager (such resolution is presently only available through PAN-sharpening, a process that can only offer pseudo-multispectral capability). Given the limited discriminatory ability of a few broad wavebands, this spatial resolution is still below that required to discern, for example, individual coral colonies (Figure 14.2), but it will increase the accuracy with which broad ecological assemblages can be identified.

14.1.3 Hyperspectral hyper-spatial satellites

It is only relatively recently that fine-resolution satellite-mounted hyperspectral imaging spectrometers have been available for civilian purposes. The first such instrument was the EO-1 Hyperion, launched in November 2000 and offering 220 spectral bands, each 10 nm wide. Designed for only a limited lifespan, this platform was primarily used for technology demonstration and was somewhat limited for work in reef environments due to a rather large 30 m² pixel size (which does, however, offer a vast increase in resolution compared to the km-scale capability of the current satellite instruments with hyperspectral capability, namely MERIS and MODIS). With such large pixels, the problem of mixed pixels for Hyperion data is pronounced. Modelling by Kutser *et al.* (2003) indicates, however, that it is at least theoretically possible to un-mix spectra of living hard and soft corals from Hyperion data in water depths of up to 15 m.

In summary, no commercially available sensors are ideally suited for reef monitoring. All could be improved with better suited temporal, spatial or spectral resolution. The most efficient band-sets for work on reefs have been investigated and settled upon (Hochberg & Atkinson, 2003; Kutser *et al.*, 2003), but a dedicated sensor with these band-sets is unlikely to be built in the near term as the cost benefits are not attractive enough.

Reef sensors need most spectral resolution in the water-penetrating blue-green spectrum. These bands could be contiguous or judicially placed to capitalize on spectral regions unique to different biota. Bands pooled in the short wavelength spectrum are, however, not well positioned for work in terrestrial environments, which require a more even distribution through the visible and into the infrared spectrum. It is on land where most remote sensing is conducted, and work on the sea floor will always have to utilize 'sensors of opportunity'.

14.2 The near future

The mainstay for remote monitoring of global change today rests with two programmes: NASA's Earth Observing System (EOS) mission, and the European Space Agency's ongoing constellation of environmentally focused satellites, ERS-1, ERS-2, and ENVISAT. Both programmes have clearly proven that satellite observation is a pivotal technology in global-scale Earth system monitoring.

In the United States, and as covered in Chapter 11 of this book, the A-Train instrument constellation comprising five satellites (Aura, PARASOL, CALIPSO, CloudSat, Aqua and, eventually, OCO-2) is already greatly expanding the data at hand to understand global change, even though OCO, which was to be the first space mission to monitor atmospheric carbon dioxide, failed upon launch in early 2009.

The A-Train constellation crosses the equator in the early afternoon local time. Preceding this is a second fleet of instruments, the morning constellation that consists of Landsat-7, EO-1, Terra and SAC-C. Terra alone measures 16 of the 24 parameters known to play a role in determining climate, including aerosols, clouds, temperature, precipitation, vegetation and radiation. The reliance on constellations rather than individual satellites is the primary concept of NASA's EOS mission. Each satellite in the constellation acquires data that is relevant to the parameters being measured by its partner instruments. Multiple platforms accrue benefits from data synergy. This is a new approach and will likely be a common theme in all future Earth observation missions.

Europe's next-generation monitoring capability is similarly synergistic and is pinned to the Sentinel satellite series. Sentinel-1 will provide day/night all-weather imaging radar of ocean and land, Sentinels 2 and 3 will be high-resolution optical for land and ocean respectively, ad Sentinels 4 and 5 will be tasked with monitoring atmospheric composition. The launch of the first sensor is tentatively scheduled for 2010/2011 and the mission collectively will be encompassed by Europe's Global Monitoring for the Environment and Security (GMES) programme.

The satellites contained within both NASA's and Europe's monitoring constellations play into a larger worldwide programme, the Global Earth Observation

System of Systems (GEOSS). This will further synergize data from many different sources and deliver them to the user through a single Internet access point, the 'GEOportal'. The GEOSS currently runs under a ten-year implementation plan that was initiated in 2005 and will expire in 2015. This plan identifies nine distinct groups of users and uses, which it calls 'Societal Benefit Areas', namely climate; water; disasters; health; energy; weather; ecosystems; agriculture; and biodiversity. Current and potential users include decision-makers in the public and private sectors, resource managers, planners, emergency responders and scientists.

The use of the Internet to rapidly disseminate raw data and information has, in the last ten years, undergone a paradigm shift with the release of *Google Earth* (www.google.com/earth) and NASA's *World Wind* (http://worldwind.arc.nasa.gov). Both, in their basic forms at least, are available to the public at no charge. We can expect further developments in the capability of the Internet as a means to archive data and to allow users to access them more intuitively and more rapidly. This trend is further discussed in Chapter 13 (Section 13.2).

In 1992, the US Department of the Interior established the National Satellite Remote Sensing Data Archive operated by USGS/EROS, at Sioux Falls, South Dakota. This archive comprises a vast collection of Landsat and related observations and provides a comprehensive permanent record of the planet's land surface derived from 45+ years of satellite remote sensing. For instance, the North American Landscape Characterization (NALC) data at EROS consists of Landsat MSS time-series triplicates acquired in 1973, 1986 and 1991 at selected locations in the USA and Mexico. These data and other imagery are available to the public through Earth Explorer, GloVis or the MRLC Consortium (Goward *et al.*, 2006; US Geological Survey, 2007).

Scientists working on global change problems are especially in need of satellite imagery covering long time periods and large regions of the globe, including areas that are frequently obscured by clouds. Web-based search-and-access tools such as the EOS Data Gateway (https://wist.echo.nasa.gov/api/) are already well advanced in such processes, but they will have to adapt to the future pressures of handling data from hundreds of EO satellites.

With a clear move to satellite systems composed of constellations of instruments as opposed to single missions, the issue of cost rapidly looms. Certainly there is truth in the fact that a multi-platform approach can deliver more than the sum of its parts. However, with individual satellites running to hundreds of millions of dollars, the long-term support of a constellation, where failed instruments must be swiftly replaced, becomes daunting. Launching satellites is extremely expensive, but lofting multiple instruments on a single rocket helps costs to become more manageable, providing that the sensors are small and light.

Several companies are already developing small satellites to be flown as fleets. A good example is the Disaster Monitoring Constellation (DMC) of micro-satellites and nano-satellites being developed by Surrey Satellite Technology Ltd (SSTL), which would provide daily global coverage. These instruments are based on advanced technology, weigh only tens of kilograms as compared to thousands for the older platforms, and cost several orders of magnitude less per satellite. Typical applications include monitoring logging and deforestation in the Amazon basin, boreal forest fire mapping in central Siberia, locust migration monitoring in Algeria, glacial calving of the Greenland ice sheet, and flooding in the Philippines. Innovative approaches to ground station networks and high-speed data flows are also important aspects of this new approach (Teillet *et al.*, 2002).

At the time of writing, SSTL is busy trying to develop a small radar satellite within the next few years. This is an ambitious aim, as radar instruments have much greater power requirements, so are much more challenging to miniaturize than visible-spectrum sensors. For this reason, radars have traditionally been housed on massive platforms such as Envisat, that weigh many tonnes and provide kilowatts of electricity. By contrast, a typical DMC satellite is a 100 kg, 60 cm cube.

The currently operational constellation of DMC satellites is managed from Guildford, England, and is composed of six small imaging instruments with spatial resolutions from 4 m to 32 m. All possess wide swath widths, exceeding 600 km, which facilitate short repeat cycles so as to quickly capture multiple images of the same point on Earth. Good temporal resolution is as important for disaster monitoring as fine spatial resolution. The DMC presently consists of satellites belonging to the UK, Spain, Algeria, China and Nigeria; a Turkish satellite is no longer operational after finishing its mission. The next planned member of the DNC is NigeriaSat-X, which is scheduled to loft in October 2010 along with another African instrument called NigeriaSat-2.

The sextet of flying DMC satellites was utilized extensively in 2005 during the aftermath of Hurricane Katrina, as well as to track the oil slick emanating from the seabed of the Gulf of Mexico after BP lost control of a wellhead there in 2010. This terrible accident followed a fire on the Deepwater Horizon rig which eventually led to its sinking (see Figure 10.13).

With the predicted rapid acceleration of the scope and capacity of EO missions comes the urgent need for ground-based measurements to keep pace with the technological advances occurring in orbit. All remote sensing data requires rigorous calibration, which becomes ever more critical as the monitored parameters become ever more subtle. A pertinent example is the satellite-based quantification of atmospheric CO_2, which demands a precision as high as 1–2 ppm against a background of 380 ppm.

The reliability and accuracy of such data relies on independent measurements carried out on (or from) the ground. Advanced field instruments and platforms are being developed for this purpose, and these include remotely controlled unmanned aircraft systems which can provide high-resolution spatial data to calibrate and validate satellite results. Some of the more advanced proposals are for drones capable of remaining aloft for stretches of months or even for several years. Power would be provided by ultra-efficient solar cells on their wings. Similarly, remotely controlled unmanned underwater robots or ocean gliders will be used more frequently to provide data over various ocean depths in order to extend and validate satellite data.

14.3 The more distant future

The logical evolution of satellite constellations populated by tens of instruments, such as synergized by the GEOSS, is the amalgamation of hundreds to thousands of craft. These will be much smaller and lighter than at present, necessitating the development of alternative sources of fuel.

Particularly important will be new technologies for linking instruments through wireless into sensor networks, allowing the information to be combined to support decisions in complex situations. In addition, the output of one or several sensors can be used to trigger observations from others, or even to rapidly reconfigure the other sensors in order to optimize the observations of a specific event.

For example, in parallel with the deployment of electronic measuring devices and developments in satellite telecommunications, the future of Earth observation satellites is exemplified by NASA's strategic Earth science vision for networks of satellite sensors, or 'sensor-webs'. Over the next several decades, webs will likely concentrate on improving predictions of Earth system changes in both the short and long term, with priority given to weather phenomena and disaster events (NASA, 2000). These orbital constellations will provide the reliability that an operational system requires by allowing many small, separate platforms to work together to accomplish what was formerly thought to require large multi-sensor platforms.

Looking perhaps a century or more into the future, some experts predict many thousands of sophisticated embedded measuring devices, each with sensors, a processor, and a transmitter, will be networked together in an extensive monitoring system for the Earth (Gross, 1999; Teillet *et al.*, 2002). Much of the research driving small inexpensive sensors is found in the area of Micro Electro-Mechanical Systems (MEMS). Scientists working with MEMS are creating tiny electronic devices from silicon, some of them smaller than a red blood cell. MEMS are common in the computer chip industry, but the technology also extends to sensor design. For instance, the University of California at Berkley is developing a sensor that is labelled 'smart dust' because it is so small that it almost floats in air. These minute devices are self-powered and contain tiny on-board sensors and a computer of just a few square millimetres in size. The objective is to use them by the thousands in interconnected networks that communicate with each other using wireless signals.

14.4 Advanced image analysis techniques

During the past several decades, major progress has been made in the analysis of satellite imagery and other remotely sensed data. However, classifying remotely sensed data into an accurate thematic map remains a challenge due to many factors, including the complexity of the landscape or water masses, the selection of the best combination of remotely sensed data and the approach used for processing and analyzing that data.

- On land, there still exist serious problems regarding the identification of the various land cover types. Confusion still occurs in the separation of the common spectral signatures of fields of crops versus grass; forest versus agriculture; forest versus forested wetland; bare soil versus impervious areas; and so on.
- In wetlands, it is difficult to distinguish between various species of plants, such as highly productive and useful *Spartina alterniflora* versus the invasive *Phragmites australis*. It is also hard to identify dry mud flats from impervious areas.
- In the ocean, the situation becomes extremely complex as one approaches coastal areas and estuaries, where the water contains a mixture of particulate and dissolved organic and inorganic substances (Schofield *et al.*, 2004).

With some confusion still existing for the classification of terrestrial targets, we face even greater challenges in mapping submerged ecosystems such as coral reefs (see Chapter 8). Here, an overlying water column serves to attenuate strongly the light available for sensing, while the corals themselves are optically inseparable except under the scrutiny of our most advanced airborne hyperspectral instruments. Never has our need been greater, though; Earth's reefs are degrading at an alarming rate and, if remote sensing is going to have a hand in their rehabilitation, then this technical limitation and many others must be solved with urgency.

It is safe to predict the further refinement of our existing classification algorithms into the future, along with the development of new strategies to identify land cover types using spectral imagery. An increased reliance on textural, spatial and context-based metrics is likely, as is the further incorporation of ancillary data into classifications (topographical models, image time-series, GIS, etc.). There is also considerable mileage in the further development of image segmentation routines, pattern recognition and object-oriented classification; such technologies have enjoyed a resurgence with the advent of powerful processors and neural-network analysis. The tried and tested parametric routines for classification that have been the mainstay for mapping over the last three decades will continue to be employed in the future, but we can expect a greater degree of automation, lessening the time required to convert a raw image into useful geographical knowledge.

14.5 Looking ahead at a changing Earth

Ever since its formation, more than four and half billion years ago, Earth's environment has undergone changes, the vast majority of which are slow and continuous. Superimposed upon this steady march appear more dramatic occurrences, where the arrangement of continental plates change or the composition of the atmosphere alters over comparatively short timescales. The geological record allows us to examine these dynamic periods and what caused them. Worryingly, it is also clear that rapid change on a global scale can profoundly depress the diversity and richness of life on Earth. The human race is not immune to these impacts, and there is no doubt that Earth's climate is presently changing at a rate far exceeding the pace of change over the last many thousands, perhaps millions, of years. At the beginning of the 21st century, climate change has become a defining issue of our time.

Though there is near-consensus among the scientific community that Earth's environment is presently in a period of rapid flux, there remain sharp divisions of opinion on what is driving the change and what impact it will have on our society over the next several decades. Answers to these questions are urgently needed in view of the mounting evidence that global change poses significant risks to human society.

The Intergovernmental Panel on Climate Change (IPCC) is the leading body for the assessment of climate change, established by the United Nations to provide the world with a clear, balanced view of the present state of understanding of climate change. The main activity of the IPCC is to provide, at regular intervals, assessment reports of the state of knowledge of climate change. At the time of writing, the IPCC has published four such reports, the most recent of which was made public in 2007 and builds upon previous releases in 1990, 1995 and 2001.

The latest offering from the IPCC contains two clear messages. First, there is no doubt that the Earth's climate is rapidly warming, and second, human activity is the most plausible driver of rapid climate change. The 2007 report finds that it is 'very likely' that emissions of heat-trapping gases from human activities have caused 'most of the observed increase in globally averaged temperatures since the mid-20th century'. Evidence that human activities are the major cause of recent climate change is even stronger than in prior assessments.

The report also makes the assertion that it is 'unequivocal' that Earth's climate is warming, 'as is now evident from observations of increases in global average air and ocean temperatures, widespread melting of snow and ice, and rising global mean sea level.' It further adds that the current atmospheric concentrations of carbon dioxide and methane, two important heat-trapping gases, 'exceeds by far the natural range over the last 650,000 years.' Since the dawn of the industrial era, concentrations of both gases have increased at a rate that is 'very likely to have been unprecedented in more than 10,000 years.'

If correct, the message from the IPCC translates to a sombre warning that the present climate, to which our society and economy are so well adapted, may be radically and irreversibly altered within a human lifetime.

Faced with this predicament, society must decide on how to prosper under an unfamiliar climatic regime. Adapt we must, because even if we were immediately to halt all releases of greenhouse gases, the emissions we have already made in the last decades are deemed sufficient to alter the climate for many years into the future (see Section 11.1, Chapter 11). Our continued emissions will serve to increase the pace and severity of climate change.

One route we could take is to place slight, or indeed no, curbs on our release of greenhouse gases, gambling that society will be able to continue to thrive under a warmer climate. This strategy is certainly easy to swallow as it demands only minor economic investment in the near term. The risk of adopting this course, however, is that the future climate may not be conducive to a vibrant and stable society, and that the economic burden of altering where and how we live will be overwhelming. Relevant to this is the fact that there is growing evidence from some climate projections that many of our traditional food sources and water supplies are likely to be jeopardized.

Certainly the stability of the world's economy and ecosystems both depend on maintaining precipitation patterns more or less as they are today. Therefore, a more conservative strategy would be immediately to implement aggressive policies that will massively reduce greenhouse gas emissions. There can be no illusions that such action would be extremely expensive, necessitating radical changes in lifestyle for much of the world's population. The developed world would be most impacted by such legislation, as the Third World consumes a smaller quantity of hydrocarbons per capita.

A middle ground is developing as the most likely scenario, with some reduction in greenhouse gas emissions over the next decade but at a level far insufficient to curb the warming trend (Matthews & Caldeira, 2008). There is no doubt that anthropogenic emissions of carbon dioxide are finite and are bounded by the certain exhaustion of Earth's fossil fuels, which will likely occur sometime between 2100 and 2300 (globally we presently consume more than one thousand barrels of oil per second). However, it is also clear that the effects of elevated carbon dioxide will manifest themselves for several centuries longer because of the huge inertia of the climate system.

Climate change will deliver warmer temperatures for much of the world. Already, 11 of the last 12 years rank among the 12 hottest years on record (since 1850, when sufficient worldwide temperature measurements began). Some, though not all, simulations point towards an increase in the frequency and intensity of storms, precipitation and drought into the future. Almost certainly, the present loss of ice from mountain glaciers and the poles will continue unabated. In turn, sea level will continue to rise as the oceans swell in response to the massive freshwater input from the melting Greenland and Antarctic ice sheets, as well as from thermal expansion, as the world's oceans absorb a large portion of the heat added to the climate. The rising sea will inundate vast tracts of low-lying coastal plains, on which many millions of people currently live and rely for food and employment. Even considering the most conservative future climate scenarios, the effects on our society are going to be profound, with the most poor and vulnerable individuals impacted most gravely.

With a shifting climate and increasing human impacts, Earth's ecosystems become degraded. An all-too-familiar pattern is a loss in species richness, leading to ecosystem collapse. Many changes are imperceptible on the scale of a human lifetime, but others, such as the present demise of the coral reefs, are clearly observable over periods of only a few decades. One simple rule of ecosystem change is almost always honoured; bad things tend to occur rapidly, good things more slowly.

If human intervention is to successfully reverse this degradation, it is prudent and ultimately less costly to intervene in the short term before ecosystem vitality is too greatly reduced. This concept is presented in Figure 14.3, where the orange line represents the decay of a hypothetical ecosystem. In the figure, the orange shaded

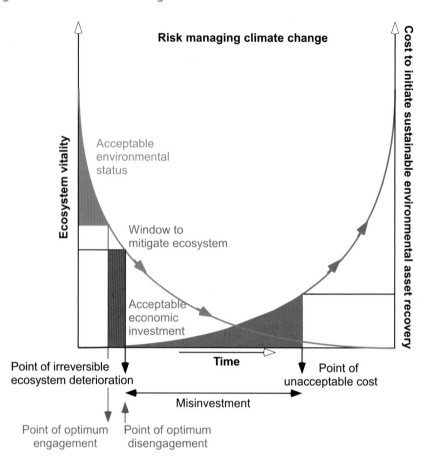

Figure 14.3 Hypothetical degradation of an ecosystem. There exists only a brief window of opportunity within which intervention will revitalize the ecosystem at acceptable cost.

area represents a level of ecosystem functioning that is acceptable. Though changes within these bounds may depress the services that this ecosystem can provide, they are deemed temporary and reversible in the short term.

Once the ecosystem is seen to break down, there exists a brief window of opportunity where intervention and investment stands a good chance of revitalizing the system and returning it to an acceptable state (depicted in magenta in the figure). If this opportunity is not seized, the cost of successful intervention rises rapidly (green line) to a point that the investment needed to repair the ecosystem becomes unpalatable and quickly unacceptable. Our present reliance on fossil fuel fits this analogy well. The sooner we switch to a more sustainable source of energy, the less the cost and the greater the chance that we will be able to reverse the climate problem.

Remote sensing is a potent technology in that it allows Earth's systems to be routinely appraised, maximizing the chance that degradation can be detected at an early stage. The choice is then ours as to how we respond. With the existence of the technologies and services discussed in this book, we can no longer say that we were not warned.

Our predictions as to the magnitude and pace of global climate perturbations remain frustratingly vague because of the lack of long-term data against which the recently observed changes can be compared. There is now significant evidence of observed changes in natural systems on every continent and most oceans. However, documentation of observed changes remains notably sparse in tropical regions and the southern hemisphere. Robust data are particularly lacking in Southeast Asia, the Indian Ocean and regions of the Pacific. Remote sensing technologies are the only means of delivering global-scale data over time periods sufficient to track the long-term trajectory of our climate.

Some targets on the Earth's surface, such as the ice caps, are relatively easy to resolve because they are large and have a distinct albedo. Others, like coral reefs, are much more challenging, demanding the isolation of a very weak signal from abundant spectral noise. It is these targets that demand our remote sensing instruments to operate at the limits of their detectability.

This book sits amid growing concern that ocean acidification, global warming, pollution and habitat loss will cause extermination, if not extinction, of much of the globe's biodiversity. Never before have we so urgently needed a comprehensive system for Earth observation to watch over our planet.

References

Abdalati, W. & Steffen, K. (2001) Greenland ice sheet melt extent: 1979–1999. *Journal of Geophysical Research*, **106**, 33983–33989.

Ackermann, F. (1999) Airborne laser scanning – present status and future expectations. *ISPRS Journal of Photogrammetry and Remote Sensing*, **54**, 64–67.

Ackleson, S.G. (2001) Ocean optics research at the start of the 21st century. *Oceanography*, **13**, 5–8.

Ackleson, S.G. (2003) Light in the shallow ocean: a brief research review. *Limnology and Oceanography*, **48**, 323–328.

Adams, J.B., Smith, M.O. & Johnson, P.E. (1986) Spectral mixture modelling: a new analysis of rock and soil types at the Viking Lander I site. *Journal of Geophysical Research*, **91**, 8098–8112.

AGU (2008) Human impacts on climate. *American Geophysical Union position statement*. *EOS*, **89**, 41.

Ahmad, W. & Neil, D.T. (1994) An evaluation of Landsat Thematic Mapper TM digital data for discriminating coral reef zonation: Heron Reef (GBR). *International Journal of Remote Sensing*, **15**, 2583–2597.

Aires, F., Prigent, C., Rossow, W.B. & Rothstein, M. (2001) A new neural network approach including first-guess for retrieval of atmospheric water vapor, cloud liquid water path, surface temperature and emissivities over land from satellite microwave observations. *Journal of Geophysical Research*, **106**, 14887–14907.

Akbari, H., Davis, S.D., Dorsano, S., Huang, J. & Winnett, S. (eds.) (1992) *Cooling Our Communities: A Guidebook On Tree Planting And Light-Colored Surfacing*. US EPA & Lawrence Berkeley Laboratory Report LBL-31587. Berkeley, CA.

Alasset, P.-J., Poncos, V., Singhroy, V. & Couture, R. (2008) InSAR monitoring of permafrost activity in the Lower Mackenzie Valley, Canada. *Geoscience and Remote Sensing Symposium, 2008. IGARSS 2008*, **3**, 530–533.

Alpers, W. (1985) Theory of radar imaging of internal waves. *Nature*, **314**, 245–247.

Alpers, W. & Hennings, I. (1984) A theory for the imaging mechanism of underwater bottom topography by real and synthetic aperture radar. *Journal of Geophysical Research*, **89(C14)**, 10529–10546.

Al-Tahir, A., Baban, S.M.J. & Ramlal, B. (2006) Utilizing emerging geo-imaging technologies for the management of tropical coastal environments. *West Indian Journal of Engineering*, **29**, 11–22.

Amelung, F., Galloway, D.L., Bell, J.W., Zebker, H.A. & Laczniak, R.J. (1999) Sensing the ups and downs of Las Vegas: InSAR reveals structural control of land subsidence and aquifer-system deformation. *Geology*, **27**, 483–486.

Anderson, D.E. (1997) Younger Dryas research and its implications for understanding abrupt climatic change. *Progress in Physical Geography*, **21**, 230–249.

Anderson, D.M., Hoagland, P., Kaoru, Y. & White, A.W. (2000) *Estimated Annual Economic Impacts from Harmful Algal Bloom (HABs) in the United States.* Technical Report WHOI-2000-11, Woods Hole Oceanographic Institute, Woods Hole, MA.

Anderson, J.R., Hardy, E.E., Roach, J.T. & Witmer, R.E. (1976) *A land use and land cover classification system for use with remote sensor data.* US Geological Survey Professional Paper, **964**, 28. Washington, DC.

Andréfouët, S. & Riegl, B.M. (2004) Remote sensing: a key tool for interdisciplinary assessment of coral reef processes. *Coral Reefs*, **23**, 1–4.

Andréfouët, S., Berkelmans, R., Odriozola, L., Done, T., Oliver, J. & Muller-Karger, F.E. (2002) Choosing the appropriate spatial resolution for monitoring coral bleaching events using remote sensing. *Coral Reefs*, **21**, 147–154.

Andréfouët, S., Gilbert, A., Yan, L., Remoissenet, G., Payri, C. & Chancerelle, Y. (2005) The remarkable population size of the endangered clam *Tridacna maxima* assessed in Fangatau Atoll (Eastern Tuamotu, French Polynesia) using *in situ* and remote sensing data. *ICES Journal of Marine Science*, **62**, 1037–1048.

Andréfouët, S., Hochberg, E.J., Chevillon, C., Muller-Karger, F.E., Brock, J.C. & Hu, C. (2005) Multi-scale remote sensing of coral reefs. In: Miller, R.L., Castillo, C.E.D. & McKee, B.A. (eds.) *Remote Sensing of Coastal Aquatic Environments: Technologies, Techniques and Applications*, 299–317. Springer, The Netherlands.

Andrews, D.G., Holton, J.R. & Leovy, C.B. (1987) *Middle Atmosphere Dynamics.* Academic Press, San Diego, CA.

Anyamba, A. & Tucker, C.J. (2005) Analysis of Sahelian vegetation dynamics using NOAA-AVHRR NDVI data from 1981–2003. *Journal of Arid Environments*, **63**, 596–614.

Apel, J.R. (2003) A new analytical model for ISWs in the ocean. *Journal of Physical Oceanography*, **33**, 2247–2269.

Arking, A. (1991) The Radiative Effects of Clouds and their Impact on Climate. *Bulletin of the American Meteorological Society*, **72**, 795–813.

Arnone, R.A. & Parsons, A.R. (2004) Real-time use of ocean color remote sensing for coastal monitoring. In: Miller, R.L., Del Castillo, C.E., McKee, B.A. (eds.) *Remote Sensing of the Coastal Aquatic Environments: Technologies, Techniques and Applications*, Chapter 14. Springer Publishing, Kluwer Academic, NY.

Arruda, W. Z., Lentini, C.A. & Campos, E.D.J. (2005) The use of satellite derived upper ocean heat content to the study of climate variability in the South Atlantic. *Revista Brasileira De Cartografia*, **57/02**, 2005. (ISSN 1808–0936).

Ashmore, M.R. (2005) Assessing the future global impacts of ozone on vegetation. *Plant, Cell, and Environment*, **28**, 949–964.

Asner, G.P. & Vitousek, P.M. (2005) Remote sensing of biological invasion and biogeochemical change. *Proceedings of the National Academy of Sciences*, **102**, 4383–4386.

Asner, G.P., Hughes, R.F., Vitousek, P.M., Knapp, D.E., Kennedy-Bowdoin, T., Boardman, J., Martin, R.E., Eastwood & M. Green, R.O. (2008) Invasive plants transform the three-dimensional structure of rain forests. *Proceedings of the National Academy of* Sciences, **105**, 4519–4523.

Auffhammer, M. & Carson, R.T. (2008) Forecasting the path of China's CO_2 emissions using province level information. *Journal of Environmental Economics and Management*, **55**, 229–247.

Avery, T.E. & Berlin, G.L. (1992) *Fundamentals of Remote Sensing and Airphoto Interpretation*, 5th edn. Macmillan Publishing Company, NY.

Avissar, R. (1996) Potential effect of vegetation on the urban thermal environment. *Atmosphere Environment*, **30**, 437–448.

Baird, A.H., Bhagooli, R., Ralph, P.J. & Takahashi, S. (2009) Coral bleaching: the role of the host. *Trends in Ecology and Evolution*, **24**, 16–20.

Baker, A.C., Glynn P.W. & Riegl, B. (2008) Climate change and coral reef bleaching: an ecological assessment of long-term impacts, recovery trends and future outlook. *Estuarine Coastal Shelf Science*, **80**, 435–471.

Baldwin, A.H. & Mendelssohn, I.A. (1998) Effects of salinity and water level on coastal marshes: an experimental test of disturbance as a catalyst for vegetation change. *Aquatic Botany*, **61**, 255–268.

Balk, D., Pozzi, F., Yetman, G., Deichmann, U. & Nelson, A. (2005) The distribution of people and the dimension of place: methodologies to improve the global estimation of urban extent. *Urban Remote Sensing Conference*, Tempe, AZ.

Banks, W.S.L. Paylor, R.L. & Hugues, W.B. (1996) Using thermal-infrared imagery to delineate ground-water discharged. *Ground Water*, **34**, 434–443.

Barber, D.G. & LeDrew, E.F. (1991) SAR sea ice discrimination using texture statistics: a multivariate approach. *Photogrammetric Engineering and Remote Sensing*, **57**, 385–395.

Barlow, J., Martin, Y. & Franklin, S.E. (2003) Detecting translational landslide scars using segmentation of Landsat ETM+ and DEM data in the northern Cascade Mountains, British Columbia. *Canadian Journal of Remote Sensing*, **29**, 510–517.

Barnsley, M.J., Barr, S.L. & Sadler, G.J. (1991) Spatial re-classification of remotely sensed images for urban land-use monitoring. *Proceedings of Spatial Data 2000, Oxford, 17–20 September* (Remote Sensing Society, Nottingham), 106–117.

Barras, J. (2006) *Land Area Change in Coastal Louisiana after the 2005 Hurricanes – A Series of Three Maps*. US Geological Survey Open-File Report 06–1274, 12p.

Barreiro, M., Fedorov, A., Pacanowski, R. & Philander, S.G. (2008) Abrupt climate changes: How freshening of the northern Atlantic affects the thermohaline and wind-driven oceanic circulations. *Annual Review of Earth and Planetary Sciences*, **36**, 33–58.

Bartholmé, E. & Belward, A.S. (2005) GLC2000: A new approach to global land cover mapping from Earth Observation data. *International Journal of Remote Sensing*, **26**, 1959.

Barton, I.J. (1995) Satellite-Derived Sea Surface Temperatures: Current Status. *Journal of Geophysical Research*, **100**, 8777.

Bathgate, J., Heron M. & Prytz, A. (2006) A method of swell parameter extraction from HF ocean surface radar spectra. *IEEE Journal of Oceanic Engineering*, **31**, 812–818.

Becker, M.W. (2006) Potential for Satellite Remote Sensing of Ground Water. *Ground Water*, **44**, 306–318.

Behrenfeld, M.J. & Falkowski, P.G. (1997) Photosynthetic rates derived from satellite-based chlorophyll concentration. *Limnology and Oceanography*, **42**, 1–20.

Behrenfeld, M.J., O'Malley, R.T., Siegel, D.A., McClain, C.R., Sarmiento, J.L., Feldman, G.C., Milligan, A.J., Falkowski, P.G., Letelier, R.M. & Boss, E.S. (2006) Climate-driven trends in contemporary ocean productivity. *Nature*, **444**, 752–755.

Bell, R.E., Studinger, M., Shuman, C.A., Fahnestock M.A. & Joughin, I. (2007) Large subglacial lakes in East Antarctica at the onset of fast-lowing ice streams. *Nature*, **445**, 904–907.

Berkelmans, R. & Oliver, J.K. (1999) Large-scale bleaching of corals on the Great Barrier Reef. *Coral Reefs*, **18**, 55–60.

Berry, B.L. (1990) Urbanization. In: Turner, B.L., Clark, W.C., Kates, R.W., Richards, J.F., Matthews, J.T., & Meyer, W.B. (eds.) *The Earth transformed by human action*. Cambridge University Press, Cambridge, UK.

Bertels, L., Vanderstraete, T., Van Coillie, S., Knaeps, E., Sterckx, S., Goossens, R. & Deronde, B. (2008) Mapping of coral reefs using hyperspectral CASI data; a case study: Fordata, Tanimbar, Indonesia. *International Journal of Remote Sensing*, **29**, 2359–2391.

Bindschadler, R., Vornberger, P., Fleming, A., Fox, A., Mullins, J., Binnie, D., Paulsen, S.J., Granneman, B. & Gorodetzky, D. (2008) The Landsat image mosaic of Antarctica. *Remote Sensing of Environment*, **112**, 4214–4226.

Bissett, W., Arnone, R.A., Davis, C.O., Dickey, T.D., Dye, D., Kohler, D.D.R. & Gould, R.W. (2004) From meters to kilometers: a look at ocean-color scales of variability, spatial coherence, and the need for fine-scale remote sensing in coastal ocean optics. *Oceanography*, June 2004.

Bjørgo, E., Johannessen, O.M. & Miles, M.W. (1997) Analysis of merged SSMR-SSMI time series of Arctic and Antarctic sea ice parameters 1978–95. *Geophysical Research Letters*, **24**, 413–416.

Black, T.A., Chen, W.J., Barr, A.G., Arain, M.A., Chen, Z., Nesic, Z., Hogg, E.H., Neumann H.H. & Yang, P.C. (2000) Increased carbon sequestration by a boreal deciduous forest in years with a warm spring. *Geophysical Research Letters*, **27**, 1271–1274.

Blackburn, G.A. (1999) Relationships between spectral reflectance and pigment concentrations in stacks of broadleaves. *Remote Sensing of Environment*, **70**, 224–237.

Blackburn, G.A. (2006) Hyperspectral remote sensing of plant pigments. *Journal of Experimental Botany*, **58**, 855–867.

Blair, J.B., Rabine, D.L. & Hofton, M.A. (1999) The laser vegetation imaging sensor: a medium-altitude, digitization-only, airborne laser altimeter for mapping vegetation and topography. *ISPRS Journal of Photogrammetry & Remote Sensing*, **54**, 115–122.

Blanchon, P., Jones, B. & Kalbfleisch, W. (1997) Anatomy of a fringing reef around Grand Cayman: storm rubble, not coral framework. *Journal of Sedimentary Research*, **67**, 1–16.

Blanchon, P., Eisenhauer, A., Fietzke, J. & Liebetrau, V. (2009) Rapid sea-level rise and reef back-stepping at the close of the last interglacial highstand. *Nature*, **458**, 881–884.

Blaschke T. & Hay, G. (2001) Object-oriented image analysis and scale-space: theory and methods for modeling and evaluating multi-scale landscape structure. *International Archives of the Photogrammetry, Remote Sensing and Spatial Information Sciences*, **34**, 22–29.

Blood, E.R., Anderson, P., Smith, P.A., Nybro, C. & Ginsberg, K.A. (1991) Effects of Hurricane Hugo on coastal soil solution chemistry in South Carolina. *Biotropica*, **23**, 348–355.

Boak, E.H. & Turner, I.L. (2005) Shoreline definition and detection: a review. *Journal of Coastal Research*, **21**, 688–703.

Bonisteel, J.M., Nayegandhi, A., Brock, J.C., Wright, C.W., Stevens, S., Yates, X. & Klipp, E.S. (2009) *EAARL Coastal Topography – Assateague Island National Seashore 2008: First Surface*: US Geological Survey Data Series 446.

Bonisteel, J.M., Nayegandhi, A., Wright, C.W., Brock, J.C. & Nagle, D.B. (2009) *Experimental Advanced Airborne Research LiDAR (EAARL) data processing manual*. US Geological Survey Open-File Report, 2009–1078.

Bornstein, R.D. (1968) Observations of the urban heat island effect in New York City. *Journal of Applied Meteorology*, **7**, 575–582.

Boss, S.K. (1996) Digital shaded relief image of a carbonate platform (northern Great Bahama Bank): scenery seen and unseen. *Geology*, **24**, 985–988.

Bosscher, H. & Schlager, W. (1992) Computer simulation of reef growth. *Sedimentology,* **39**, 503–512.

Bradley, R.S. & England, J. (1978) Volcanic dust influence on glacier mass balance at high latitudes. *Nature*, **271**, 736–738.

Braithwaite, C.J.R., Montaggioni, L.F., Camoin, G.F., Dalmasso, H., Dullo, W.C. & Mangini, A. (2000) Origins and development of Holocene coral reefs: a revisited model based on reef boreholes in the Seychelles, Indian Ocean. *International Journal of Earth Sciences*, **89**, 431–445.

Brakke, T.W., Kanemasu, E.T., Steiner, J.L., Ulaby, F.T. & Wilson, E. (1981) Microwave radar response to canopy moisture, leaf area index and dry weight of wheat, corn, and sorghum. *Remote Sensing of Environment*, **11**, 207–220.

Brandt, O., Langley, K., Kohler J. & Hamran, S-E. (2007) Detection of buried ice and sediment layers in permafrost using multi-frequency ground penetrating radar: A case examination on Svalbard. *Remote Sensing of Environment*, **111**, 212–227.

Breaker, L.C., Krasnopolsky,V.M., Rao, D.B. & Yan, X.-H. (1994) The feasibility of estimating ocean surface currents on an operational basis using satellite feature tracking methods. *Bulletin of the American Meteorological Society*, **75**: 2085–2090.

Brecke, C. & Solberg, A.H.S. (2005) Oil spill detection by satellite remote sensing. *Remote Sensing of Environment*, **95**, 1–13.

Breecker, D.O., Sharp, Z.D. & McFadden, L.D. (2010) Atmospheric CO_2 concentrations during ancient greenhouse climates were similar to those predicted for AD 2100. *Proceedings of the National Academy of Sciences*, **107**, 576–580.

Brenner, A.R. & Roessing, L. (2008) Radar imaging of urban areas by means of very high-resolution SAR and interferometric SAR. *IEEE Transactions on Geoscience and Remote Sensing*, **46**, 2971–2982.

Breshears, D.D., Cobb, N.S., Rich, P.M., Price, K.P., Allen, C.D., Balice, R.G., Romme, W.H., Kastens, J.H., Floyd, M.L., Belnap, J., Anderson, J.J., Myers, O.B. & Meyer, C.W. (2005) Regional vegetation die-off in response to global-change-type drought. *Proceedings of the National Academy of Sciences*, **102**, 15144–48.

Bricaud, A., Roesler, C., Zaneveld, J.R.V. (1995) In situ methods for measuring the inherent optical properties of ocean waters. *Limnology and Oceanography*, **40**, 393–410.

Brierley, A.S. & Kingsford, M.J. (2009) Impacts of climate change on marine organisms and ecosystems. *Current Biology*, **19**, 602–614.

Briffa, K., Jones, P., Schweingruber, P. & Osborn, T. (1998) Influence of volcanic eruptions on northern hemisphere summer temperature over the past 600 years. *Nature*, **393**, 450–454.

Brock, J. & Sallenger, A. (2000) *Airborne Topographic Mapping for Coastal Science and Resource Management*. USGS Open-File Report 01–46.

Brock, J., Wright, C.W., Hernandez, R. & Thompson, P. (2006) Airborne LiDAR sensing of massive stony coral colonies on patch reefs in the northern Florida reef tract. *Remote Sensing of Environment*, **104**, 31–42.

Brock, J.C. & Purkis, S.J. (2009) The emerging role of LiDAR remote sensing in coastal research and resource management. *Journal of Coastal Research*, **53**, 1–5.

Brock, J.C., Sallenger, A.H., Krabill, W.B., Swift, R.N. & Wright, C.W. (2001) Recognition of fiducial surfaces in LiDAR surveys of coastal topography. *Photogrammetric Engineering and Remote Sensing*, **67**, 1245–1258.

Brock, J.C., Wright, C.W., Clayton, T.D., Nayegandhi, A. (2004) LIDAR optical rugosity of coral reefs in Biscayne National Park, Florida. *Coral Reefs*, **23**, 48–59.

Brock, J.C., Palaseanu-Lovejoy, M., Wright, C.W. & Nayegandhi, A. (2008) Patch-reef morphology as a proxy for Holocene sea-level variability, Northern Florida Keys, USA. *Coral Reefs*, **27**, 555–568.

Broecker, W.S. (1997) Thermohaline circulation, the Achilles heel of our climate system: will man-made CO_2 upset the current balance? *Science*, **278**, 1582–1588.

Broge, N.H. & Mortensen, J.V. (2002) Deriving green crop area index and canopy chlorophyll density of winter wheat from spectral reflectance data. *Remote Sensing of Environment*, **81**, 45–57.

Browell, E.V., Gregory, G.L., Harriss, R.C. & Kirchhoff, V.W.J.H. (1988) Troposheric ozone and aerosol distribution across the Amazon Basin. *Journal of Geophysical Research*, **93**, 1431–1451.

Brown, J., Ferrians, O.J.J., Heginbottom, J.A., Melnikov, E.S. (1997) International Permafrost Association Circum-Arctic Map of Permafrost and Ground Ice Conditions, Scale 1:10,000,000. *US Geological Survey*.

Bruckner, A.W., Bruckner, R.J., (2006) The recent decline of *Montastrea annularis* (complex) coral populations in western Curaçao: a cause for concern? *Revista de Biología Tropical*, **54**, 45–58.

Bruno, J.F., Selig, E.R., Casey, K.S., Page, C.A., Willis, B.L., Harvell, C.D., Sweatman, H., Melendy, A.M. (2007) Thermal stress and coral cover as drivers of coral disease outbreaks. *PLoS Biology*, **5**, e124, doi:10.1371/journal.pbio.0050124.

Bryant, D., Burke, L., McManus, J. & Spalding, M. (1998) *Reefs at Risk: A map-based indicator of potential threats to the world's coral reefs*. World Resources Institute, Washington, DC; International Center for Living Aquatic Resource Management, Manila; and United Nations Environment Programme – World Conservation Monitoring Centre, Cambridge, MA.

Bryden, H.L., Longworth, H.R. & Cunningham, S.A. (2005) Slowing of the Atlantic meridional overturning circulation at 25° N. *Nature*, **438**, 655–657.

Burnett, C. & Blaschke, T. (2003) A multi-scale segmentation/object relationship modeling methodology for landscape analysis. *Ecological Modelling*, **168**, 233–249.

Burrage, D., Wesson, J., Martinez, C., Perez, T. Moller Jr., O. & Piola, A. (2008) Patos lagoon overflow within the Rio de la Plata plume using an airborne salinity mapper: observing an embedded plume. *Continental Shelf Research*, **28**, 1625–1638.

Burrage, D.M., Heron, M.L., Hacker, J.M., Miller, J.L., Stieglitz, T.C., Steinberg, C.R. & Prytz, A. (2003) Structure and influence of tropical river plumes in the Great Barrier reef: application and performance of an airborne sea surface salinity mapping system. *Remote Sensing of Environment*, **85**, 204–220.

Cabanes, C., Cazenave, A. & Le Provost, C. (2001) Sea level rise during past 40 years determined from satellite and *in situ* observations. *Science*, **26**, 840–842.

Caldeira, K. & Wickett, M.E. (2003) Anthropogenic carbon and ocean pH. *Nature*, **425**, 365.

Caldeira, K., Jain, A.K. & Hoffert, M.I. (2003) Climate sensitivity uncertainty and the need for energy without CO_2 emission. *Science*, **299**, 2052–2054.

Campbell, J.B. (2007) *Introduction to Remote Sensing*, 4th edn. The Guilford Press, NY.

Canton-Garbin, M. (2008) Satellite ocean observation in relation to global change. In: E.Chuvieco (Ed.), *Earth Observation of Global Change*. Springer-Verlag, Berlin.

Carlson, T.N. (1979) Atmospheric turbidity in Saharan dust outbreaks as determined by analysis of satellite brightness data. *Monthly Weather Review*, **107**, 322–335.

Carslaw, K.S., Harrison, R.G. & Kirkby, J. (2002) Cosmic rays, clouds, and climate. *Science*, **298**, 1732–1737.

Casey, K.S. & Cornillon, P. (1999) A comparison of satellite and *in situ* based sea surface temperature climatologies. *Journal of Climate*, **12**, 1848–1863.

Cavalieri, D.J., Gloersen, P., Parkinson, C.L., Comiso, J.C., & Zwally, H.J. (1997) Observed hemispheric asymmetry in global sea ice changes. *Science*, **278**, 1104–1106.

Cavalieri, D.J., Parkinson, C.L., Gloersen, P., Comiso, J.C. & Zwally H.J. (1999) Deriving long-term time series of sea ice cover from satellite passive-microwave multisensor data sets. *Journal of Geophysical Research*, **104**, 803–814.

Chagnon S. (ed.) (2000) *El Niño 1997–1998: The Climate Event of the Century*. Oxford Press.

Chang, A.T.C., Foster, J.L. & Hall, D.K. (1987) Nimbus-7 SSMR derived global snow cover parameters. *Annals of Glaciology*, **9**, 39–44.

Chang, K. (2008) *Introduction to Geographic Information Systems*. McGraw-Hill, NY.

Chant, R.J., Glenn, S.M., Hunter, E., Kohut, J., Chen, R.F., Houghton, R.W., Bosch, J. & Schofield, O. (2008) Bulge formation of a buoyant river outflow. *Journal of Geophysical Research*, **113**, C01017.

Chappell, J. & Polach, H. (1991) Post-glacial sea-level rise from a coral record at Huon Peninsula, Papua New Guinea. *Nature*, **349**, 147–149.

Chen, J.L., Wilson, C.R., Tapley, B.D., Yang, Z.L. & Niu, G.Y. (2009) 2005 drought event in the Amazon River basin as measured by GRACE and estimated by climate models. *Journal of Geophysical Research*, **114**, B05404.

Chen, J.M. & Black, T.A. (1992) Defining leaf area index for non-flat leaves. *Agricultural and Forest Meteorology*, **57**, 1–12.

Chen, K. (2001) An approach to linking remotely sensed data and areal census data. *International Journal of Remote Sensing*, **23(1)**, 37–48.

Chen, X. & Vierling, L. (2006) Spectral mixture analyses of hyperspectral data acquired using a tethered balloon. *Remote Sensing of Environment*, **103**, 338–350.

Chevallier, F., Maksyutov, S., Bousquet, P., Breon, F-M., Saito, R., Yoshida, Y. & Yokota, T. (2009) On the accuracy of the CO_2 surface fluxes to be estimated from the GOSAT observations. *Geophysical Research Letters*, **36**, L19807.

Chipman, J.W., Lillesand, T.M., Schmaltz, J.E., Leale, J.E. & Nordheim, M.J. (2004) Mapping lake water clarity with Landsat images in Wisconsin, USA. *Canada Journal of Remote Sensing*, **30**, 1–7.

Chomentowski, W., Salas, W. & Skole, D. (1994) Landsat Pathfinder Project advances deforestation mapping. *GIS World*, April 1994, 34–38.

Church, J.A., Gregory, J.M., Huybrechts, P., Kuhn, M., Lambeck, K., Nhuan, M.T., Qin, D. & Woodworth, P.L. (2001). Changes in Sea Level. In: Houghton, J.T, Ding, Y., Griggs, D.J., Noguer, M., Van der Linden, P.J., Dai, X., Maskell, K. & Johnson, C.A. (eds.) *Climate Change 2001: The Scientific Basis: Contribution of Working Group I to the Third Assessment Report of the Intergovernmental Panel on Climate Change*, 639–694. Cambridge University Press, Cambridge, NY.

Chuvieco, E. (2008) *Earth observation of global change: The role of satellite remote sensing in monitoring the global environment*. Springer-Verlag, Berlin.

Clerbaux, C., Edwards, D,P., Deeter, M., Emmons, L., Lamarque, J.-F., Tie, X.X., Massie, S.T. & Gille, J. (2008) Carbon monoxide pollution from cities and urban areas observed by the Terra/MOPITT mission. *Geophysical Research Letters*, **35**, L03817.

Cohen, A.L., McCorkle, D.C., de Putron, S., Gaetani, G.A. & Rose, K.A. (2009) Morphological and compositional changes in the skeletons of new coral recruits reared in acidified seawater: Insights into the biomineralization response to ocean acidification. *Geochemistry Geophysics Geosystems*, **10**, Q07005.

Collier, J.S. & Humber, S.R. (2007) Time-lapse side-scan sonar imaging of bleached coral reefs: A case study from the Seychelles. *Remote Sensing of Environment*, **108**, 339–356.

Collin, A., Archambault, P. & Long, B. (2008) Mapping the shallow water seabed habitat with SHOALS. *IEEE Transactions on Geoscience and Remote Sensing*, **46**, 2947–2955.

Collins, M.J., Livingstone C.E. & Raney R.K. (1997) Discrimination of sea ice in the Labrador marginal ice zone from synthetic aperture radar image texture. *International Journal of Remote Sensing*, **18(3),** 535–571.

Collins, W., Colman, R., Haywood, J., Manning, M.R. & Mote, P. (2007) The physical science behind climate change. *Scientific American*, **Aug. 2007**, 64–73.

Combal, B., Baret, F., Weiss, M., Trubuil, A., Mace, D., Pragnere, A., Myneni, R., Knyazikhin, Y. & Wang, L. (2002) Retrieval of canopy biophysical variables from bidirectional reflectance: using prior information to solve the ill-posed inverse problem. *Remote Sensing of Environment*, **84**, 1–15.

Comiso, J. C. (2003) Warming Trends in the Arctic. *Journal of Climate*, **16(21)**, 3498–3510.

Comiso, J.C. (2006) Abrupt decline in the Arctic winter sea ice cover. *Geophysical Research Letters*, **33**, L18504.

Comiso, J.C., & Nishio, F. (2008) Trends in the sea ice cover using enhanced and compatible AMSR-E, SSM/I, and SMMR data. *Journal of Geophysical Research*, **113**, C02S07.

Comiso, J. C. & Parkinson, C. L. (2004) Satellite observed changes in the Arctic. *Physics Today*, **57(8)**, 38–44.

Congalton, R. (1991) A review of assessing the accuracy of classifications of remotely sensed data. *Remote Sensing of Environment*, **37**, 35–46.

Conger, C.L., Hochberg, E.J., Fletcher, C.H. & Atkinson, M.J. (2006) Decorrelating remote sensing color bands from bathymetry in optically shallow waters. *IEEE Transactions on Geoscience and Remote Sensing*, **44**, 1655–1660.

Connell, J.H. (1973) Population ecology of reef-building corals. In: Jones, O.A. & Endean, R. (eds.) *Biology and Geology of Coral Reefs, Vol.2, Biology*. Academic Press, NY.

Coops, N.C. & Stone, C. (2005) A comparison of field-based and modelled reflectance spectra from damaged *Pinus radiata* foliage. *Australian Journal of Botany*, **53**, 417–429.

Coppin, P., Jonckhere, I., Nackaerts, K., Mays, B. & Lambin, E. (2004) Digital change detection methods in ecosystem monitoring: a review. *International Journal of Remote Sensing*, **25**, 1565–1596.

Costa, B.M., Battista, T.A. & Pittman, S.J. (2009) Comparative evaluation of airborne LiDAR and ship-based multibeam SoNAR bathymetry and intensity for mapping coral reef ecosystems. *Remote Sensing of Environment*, **113**, 1082–1100.

Costa, M.J., Cervino, M., Cattani, E., Torricella, F., Levizzani, V., Silva, A.M. & Melani, S. (2002) Aerosol characterization and optical thickness retrievals using GOME and METEOSAT satellite data. *Meteorology and Atmospheric Physics*, **81**, 289–298.

Cowardin, L., Carter, V., Golet, F. & LaRoe, E. (1979) *Classification of wetlands and deep water habitats of the United States*. US Department of the Interior, Fish and Wildlife Service, Office of Biological Services, FWS/OBS-79/31. Washington, DC, p.131.

Cowardin, L.M. (1978) Wetland Classification in the United States. *Journal of Forestry*, **76(10)**, 666–668.

Cowen, D.J., Jensen, J.R., Bresnahan, P.J., Ehler, G.B., Graves, D., Huang, X., Wiesner, C., & Mackey, Jr. H.E. (1995) The design and implementation of an integrated GIS for environmental applications. *Photogrammetric Engineering and Remote Sensing*, **61**, 1393–1404.

Cowles, T. & Donaghay, P. (1998) Thin Layers: Observations of small-scale patterns and processes in the upper ocean. (Editorial) *Oceanography*, **11**, 1.

Cracknell, A. P. & Hayes, L. (2007) *Introduction to Remote Sensing*. CRC Press, NY.

Crane, R. G., Barry, R. G. & Zwally, H. J. (1982) Analysis of atmosphere-sea ice interactions in the Arctic Basin using ESMR microwave data. *International Journal of Remote Sensing*, **3**, 259–276.

Crosetto, M., Castillo, M. & Arbiol, R. (2003) Urban subsidence monitoring using radar interferometry: algorithms and validation. *Photogrammetric Engineering and Remote Sensing*, **69**, 775–783.

Crowley, T.J. & Berner, R.A. (2001) CO_2 and climate change. *Science*, **292**, 870–872.

Cullen, J.J., Ciotti, A.M., Davis, A.F. & Lewis, M.R. (1997) Optical detection and assessment of algal blooms. *Limnology and Oceanography*, **42** (5, part 2), 1223–1229.

Cunningham, S.A., Kanzow, T., Rayner, D., Baringer, M.O., Johns, W.E., Marotzke, J., Longworth, H.R., Grant, E.M., Hirschi, J. J.-M., Beal, L.M., Meinen, C.S. & Bryden, H.L. (2007) Temporal variability of the Atlantic meridional overturning circulation at 26.5°N. *Science*, **317**, 935–938.

Curry, J.A., Schramm, J.L. & Ebert, E.E. (1995) Sea ice-albedo climate feedback mechanism. *Journal of Climate*, 8, 240–247.

D'Souza, G., Belward, A.S., Malingreau, J.-P. (1996) *Advances in the Use of NOAA AVHRR Data for Land Applications*. Kluwer, Dordrecht, The Netherlands.

Dahl, T.E. (2006) Status and trends of wetlands in the conterminous United States 1998 to 2004. Washington, DC. *US Department of the Interior, Fish and Wildlife Service Publication*, 116.

Daiber, F.C. (1986) *Conservation of Tidal Marshes*. Van Nostrand Reinhold, NY.

Dall, J., Madsen, S.N., Keller, K. & Forsberg, R. (2001) Topography and penetration of the Greenland Ice Sheet measured with airborne SAR interferometry. *Geophysical Research Letters*, **28**, 1703–1706.

Davis, C.H., Li, Y., McConnell, J.R., Frey, M.M. & Hanna, E. (2005) Snowfall-Driven Growth in East Antarctic Ice Sheet Mitigates Recent Sea-Level Rise. *Science*, published online, doi:10.1126/science.1110662.

Davis, C.O., Lamela, G.M., Donato, T.F. & Bachmann, C.M. (2004) Coastal Margins and Estuaries. American Society of Photogrammetry and Remote Sensing. *Manual of Remote Sensing*, **4**, 401–446.

Davis, R.E. (1985) Drifter observations of coastal surface currents during CODE: the statistical and dynamical view. *Journal of Geophysical Research*, **90**, 4756–4772.

De Angelis H. & Skvarca. P. (2003) Glacier surge after ice shelf collapse. *Science*, **299**, 1560–1562.

Dekker, A.G. (1993) *Detection of optical water quality parameters for eutrophic waters by high resolution remote sensing*. PhD thesis, Vrije Universiteit Amsterdam.

Dekker, A.G., Vos, R.J. & Peters, S.W.M. (2001a) Comparison of remote sensing data, model results and *in situ* data for the Southern Frisian Lakes. *The Science of the Total Environment*, **268**, 197–214.

Dekker, A.G., Brando, V.E., Anstee, J.M., Pinnel, N., Kutser, T., Hoogenboom, E.J., Peters, S., Pasterkamp, R., Vos, R., Olbert, C. & Malthus, T.J.M. (2001b) Imaging spectrometry of water. In: Van der Meer, F.D. & de Jong, S.M. (eds.) *Imaging Spectrometry*, 307–359.

Dekker, A.G., Vos, R.J. & Peters, S.W.M. (2002). Analytical algorithms for lake water TSM estimation for retrospective analysis of TM and SPOT sensor data. *International Journal of Remote Sensing*, **23**, 15–35.

DeMott, P.J., Sassen, K., Poellot, M.R., Baumgardner, D., Rogers, D.C., Brooks, S.D., Prenni, A.J. & Kreidenweis, S.M. (2003) African dust aerosols as atmospheric ice nuclei. *Geophysical Research Letters*, **30**, 1732.

Dentener, F., Carmichael, G., Zhang, Y., Crutsen, P. & Lelifeld, J. (1996) The role of mineral aerosols as a reactive surface in the global troposphere. *Journal of Geophysical Research*, **101**, 22869–22890.

Deronde, B., Houthuys, R., Debruyn, W., Fransaer, D., Lancker, Vera V. & Hernriet, J.-P. (2006) Use of airborne hyperspectral data and Laserscan data to study beach morphodynamics along the Belgian coast. *Journal of Coastal Research*, **22**, 1108–1117.

Diaz-Pulido, G., McCook, L.J., Dove, S., Berkelmans, R., Roff, G., Kline, D.I., Weeks, S., Evans, R.D., Williamson, D.H. & Hoegh-Guldberg, O. (2009) Doom and boom on a resilient reef: climate change, algal overgrowth and coral recovery. *PLoS ONE*, **4**, e5239.

DigitalGlobe (2003) *Quickbird Imagery Products and Product Guide* (revision 4). DigitalGlobe Inc., CO.

Diner, D.J., Abdou, W.A., Bruegge, C.J., Conel, J.E., Crean, K.A., Gaitley, B.J., Helmlinger, M.C., Kahn, R.A., Martonchik, J.V., Pilorz, S.H. & Holben, B.N. (2001) MISR aerosol optical depth retrievals over southern Africa during the SAFARI-2000 dry season campaign. *Geophysical Research Letters*, **28**, 3127–3130.

Dixon, T.H., Amelung, F., Ferretti, A., Novali, F., Rocca, F., Dokka, R., Sellall, G., Kim, S.-W., Wdowinski, S. & Whitman, D. (2006) Space geodesy: subsidence and flooding in New Orleans. *Nature*, **441**, 587–588.

Doake, C.S.M., Corr, H.F.J., Rott, H., Skvarca, P. & Young, N.W. (1998) Breakup and conditions for stability of the northern Larsen Ice Shelf, Antarctica. *Nature*, **391**, 778–780.

Dobson, J.E., Bright, E.A., Ferguson, R.L., Field, D.W., Wood, L.L., Haddad, K.D., Iredale, III, H., Jensen, J.R., Klemas, V, Orth, R.J. & Thomas, J.P. (1995). *NOAA Coastal Change Analysis Program (C-CAP): guidance for regional implementation*, 92. NOAA Technical Report NMFS-123. US Department of Commerce, Washington, DC.

Domack, E.W., Duran, D., McMullen, K., Gilbert, R. & Leventer, A. (2002) Sediment lithofacies from beneath the Larsen B Ice Shelf: can we detect ice shelf fluctuations? *EOS*, **83**, F301.

Donato, T.F. & Klemas, V. (2001) Remote sensing and modeling applications for coastal resource management. *Geocarto International*, **16**, 23–29.

Done, T.J. (1999) Coral community adaptability to environmental change at the scales of regions, reefs and dead zones. *American Zoologist*, **39**, 66–79.

Donner, S.D. (2009) Coping with commitment: Projected thermal stress on coral reefs under different future scenarios. *PLoS ONE*, **4**, e5712.

Drew T. Shindell, D.T., Rind, D. & Lonergan, P. (1998) Increased polar stratospheric ozone losses and delayed eventual recovery owing to increasing greenhouse-gas concentrations. *Nature*, **392**, 589–592.

Dubayah, R.O. & Drake, J.B. (2000) LiDAR remote sensing of forestry. *Journal of Forestry*, **98**, 44–46.

Duguay, C.R., Zhang, T., Leverington, D.W. & Romanovsky, V.E. (2005) Satellite remote sensing of permafrost and seasonally frozen ground. In: Duguay, C.R. & Pietroniro, A. (eds.) *Remote Sensing in Northern Hydrology: Measuring Environmental Change*, 91–118. Geophysical Monograph 163, American Geophysical Union.

Durand, D., Bijaoui, J. & Cauneau, F. (2000) Optical remote sensing of shallow-water environmental parameters: A feasibility study. *Remote Sensing of Environment*, **73**, 152–161.

Durden, S.L., Morrissey, L.A. & Livingston, G.P. (1995) Microwave backscatter and attenuation dependence on leaf area index for flooded rice fields. *IEEE Geoscience and Remote Sensing Letters*, **33**, 807–810.

Dustan, P., Dobson, E. & Nelson, G. (2001) Landsat TM: detection of shifts in community composition of coral reefs. *Conservation Biology*, **15**, 892–902.

Dzwonkowski, B. & Yan, X.-H. (2005) Tracking of a Chesapeake Bay estuarine outflow plume with satellite-based ocean color data. *Continental Shelf Research*, **25**, 1942–1958.

Easterling, D.R., Horton, B., Jones, P.D., Peterson, T.C., Karl, T.R., Parker, D.E., Salinger, M.J., Razuvayev, V., Plummer, N., Jamason, P. & Folland, C.K. (1997) Maximum and minimum temperature trends for the globe. *Science*, **277**, 364–367.

Eastwood, J.A., Yates, M.G., Thomson, A.G. & Fuller, R.M. (1997) The reliability of vegetation indices for monitoring salt marsh cover. *International Journal of Remote Sensing*, **18**, 3901–3907.

Ehlers, M. (2007) New developments and trends for urban remote sensing. In: Weng, Q. & Quattrochi, D.A. (eds.) *Urban Remote Sensing*, 412. CRC Press, Boca Raton.

Ekercin, S. (2007) Water quality retrievals from high resolution IKONOS multispectral imagery: a case study in Istanbul, Turkey. *Water, Air & Soil Pollution*, **183**, 239–251

Elachi, C. & van Ziel, Z. (2006) *Introduction to the physics and techniques of remote sensing* (2nd ed.). John Wiley and Sons, NJ.

Ellis, J.M. & Dodd, H.S. (2000) *Applications and lessons learned with airborne multispectral imaging.* Fourteenth International Conference on Applied Geologic Remote Sensing, Las Vegas, NV.

Ellis, R.P., Bersey, J., Rundle, S.D., Hall-Spencer, J.M. & Spicer, J.I. (2009) Subtle but significant effects of CO_2 acidified seawater on embryos of the intertidal snail, *Littorina obtusata. Aquatic Biology*, **5**, 41–48.

Elvidge, C.D., Dietz, J.B., Berkelmans, R., Andréfouët, S., Skirving, W.J., Strong, A.E. & Tuttle, B.T. (2004) Satellite observation of Keppel Islands (Great Barrier Reef) 2002 coral reef bleaching using IKONOS data. *Coral Reefs*, **23**, 123–132.

Elvidge, C.D., Tuttle, B.T., Sutton, P.S., Baugh, K.E., Howard, A.T., Milesi, C., Bhaduri, B.L. & Nemani, R. (2007) Global distribution and density of constructed impervious surfaces. *Sensors*, **7**, 1962–1979.

Emanuel, K., Sundararajan, R. & Williams, J. (2008) Hurricanes and global warming: Results from downscaling IPCC AR4 simulations. *Bulletin of the American Meteorological Society*, **89**, 347–367.

Emanuel, K.A. (1987) The dependence of hurricane intensity on climate. *Nature*, **326**, 483–485.

Emanuel, K.A. (2005) Increasing destructiveness of tropical cyclones over the past 30 years. *Nature*, **436**, 686–688.

England, A.W. (1990) Radiobrightness of diurnally heated, freezing soil. *IEEE Transactions on Geoscience and Remote Sensing*, **28**, 464–476.

England, A.W., Galantowicz, J.F. & Schretter, M.S. (1992) The radiobrightness thermal inertia measure of soil moisture. *IEEE Transactions on Geoscience and Remote Sensing*, **30**, 132–139.

Everitt, J.H., Yang, C., Fletcher, R.S., Davis, M.R. & Drawe, D.L. (2004) Using aerial color-infrared photography and QuickBird satellite imagery for mapping wetland vegetation. *Geocarto International*, **19**, 15–22.

Fahnestock, M., Bindschadler, R., Kwok, R. & Jezek, K (1993) Greenland ice sheet surface properties and ice dynamics from ERS-1 SAR imagery. *Science*, **262**, 1530–1534.

Fahnestock, M.A., Abdalati, W. & Shuman, C.A. (2002) Long melt seasons on ice shelves of the Antarctic Peninsula: an analysis using satellite-based microwave emission measurements. *Annals of Glaciology*, **34**, 127–133.

Fairbanks, R.G. & Dodge, R.E. (1979) Annual periodicity of $^{18}O/^{16}O$ and $^{13}C/^{12}C$ ratios in the coral *Montastrea annularis. Geochimica et Cosmochimica Acta*, **43**, 1009–1020.

Farman, J.C., Gardiner, B.G. & Shanklin, J.D. (1985) Large losses of total ozone in Antarctica reveal seasonal CLOx/NOx interaction. *Nature*, **315**, 207–210.

Farr, T.G. & Kobrick, M. (2000) Shuttle Radar Topography Mission produces a wealth of data. *Eos, Transactions, AGU*, **81**, 583–585.

Feely, R.A., Sabine, C.L., Lee, K., Berelson, W., Kleypas, J., Fabry, V.J. & Millero, F.J. (2004) Impact of anthropogenic CO_2 on the $CaCO_3$ system in the ocean. *Science*, **305**, 362–366.

Fettweis, X., van Ypersele, J.-P., Gallée, H., Lefebre, F. & Lefebvre, W. (2007) The 1979–2005 Greenland ice sheet melt extent from passive microwave data using an improved version of the melt retrieval XPGR algorithm. *Geophysical Research Letters*, **34**, L05502, doi:10.1029/2006GL028787.

Field, R.T. & Philipp, K.R. (2000a) Vegetation changes in the freshwater tidal marsh of the Delaware estuary. *Wetlands Ecology and Management*, **8**, 79–88.

Field, R.T. & Philipp, K.R. (2000b) *Tidal inundation, vegetation type, and elevation at Milford Neck Wildlife Conservation Area: An exploratory analysis.* Report prepared for Delaware

Division of Fish and Wildlife, under contract AGR 199990726 and the Nature Conservancy under contract DEFO–0215000–01.

Fielding, E.J., Blom, R.G. & Goldstein, R.M. (1998) Rapid subsidence over oil fields measured by SAR interferometry. *Geophysical Research Letters*, **25**, 3215–3218.

Fily, M., Royer, A., Goïta, K. & Prigent, C. (2003) A simple retrieval method for land surface temperature and fraction of water surface determination from satellite microwave brightness temperatures in sub-arctic areas. *Remote Sensing of Environment*, **85**, 328–338.

Fishman, J. & Brackett, V.G. (1997) The climatological distribution of tropospheric ozone derived from satellite measurements using version 7 Total Ozone Mapping Spectrometer and Stratospheric Aerosol and Gas Experiment data sets. *Journal of Geophysical Research*, **102** (D15), 19275–19278.

Fishman, J., Brackett, V.G., Browell, E.V. & Grant, W.B. (1996) Tropospheric ozone derived from TOMS/SBUV measurements during TRACE A. *Journal of Geophysical Research*, **101**, 69–82.

Fitt, W.K., McFarland, F.K., Warner, M.E. & Chilcoat, G.C. (2000) Seasonal patterns of tissue biomass and densities of symbiotic dinoflagellates in reef corals and relation to coral bleaching. *Limnology and Oceanography*, **45**, 677–685.

Fonstad, M.A. & Marcus, W.A. (2005) Remote sensing of stream depths with hydraulically assisted bathymetry (HAB) models. *Geomorphology*, **72**, 320–339.

Foody, G.M. & Cox, D.P. (1994) Sub-pixel land cover composition estimation using a linear mixture model and fuzzy membership functions. *International Journal of Remote Sensing*, **15**, 619–631.

Foster, G., Walker, B.K. & Riegl, B.M. (2009) Interpretation of Single-Beam Acoustic Backscatter Using LiDAR-Derived Topographic Complexity and Benthic Habitat Classifications in a Coral Reef Environment. *Journal of Coastal Research*, **53**, 16–26.

Fraser, C., Baltsavias, E. & Gruen, A. (2002) Processing of IKONOS imagery for submetre 3-D positioning and building extraction. *ISPRS Journal of Photogrammetry and Remote Sensing*, **56**, 177–194.

Freidlingstein, P., Cox, P., Betts, R., Bopp, L., von Bloh, W., Brovkin, V., Cadule, P., Doney, S., Eby, M., Fung, I., Bala, G., John, J., Jones, C., Joos, F., Kato, T., Kawamiya, Knorr, M., Lindsay, W., Matthews, K., Raddatx, H.D., Rayner, T., Reick, P., Roeckner, C., Schnitzier, E., Schnur, K.-G., Strassmann, R., Weaver, K., Yoshikawa, A.J. & Zeng, C. (2006) Climate-carbon cycle feedback analysis: results from the C4MIP model intercomparison. *Journal of Climate*, **19**, 3337–3353.

Frietas, R., Rodrigues, A.M. & Quintino, V. (2003). Benthic biotopes remote sensing using acoustics. *Journal of Experimental Marine Biology and Ecology*, **285**, 339–353.

Freitas, R., Sampaio, L., Oliveira, J., Rodrigues, A.M. & Quintino, V. (2006) Validation of soft bottom benthic habitats identified by single-beam acoustics. *Marine Pollution Bulletin*, **53**, 72–79.

Freitas, R., Rodrigues, A.M., Morris, E., Perez-Llorens, J.L. & Quintino, V. (2008) Single-beam acoustic ground discrimination of shallow water habitats: 50 kHz or 200 kHz frequency survey? *Estuarine, Coastal and Shelf Science*, **78**, 613–622.

Friedl, M.A., McIver D.K., Hodges, J.C.F., Zhang, X.Y., Muchoney, D., Strahler, A.H., Woodcock, C.E., Gopal, A., Schneider, A., Cooper, A., Baccini, A., Gao, F. & Schaaf, C. (2002) Global land cover from MODIS: Algorithms and early results. *Remote Sensing of Environment*, **83**, 135–148.

Fu, F.X., Zhang, Y.H., Warner, M.E., Feng, Y.Y., Sun, J. & Hutchins, D.A. (2008) A comparison of future increased CO_2 and temperature effects on sympatric *Heterosigma akashiwo* and *Prorocentrum minimum*. *Harmful Algae*, **7**, 76–90.

Fuglestvedt, J., Berntsen, T., Myhre, G., Rypdal, K. & Skeie, R.B. (2008) Climate forcing from the transport sectors. *Proceedings of the National Academy of Sciences [USA]*, **105**, 454–458.

Fung, I.Y., Meyn, S.K., Tegen, I., Doney, S.C., John, J.G. & Bishop J.K.B. (2000) Iron supply and demand in the upper ocean. *Global Biogeochemical Cycles*, **14**, 697–700.

Gallo, K., Ji, L., Reed, B., Eidenshink, J., & Dwyer, J. (2005) Multiplatform comparisons of MODIS and AVHRR normalized difference vegetation index data. *Remote Sensing of Environment*, **99**, 221–231.

Gallo, K.P. & Owen, T.W. (1999) Satellite-based adjustments for urban heat island temperature bias. *Journal of Applied Meteorology*, **38**, 806–813.

Gallo, K.P., Easterling, D.R. & Peterson, T.C. (1996) The Influence of Land Use/Land Cover on Climatological Values of the Diurnal Temperature Range. *Journal of Climate*, **9**, 2941–2944.

Gallo, K.P., Owen, T.W., Easterling, D.R. & Jamason, P.F. (1999) Temperature trends of the US historical climatology network based on satellite-designated land use/land cover. *Journal of Climate*, **12**, 1344–1348.

Gao, Y., Kaufman, Y.J., Tanré, D., Kolber, D. & Falkowski, P.G. (2001) Seasonal distributions of aeolian iron fluxes to the global ocean. *Geophysical Research Letters*, **28**, 29–32.

Garono, R.J., Simenstad, C.A., Robinson, R. & Ripley, H. (2004) Using high spatial resolution hyperspectral imagery to map intertidal habitat structure in Hood Canal Washington USA. *Canadian Journal of Remote Sensing*, **30**, 54–63.

Gates, D.M. (1993) *Climate Change and Its Biological Consequences*. Sinauer Associates Inc, Sunderland, MA.

Gege, P. (2004) The water colour simulator WASI: an integrating software tool for analysis and simulation of optical in-situ spectra. *Computers and Geosciences*, **30**, 523–532.

Giardino, C., Brando, V.E., Dekker, A.G., Strömbeck, N. & Candianig, G. (2007) Assessment of water quality in Lake Garda (Italy) using Hyperion. *Remote Sensing of Environment*, **109**, 83–195.

Giovinetto, M.B. & Zwally, H.J. (2000) Spatial distribution of net surface accumulation on the Antarctic ice sheet. *Annals of Glaciology*, **31**, 171–178.

Gitelson, A. (1993) Quantitative remote sensing methods for real-time monitoring of inland water quality. *International Journal of Remote Sensing*, **14**, 1269–1295.

Gleason, A.C.R., Eklund, A.-M., Reid, R.P. & Koch, V. (2006) Acoustic Signatures of the Seafloor: Tools for Predicting Grouper Habitat. *NOAA Professional Papers NMFS*, **5**, 38–47.

Gleick, P.H. (1993) Water and conflict: fresh water resources and international security. *International Security*, **18**, 79–112.

Gloersen, P. & Barath, F.T. (1977) A scanning multichannel microwave radiometer for Nimbus-G and Seasat-A. *IEEE Journal of Oceanic Engineering*, **2**, 172–178.

Glynn, P.W. (1993) Coral reef bleaching: facts, hypotheses and implications. *Coral Reefs*, **12**, 1–7.

Glynn, P.W. (1996) Coral reef bleaching: facts, hypotheses and implications. *Global Change Biology*, **2**, 495–509.

Gobbi, G.P., Barnaba, F., Giorgi, R. & Santacasa, A. (2000) Altitude-resolved properties of a Saharan-Dust event over the Mediterranean. *Atmospheric Environment*, **34**, 5119–5127.

Gobbi, G.P., Barnaba, F., Blumthaler, M., Labow, G. & Herman, J.R. (2002) Observed effects of particles nonsphericity on the retrieval of marine and desert dust aerosol optical depth by LiDAR. *Atmospheric research*, **61**, 1–114.

Goetz, S. J., Jantz, C. A., Prince, S. D., Smith, A. J., Wright, R. & Varlyguin, D. (2004) Integrated analysis of ecosystem interactions with land use change: the Chesapeake Bay watershed.

In: Defries, R.S., Asner, G.P. & Houghton, R.A. (eds.) *Ecosystems and Land Use Change*, 263–275. American Geophysical Union, Washington, DC.

Goff, J.A., Olson, H.C. & Duncan, C.S. (2000) Correlation of side-scan backscatter intensity with grain-size distribution of shelf sediments, New Jersey margin, *Geo-Marine Letters*, **20**, 43–49.

Goldstein, R.M., Engelhardt, H., Kamb, B. & Frolich, R.M. (1993) Satellite radar interferometry for monitoring ice sheet motion: application to an Antarctic ice stream, *Science*, **262**, 1525–1530.

González, J.E., Luvall, J.C., Rickman, D.L., Comarazamy, D. & Picon, A.J. (2007) Urban heat island identification and climatological analysis in a coastal, tropical city: San Juan, Puerto Rico. In: Weng, Q., Quattrochi, D.A. (eds.) *Urban Remote Sensing*, 412. CRC Press, Boca Raton, FL.

Goodchild, M.F. (2003) Geographic Information Science and Systems for Environmental Management. *Annual Review of Environment and Resources*, **28**, 493–519.

Gooding, R.A., Harley, C.D.G. & Tang, E. (2009) Elevated water temperature and carbon dioxide concentration increase the growth of a keystone echinoderm. *Proceedings of the National Academy of Sciences of the United States of America*, **106**, 9316–9321.

Goodman, J.A. & Ustin, S.L. (2007) Classification of benthic composition in a coral reef environment using spectral unmixing. *Journal of Applied Remote Sensing*, **1**, 011501.

Gordon, H.R. & Morel, A.Y. (1983) Remote assessment of ocean color for interpretation of satellite visible imagery: a review. *Lecture Notes on Coastal and Estuarine Studies*, **4**. Springer Verlag, NY.

Goward, S., Arvidson, T., Williams, D., Faundeen, J., Irons, J. & Franks, S. (2006) Historical record of Landsat global coverage: Mission operations, NSLRSDA, and international cooperator stations. *Photogrammetric Engineering and Remote Sensing*, **72**, 1155–1169.

Goward, S.N., Markham, B., Dye, D.G., Dulaney W. & Yang, J. (1991) Normalized Difference Vegetation Index measurements from the Advanced Very High Resolution Radiometer. *Remote Sensing of Environment*, **35**, 257–277.

Gower, S.T. & Norman, J.M. (1991) Rapid estimation of leaf area index in conifer and broad-leaf plantations. *Ecology*, **72**, 1896–1900.

Goyet, C., Gonçalves, R.I. & Touratier, F. (2009) Anthropogenic carbon distribution in the eastern South Pacific Ocean. *Biogeosciences*, **6**, 149–156.

Graber, H., Haus, B., Chapman, R. & Shay, L. (1997) HF radar comparisons with moored estimates of current speed and direction: Expected differences and implications. *Journal of Geophysical Research*, **102**, 18749–18766.

Graham, N.A.J., Wilson, S.K., Jennings, S., Polunin, N.V.C., Bijoux, J.P. & Robinson, J. (2006) Dynamic fragility of oceanic coral reef ecosystems. *Proceedings of the National Academy of Sciences for the United States of America*, **103**, 8425–8429.

Green, E.P., Mumby, P.J., Edwards, A.J. & Clark, C.D. (2000) Remote sensing handbook for tropical coastal management. In: Edwards, A.J. (ed.) *Coastal Management Sourcebooks*, **3**, 316. UNESCO, Paris.

Green, K., Finney, M., Campbell, J., Weinstein, D. & Lanndrum, V. (1995) Using GIS to predict fire behaviour. *Journal of Forestry*, **93**, 21–25.

Greening, H., Doering, P. & Corbett, C. (2006) Hurricane impacts on coastal ecosystems. *Estuaries and Coasts*, **29**, 877–879.

Greenstreet, S.P.R., Tuck, I.D., Grewar, G.N., Armstrong, E., Reid, D.G. & Wright, P.J. (1997) An assessment of the acoustic survey technique, RoxAnn, as a means of mapping seabed habitat. *ICES Journal of Marine Science*, **54**, 939–959.

Gregg, M. (2007) NOAA's Climate Goal. US Department of Commerce. *NOAA Earth System Monitor*, **16**, 3–4.

Grody, N.C. & Basist, A.N. (1996) Global identification of snowcover using SSM/I measurements. *IEEE Transactions on Geoscience and Remote Sensing*, **34**, 237–249.

Gross, M.F., Hardisky, M.A., Klemas, V. & Wolf, P.L. (1987) Quantification of biomass of the marsh grass *Spartina alterniflora loisel* using Landsat Thematic Mapper imagery. *Photogrammetric Engineering and Remote Sensing*, **53**, 1577–1583.

Gross, N. (1999) The Earth will don an electronic skin (cover story: 21 Ideas for the 21st Century). *Business Week Online*, August 30.

Gruhier, C., de Rosnay, P., Kerr, Y., Mougin, E. Ceschia, E., Calvet, J.-C. & Richaume, P. (2008) Evaluation of AMSR-E soil moisture product based on ground measurements over temperate and semi-arid regions. *Geophysical Research Letters*, **35**, (L10405).

Gu, G., Alder, R.F., Huffman, G.J. & Curtis, S. (2007) Tropical rainfall variability on interannual-to-interdecadal and longer time scales derived from the GPCP monthly product. *Journal of Climate*, **20**, 4033–4046.

Guenther, G. (2007) Airborne LiDAR bathymetry digital elevation. Model technologies and Applications: In: Maune, D. (ed.) *The DEM users manual*, 253–320. American Society for Photogrammetry and Remote Sensing.

Guenther, G., LaRocque, P. & Lillycrop, W. (1994) Multiple surface channels in SHOALS airborne LiDAR. *SPIE: Ocean Optics,* **XII 2258**, 422–430.

Guenther, G.C., Tomas, R.W.L. & LaRocque, P.E. (1996) Design considerations for achieving high accuracy with the SHOALS bathymetric LiDAR system. *SPIE: Laser Remote Sensing of Natural Waters: From Theory to Practice*, **15**, 54–71.

Guild, L., Lobitz, B., Armstrong, R., Gilbes, F., Gleason, A., Goodman, J., Hochberg, E., Monaco, M., Berthold, R. & Kerr, J. (2008) NASA Airborne AVIRIS and DCS Remote Sensing of Coral Reefs, 11th International Coral Reef Symposium, Fort Lauderdale, FL.

Guindon, B., Zhang, Y. & Dillabaugh, C. (2004) Landsat urban mapping based on a combined spectral-spatial methodology. *Remote Sensing of Environment*, **92**, 218–232.

Guinotte, J.M. & Fabry, V.J. (2008) Ocean acidification and its potential effects on marine ecosystems. *Annals of the New York Academy of Sciences*, **1134**, 320–342.

Gurney, K.R., Mendoza, D.L., Zhou, Y., Fischer, M.L., Miller, C.C., Geethakumar, S. & de La Rue du Can, S. (2009) High resolution fossil fuel combustion CO_2 emission fluxes for the United States. *Environmental Science and Technology*, **43**, 5535–5541.

Gutierrez, R., Gibeaut, J.C., Crawford, M.M., Mahoney, M.P., Smith, S., Gutelius, W., Carswell, D. & MacPherson, E. (1998) Airborne laser swath mapping of Galveston Island and Bolivar Peninsula, Texas. *Proc. Fifth International Conference on Remote Sensing for Marine and Coastal Environments, San Diego*, **1**, 236–243.

Hahn, C. J., Warren, S.G. & London, J. (1995) The effect of moonlight on observation of cloud cover at night, and application to cloud climatology. *Journal of Climate*, **8**, 1429–1446.

Hall, D.K. & Martinec, J. (1985) *Remote Sensing of Ice and* Snow, 189. Chapman & Hall, London.

Hall, D.K., Williams, R.S., Luthcke, S.B. & Digirolamo, N.E. (2008) Greenland ice sheet surface temperature, melt and mass loss: 2000–06. *Journal of Glaciology*, **54**, 81–93.

Hallegraeff, G.M. (1993) A review of harmful algal blooms and their apparent global increase. *Phycologia*, **32**, 79–99.

Hamilton, L.J., Mulhearn, P.J., Poeckert, R. (1999) Comparison of RoxAnn and QTC-View acoustic bottom classification system performance for the Cairns area, Great Barrier Reef, Australia. *Continental Shelf Research*, **19**, 1577–1597.

Hamylton, S. (2009) Determination of the separability of coastal community assemblages of the Al Wajh Barrier Reef, Red Sea, from hyperspectral data. *European Journal of Geosciences*, 1–11.

Han, G., (2005) Altimeter surveys, coastal tides and shelf circulation. In: Schwartz, M.L. (ed.) *Encyclopedia of Coastal Science*, 27–28. Springer, Dordrecht, The Netherlands.

Hand, J.L., Mahowald, N., Chen, Y., Siefert, R., Luo, C., Subramaniam A. & Fung, I. (2004) Estimates of soluble iron from observations and a global mineral aerosol model: Biogeochemical implications. *Journal of Geophysical Research*, **109**, D17205, doi: 10.1029/2004JD004574.

Handcock, R.N., Gillespie, A.R., Cherkauer, K.A., Kay, J.E., Burges, S.J. & Kampf, S.K. (2006) Accuracy and uncertainty of thermal-infrared remote sensing of stream temperatures at multiple spatial scales. *Remote Sensing of Environment*, **100**, 427–440.

Hanebuth, T.J.J., Stattegger, K. & Bojanowski, A. (2009) Termination of the Last Glacial Maximum sea-level lowstand: the Sunda-Shelf data revisited. *Global and Planetary Change*, **66**, 76–84.

Hankin, S., Malone, T.C., Lindstrom, E. & Cohen, R. (2004) US Integrated Ocean Observing System: Data and communications infrastructure. US Department of Commerce, NOAA. *Earth System Monitor*, **14**, 1–6.

Hansen, J., Lacis, A., Ruedy, R. & Sato, M. (1992) Potential climate impact of Mount-Pinatubo eruption. *Geophysical Research Letters*, **19**, 215–218.

Hansen, J.E., Ruedy, R., Sato, M., Imhoff, M., Lawrence, W., Easterling, D., Peterson, T. & Karl, T. (2001) A closer look at United States and global surface temperature change. *Journal of Geophysical Research*, **106**, 23947–23963.

Hansen, M.C. & Reed, B. (2000) A comparison of the IGBP DISCover and University of Maryland 1km Global Land Cover products. *International Journal of Remote Sensing*, **21**, 1365.

Hansen, M.C., DeFries, R.S., Townshend, J.R.G., Carroll, M., Dimiceli, C. & Sohlberg, R.A. (2003) Global percent tree cover at a spatial resolution of 500 meters: first results of the MODIS vegetation continuous fields algorithm. *Earth Interactions*, **7**, 1–15.

Harding, D.J., Lefsky, M.A., Parker, G.G. & Blair, J.B. (2001) Laser altimetry height profiles methods and validation for closed-canopy, broadleaf forests. *Remote Sensing of Environment* **76**, 283–297.

Hardisky, M.A., Daiber, F.C., Roman, C.T. & Klemas, V. (1984) Remote sensing of biomass and annual net aerial productivity of a salt marsh. *Remote Sensing of Environment*, **16**, 91–106.

Hare, C.E., Leblanc, K., DiTullio, G.R., Kudela, R.M., Zhang, Y., Lee, P.A., Riseman, S. & Hutchins, D.A. (2007) Consequences of increased temperature and CO_2 for phytoplankton community structure in the Bering Sea. *Marine Ecology Progress Series*, **352**, 9–16.

Harvell, C.D., Kim, K., Burkholder, J.M., Colwell, R.R., Epstein, P.R., Grimes, D.J., Hofman, E.E., Lipp, E.K., Osterhaus, A.D.M.E., Overstreet, R.M., Porter, J.W., Smith, G.W. & Vasta, G.R. (1999) Emerging marine diseases – climate links and anthropogenic factors. *Science*, **285**, 1505–1510.

Haywood, J. & Boucher, O. (2000) Estimates of the direct and indirect radiative forcing due to Tropospheric aerosols: a review. *Reviews of Geophysics*, **38**, 513–543.

Haywood, J.M. & Ramaswamy, V. (1998) Global sensitivity studies of the direct radiative forcing due to anthropogenic sulphate and black carbon aerosols, *Journal of Geophysical Research*, **103**, 6043–6058.

Hedley, J.D. & Mumby, P.J. (2003) A remote sensing method for resolving depth and sub-pixel composition of aquatic benthos. *Limnology and Oceanography*, **48**, 480–488.

Hedley, J.D., Mumby, P.J., Joyce, K.E. & Phinn, S.R. (2004) Spectral unmixing of coral reef benthos under ideal conditions. *Coral Reefs*, **23**, 60–73.

Hedley, J.D., Harborne, A.R. & Mumby, P.J. (2005) Simple and robust removal of sun glint for mapping shallow-water benthos. *International Journal of Remote Sensing*, **26**, 2107–2112.

Heinrichs, J.F., Cavalieri, D.J. & Markus, T. (2006) Assessment of the AMSR-E sea ice-concentration product at the ice edge using RADARSAT-1 and MODIS imagery. *IEEE Transactions on Geoscience and Remote Sensing*, **44**, 3070–3080.

Herman, J.R., Bhartia, P.K., Torres, O., Hsu, C., Seftor, C. & Celarier, E. (1997) Global distribution of UV-absorbing aerosol from Nimbus-7/TOMS data. *Geophysical research Letters*, **102**, 16911–16922.

Herzfeld, U.C., Lingle, C.S., Freeman, C., Higginson, C.A., Lambert, M.P., Lee, L.-H. & Voronina, V.A. (1997) Monitoring Changes of Ice Streams Using Time-series of Satellite-Altimetry-Based Digital Terrain Models. *Journal Mathematical Geology*, **29**, 859–890.

Hess, L., Melack, J. & Simonett, D. (1990) Radar detection of flooding beneath the forest canopy: A review. *International Journal of Remote Sensing*, **11**, 1313–1325.

Hilldale, R.C. & Raff, D. (2008) Assessing the ability of airborne LiDAR to map river bathymetry. *Earth Surface Processes and Landforms*, **33**, 773–783.

Hinz, A., Doerstel, C. & Meier, H. (2001) DMC – The Digital Sensor Technology of Z/I Imaging. In: Fritsch, D. & Spiller, R. (eds.), *Photogrammetric Week 2001*, 93–103. Wichmann, Heidelberg.

Hipple, J.D., Drazkowski, B. & Thorsell, P.M. (2005) Development in the Upper Mississippi Basin: 10 years after the Great Flood of 1993. *Landscape and Urban Planning*, **72**, 313–323.

Hirofumi, H., Melton, F., Ichii, K., Milesi, C., Wang, W. & Nemani, R.R. (2009) Evaluating the impacts of climate and elevated carbon dioxide on tropical rainforests of the western Amazon basin using ecosystem models and satellite data. *Global Change Biology*, 10.1111/j.1365–2486.2009.01921.

Hochberg, E.J. & Atkinson, M.J. (2000) Spectral discrimination of coral reef benthic communities. *Coral Reefs*, **19**, 164–171.

Hochberg, E.J. & Atkinson, M.J. (2003) Capabilities of remote sensors to classify coral, algae, and sand as pure and mixed spectra. *Remote Sensing of Environment*, **85**, 174–189.

Hochberg, E.J., Andréfouët, S. & Tyler, M.R. (2003) Sea surface correction of high spatial resolution IKONOS images to improve bottom mapping in near-shore environments. *IEEE Transactions on Geoscience and Remote Sensing*, **41(7)**, 1724–1729.

Hochberg, E.J., Apprill, A.M., Atkinson, M.J. & Bidigare, R.R. (2006) Bio-optical modeling of photosynthetic pigments in corals. *Coral Reefs*, **25**, 99–109.

Hocking, W.K., Carey-Smith, T., Tarasick, D.W., Argall, P.S., Strong, K., Rochon, Y., Zawadzki, I. & Taylor, P.A. (2007) Detection of stratospheric ozone intrusions by windprofiler radars. *Nature*, **450**, 281–284.

Hoegh-Guldberg, O. (1999) Coral bleaching, climate change and the future of the world's coral reefs. *Marine and Freshwater Research*, **50**, 839–866.

Hoegh-Guldberg, O., Mumby, P.J., Hooten, A.J., Steneck, R.S., Greenfield, P., Gomez, E., Harvell, C.D., Sale, P.F., Edwards, A.J., Caldeira, K., Knowlton, N., Eakin, C.M., Iglesias-Prieto, R., Muthiga, N., Bradbury, R.H., Dubi, A. & Hatziolos, M.E. (2007) Coral Reefs Under Rapid Climate Change and Ocean Acidification. *Science*, **318**, 1737–1742.

Hoegh-Guldberg, O., Hoegh-Guldberg, H., Veron, J.E.N., Green, A., Gomez, E.D., Lough, J., King, M., Ambariyanto, Hansen, L., Cinner, J., Dews, G., Russ, G., Schuttenberg, H.Z., Peñaflor, E.L., Eakin, C.M., Christensen, T.R.L., Abbey, M., Areki, F., Kosaka, R.A., Tewfik, A. & Oliver, J. (2009) *The Coral Triangle and Climate Change: Ecosystems, People and Societies at Risk*. WWF Australia, Brisbane.

Hofmann, D.J., Butler, J.H. & Tans, P.P. (2009) A new look at atmospheric carbon dioxide. *Atmospheric Environment*, **43**, 2084–2086.

Hoffman, P.F., Kaufman, A.J., Halverson, G.P. & Schrag, D.P. (1998) A Neoproterozoic snowball Earth. *Science*, **281**, 1342–1346.

Holben, B., Vermote, E., Kaufman, Y.J., Tanré, D. & Kalb, V. (1992) Aerosol retrieval over-land from AVHRR data – application for atmospheric correction. *IEEE Transaction on Geoscience and Remote Sensing*, **30**, 212–222.

Holben, B.N., Kandell, J. & Kaufman, Y.J. (1990) Calibration of NOAA-11 AVHRR Visible and Near-IR bands. *International Journal of Remote Sensing*, **11**, 1511–1519.

Hollinger, J.P., Pierce, J.L. & Poe, G.A. (1990) SSM/I instrument evaluation. *IEEE Transactions of Geoscience and Remote Sensing*, **28**, *781–790*.

Holton, J.R., Haynes, P.H., McIntyre, M.E., Douglass, A.R., Rood, R.B. & Pfister, L. (1995) Stratosphere-troposphere exchange. *Reviews of Geophysics*, **33**, 403–439.

Hoogenboom, H.J., Dekker, A.G. & de Haan, J.F. (1998a) Retrieval of chlorophyll and sus-pended matter in inland waters from CASI data by matrix inversion. *Canadian Journal of Remote Sensing*, **24**, 144–152.

Hoogenboom, H.J., Dekker, A.G. & Althuis, I.J.A. (1998b) Simulation of AVIRIS sensitivity for detecting chlorophyll over coastal and inland waters. *Remote Sensing of Environment*, **65**, 333–340.

Hooker, S.B. & Esaias, W.E. (1993) An overview of the SeaWiFS project. *American Geophysical Union, EOS Transactions*, **74**, 241–246.

Hooker, S.B., McLain, C.R. & Mannino, A. (2007) *A comprehensive plan for the long-term calibration and validation of oceanic biogeochemical satellite data.* NASA strategic plan-ning document: NASA/SP–2007–214152, July 2007.

Hsu, N.C., Herman, J.R., Bhartia, P.K., Seftor, C.J., Torres, O., Thompson, A.M., Gleason, J.F., Eck, T.F. & Holben, B.N. (1996) Detection of biomass burning smoke from TOMS measurements. *Geophysical research Letters*, **23**, 745–748.

Hudson, R.D. & Thompson, A.M. (1998) Tropical tropospheric ozone from total ozone mapping spectrometer by a modified residual method. *Journal of Geophysical Research*, **103**, 129–145.

Hughes, T.P., Baird, A.H., Bellwood, D.R., Card, M., Connolly, S.R., Folke, C., Grosberg, R., Hoegh-Guldberg, O., Jackson, J.B.C., Kleypas, J., Lough, J.M., Marshall, P., Nyström, M., Palumbi, S.R., Pandolfi, J.M., Rosen, B. & Roughgarden, J. (2003) Climate change, human impacts, and the resilience of coral reefs. *Science*, **310**, 929–933.

Hughes, T.P., Rodrigues, M.J., Bellwood, D.R., Ceccarelli, D., Hoegh-Guldberg, O., McCook, L., Moltschaniwskyj, N., Pratchett, M.S., Steneck, R.S. & Willis, B. (2007) Phase shifts, herbivory, and the resilience of coral reefs to climate change. *Current Biology*, **17**, 360–365.

Hwang, Y. & Southwell, B.G. (2009) Science TV news exposure predicts science beliefs: real world effects among a national sample. *Communication Research*, **36**, 724–742.

Idso, S.B., Balling, Jr., R.C. & Cerveny, R.S. (1990) Carbon dioxide and hurricanes: implica-tions of northern hemispheric warming for Atlantic/Caribbean storms. *Meteorology and Atmospheric Physics*, **42**, 259–263.

Ignatov, A. & Stowe, L. (2002) Aerosol retrievals from individual AVHRR channels. Part I: Retrieval algorithm and transition from Dave to 6S radiative transfer model. *Journal of the Atmospheric Sciences*, **59**, 313–334.

Ikeda, M. & Dobson, F.W. (1995) *Oceanographic Applications of Remote Sensing.* CRC Press, NY.

Imhoff, M.L., Lawrence, W.T., Stutzer, D.C. & Elvidge, C.D. (1997) A technique for using composite DMSP/OLS 'City Lights' satellite data to map urban area. *Remote Sensing of the Environment*, **61**, 361–370.

Intergovernmental Panel on Climate Changes (2007) *Climate Change 2007: The Physical Science Basis.* WMO/UNEP, Paris (www.ipcc.ch).

IOCCG (2000) *Remote sensing of ocean colour in coastal, and other optically-complex, waters.* Volume 3 of Reports of the International Ocean-Colour Coordinating Group. IOCCG, Dartmouth, Canada.

IPCC (2007) *Climate change 2007: The physical science basis. Summary for policymakers. Contribution of Working Group 1 to the fourth assessment report.* The Intergovernmental Panel on Climate Change.

Irish, J.L. & Lillycrop, W.J. (1999) Scanning laser mapping of the coastal zone: the SHOALS system. *ISPRS J of Photogram and Remote Sensing*, **54**, 123–129.

Island Press (2000) *The Hidden Costs of Coastal Hazards: Implications for Risk Assessment and Mitigation.* Island Press, Washington, DC.

Jackson, J.B., Kirby, M.X., Berger, W.H., Bjorndal, K.A., Botsford, L.W., Bourque, B.J., Bradbury, R., Cooke, R., Erlandson, J, Estes, J.A., Hughes, T.P., Kidwell, S., Lange, C.B., Lenihan, H.S., Pandolfi, J.M., Peterson, C.H., Steneck, R.S., Tegner, M.J. & Warner, R. (2001) Historical over-fishing and the recent collapse of coastal ecosystems. *Science,* **293**, 629–639.

Jacobs, S.S., Giulivi, C.F. & Mele, P.A. (2002) Freshening of the Ross Sea during the late 20th Century. *Science*, **297**, 386–389.

Jago, R.A., Cutler, M.E.J. & Curran, P.J. (1999) Estimating canopy chlorophyll concentration from field and airborne spectra. *Remote Sensing of Environment*, **68**, 217–224.

Jensen, J.R. (1996) *Introductory Digital Image Processing: A Remote Sensing Perspective*, 2nd edn. Prentice-Hall, NJ.

Jensen, J.R. (2007) *Remote Sensing of the Environment: An Earth Resource Perspective.* Prentice-Hall, NJ.

Jensen, J.R. & Cowen, D.C. (1999) Remote sensing of urban/suburban infrastructures and socio-economic attributes. *Photogrammetric Engineering and Remote Sensing*, **65**, 611–622.

Jensen, J.R., Olson, G., Schill, S.R., Porter, D.E. & Morris, J. (2002) Remote sensing of biomass, leaf-area-index, and chlorophyll a and b content in the ACE Basin National Estuarine Research Reserve using sub-meter digital camera imagery. *Geocarto International*, **17**, 25–34.

Jerlov, N.G. (1976) *Marine Optics.* Elsevier, Amsterdam.

Jickells, T.D., An, Z. S., Andersen, K.K., Baker, A.R., Bergametti, G., Brooks, N., Cao, J.J., Boyd, P.W., Duce, R.A., Hunter, K.A., Kawahata, H., Kubilay, N., laRoche, J., Liss, Mahowald, P.S., Prospero, N., Ridgwell, J.M., Tegen, A.J. & Torres, I. (2005) Global iron connections between desert dust, ocean biogeochemistry, and climate. *Science*, **308**, 67–71.

Johannessen, O.M., Khvorostovsky, K., Miles, M.W. & Bobyley, L.P. (2005) Recent ice-sheet growth in the interior of Greenland. *Science*, **310**, 1013–1016.

Jones, A.T., Thankappan, M., Logan, G.A., Kennard, J.M., Smith, C.J., Williams, A.K. & Lawrence, G.M. (2006) Coral spawn and bathymetric slicks in Synthetic Aperture Radar (SAR) data from the Timor Sea, north west Australia. *International Journal of Remote Sensing*, **27**, 2063–2069.

Jones, T.A. & Christopher, S.A. (2007) MODIS derived fine mode fraction characteristics of marine, dust, and anthropogenic aerosols over the ocean. *Journal of geophysical research*, **112**, D22204.1–D22204.

Jung, M., Churkina, G. & Henkel, K. (2006) Exploiting synergies of global land cover products for carbon cycle modeling. *Remote Sensing of Environment*, **101**, 534–553.

Justice, C.O. (ed.) (1986) Monitoring the grasslands of semi-arid Africa using NOAA-AVHRR data. *International Journal of Remote Sensing*, **7**, 1383–1622.

Kalnay, E. & Cai, M. (2003) Impact of urbanization and land-use change on climate. *Nature*, **423**, 528–531.

Kampe, T.U., Johnson, B.R., Kuester M. & Keller, M. (2010) NEON: the first continental-scale ecological observatory with airborne remote sensing of vegetation canopy biochemistry and structure. *Journal of Applied Remote Sensing*, **4**, 1–25.

Kattawar, G.W. & Plass, G.N. (1972) Radiative transfer in the Earth's atmosphere-II. Radiance in the atmosphere and ocean. *Journal of Physical Oceanography*, **2**, 148–156.

Katz, M.E., Cramer, B.S., Mountain, G.S., Katz, S. & Miller, K.G. (2001) Uncorking the bottle: what triggered the Paleocene/Eocene thermal maximum methane release. *Paleoceanography*, **16**, 549–562.

Kaufman, Y.J., Fraser, R.S. & Ferrare, R.A. (1990) Satellite remote sensing of large-scale air pollution: Method. *Journal of Geophysical Research*, **95**, 9895–9909.

Kaufman, Y.J., Tanré, D. & Remer, L. (1997) Remote Sensing of Tropospheric aerosol from EOS-MODIS over the land using dark targets and dynamic aerosol models. *Journal of Geophysical Research*, **102**, 17051–17067.

Kaufman, Y.J., Koren, I., Remer, L.A., Tanré, D., Ginoux, P. & Fan, S. (2005) Dust transport and deposition observed from the Terra-MODIS spacecraft over the Atlantic Ocean. *Journal of Geophysical Research*, **110**, D10S12.1–D10S12.16.

Kaufmann, R.K., Seto, K.C., Schneider, A., Zhou, L. & Liu, Z. (2007) Climate Response to Rapid Urban Growth: Evidence of a Human-Induced Precipitation Deficit. *Journal of Climate*, **20**, 2299–2306.

Kay, S., Hedley, J.D. & Lavender, S. (2009) Sun glint correction of high and low spatial resolution images of aquatic scenes: A review of methods for visible and near-infrared wavelengths. *Remote Sensing*, **1**, 697–730.

Keafer, B.A., Anderson, D.M. (1993) Use of remotely-sensed sea surface temperatures in studies of *Alexandrium tamarense* bloom dynamics. In: Smayda, T.M. & Shimizu, Y. (eds.) *Toxic Phytoplankton in the Sea*, 763–768. Proc. 5th Int'l Conf., Elsevier, Amsterdam.

Keck, J., Houston, R.S., Purkis, S.J., & Riegl, B. (2005) Unexpectedly high cover of *Acropora cervicornis* on offshore reefs in Roatán (Honduras). *Coral Reefs*, **24**, 509–509.

Kelly, M. & Tuxen, M. (2009) Remote sensing support for tidal vegetation research and management. In: Yang, X. (Ed.) *Remote sensing and geospatial technologies for coastal ecosystem assessment and managemant*, 341–362. Springer Verlag, Berlin.

Kempeneers, P., Deronde, B., Provoost, S. & Houthuys, R. (2009) Synergy of airborne digital camera and LiDAR data to map coastal dune vegetation. *Journal of Coastal Research*, **53**, 73–82.

Kenyi, L. & Kaufmann, V. (2003) Estimation of rock glacier surface deformation using SAR interferometry data. *IEEE Transactions on Geoscience and Remote Sensing*, **41**, 1512–1515.

Key, J. & Haefliger, M. (1992) Arctic ice surface temperature retrieval from AVHRR thermal channel. *Journal of Geophysical Research*, **97** (D5), 5885–5893.

Key, J.R., Collins, J.B., Fowler, C. & Stone, R.S. (1997) High-latitude surface temperature estimates from thermal satellite data. *Remote Sensing of Environment*, **61**, 302–309.

Kheshgi, H.S., Smith, S.J. & Edmonds, J.A. (2005) Emissions and atmospheric CO_2 stabilization: long-term limits and paths. *Journal Mitigation and Adaptation Strategies for Global Change*, **10**, 213–220.

Kidder, S.Q. & Vonder Haar, T.H. (1995) *An Introduction to Satellite Meteorology*. Academic Press, San Diego, CA.

Kimball, J., McDonald K.C., Frolking, S. & Running, S.W. (2004) Radar remote sensing of the spring thaw transition across a boreal landscape. *Remote Sensing of Environment*, **89**, 163–175.

Kimball, J.S., Keyser, A.R., Running, S.W. & Saatchi, S.S. (2000) Regional assessment of boreal forest productivity using an ecological process model and remote sensing parameter maps. *Tree Physiology*, **20**, 761–775.

Kinsey, D.W. & Hopley, D. (1991) The significance of coral reefs as global carbon sinks – Response to greenhouse. *Palaeogeography, Palaeoclimatology, Palaeoecology*, **89**, 363–377.

Kinzel, P.J., Wright, C.W., Nelson, J.M. & Burman, A.R. (2007) Evaluation of an experimental LiDAR for surveying a shallow, braided, sand-bedded river. *Journal of Hydraulic Engineering*, **133**, 838–842.

Kiraly, S.J., Cross, F.A. & Buffington, J.D. (eds.) (1990) Federal coastal wetland mapping programs: A report by the National Ocean Pollution Policy Board's Habitat Loss and Modification Working Group. *Biological Report*, **90(18),** 174. US Department of Interior, Fish and Wildlife Service, Washington, DC.

Kirchner, I., Stenchikov, G., Graf, H.-F., Robock, A. & Antuna, J. (1999) Climate model simulation of winter warming and summer cooling following the 1991 Mount Pinatubo volcanic eruption. *Journal of Geophysical Research*, **104**, 19039–19055.

Kirk, J.T.O. (1981) Monte Carlo procedure for simulating the penetration of light into natural waters. *CSIRO Division of Plant Industry, Technical Paper*, **36**, 1–16.

Klemas, V. (2000) Remote sensing of landscape level environmental indicators. *Environmental Management*, **27**, 47–57.

Klemas, V. (2005) Remote sensing: Wetlands classification. In: Schwartz, M.L. (ed.) *Encyclopedia of Coastal Science*, 804–807. Springer, Dordrecht, The Netherlands.

Klemas, V. (2007) Remote sensing of coastal wetlands and estuaries. *Proceedings of Coastal Zone 07*. NOAA Coastal Services Center, Charleston, SC.

Klemas, V. (2009) Sensors and Techniques for Observing Coastal Ecosystems. In: Yang, X. (ed.) *Remote Sensing and Geospatial Technologies for Coastal Ecosystem Assessment and Management*. Springer-Verlag, Berlin.

Klemas, V. (2009) The role of remote sensing in predicting and determining coastal storm impacts. *Journal of Coastal Research*, **25**, 1264–1275.

Klemas, V. & Philpot, W.D. (1981) Drift and dispersion studies of ocean-dumped waste using Landsat imagery and current drogues. *Photogrammetric Engineering and Remote Sensing*, **47**, 533–542.

Klemas, V., Dobson, J. E., Ferguson, R. L. & Haddad, K.D. (1993) A coastal land cover classification system for the NOAA Coastwatch Change Analysis Project. *Journal of Coastal Research*, **9(3),** 862–872.

Kleypas, J.A. & Langdon, C. (2006) Coral reefs and changing seawater chemistry. In: Phinney, J.T., Hoegh-Guldberg, O., Kleypas, J., Skirving, W. & Strong, A. (eds.) *Coral reefs and climate change: science and management*, **61**, 73–110. AGU Monograph Series, Coastal and Estuarine Studies, American Geophysical Union. Washington, DC.

Kleypas, J.A., Buddemeier, R.W., Archer, D., Gattuso, J.-P., Langdon, C. & Opdyke, B.N. (1999) Geochemical consequences of increased atmospheric carbon dioxide on coral reefs. *Science*, **284**, 118–120.

Kleypas, J.A., Buddemeier, R.W. & Gattuso, J.-P. (2001) The future of coral reefs in an age of global change. *International Journal of Earth Sciences*, **90**, 426–437.

Kloiber, S.M., Brezonik, P.L. & Bauer, M.E. (2002a) Application of Landsat imagery to regional-scale assessments of lake clarity. *Water research*, **36**, 4330–4340.

Knoll, A.H., Bambach, R.K., Canfield, D.E. & Grotzinger, J.P. (1996) Comparative earth history and Late Permian mass extinction. *Science*, **273**, 452–457.

Knuston, T.R., Tuleya, R.E. & Kurihara, Y. (1998) Simulated increase of hurricane intensities in a CO_2-warmed climate. *Science*, **279**, 1018–1020.

Knutti, R., Stocker, T.F., Joos, F & Plattner, G-K. (2002) Constraints on radiative forcing and future climate change from observations and climate model ensembles. *Nature*, **416**, 719–723.

Kogan, F.N. (1997) Global drought watch from space. *Bulletin of the American Meteorological Society*, **78**, 621–636.

Kogan, F.N. (2001) Operational space technology for global vegetation assessment. *Bulletin of the American Meteorological Society*, **82**, 1949–1964.

Kogan, F.N. (2002) World Droughts in the New Millennium from AVHRR-based Vegetation Health Indices. *American Geophysical Union, EOS, Transactions*, **83**, 557–563.

Kohler, K.E., Purkis, S.J., Rohmann, S.O. & Riegl, B.M. (2006) Development of a Hybrid Mapping Tool (HMT) for the characterisation of coral reef landscapes. *Proceedings of the International Society for Reef Studies European Meeting*, 19–22 September. Bremen, Germany.

Kondratyev, K.Y. & Cracknell, A.P. (1998) *Observing Global Climate Change*. Taylor and Francis, London.

Kondratyev, K.Y. & Varotsos, C. (2002) Remote sensing and global Tropospheric ozone observed dynamics. *International Journal of Remote Sensing*, **23**, 159–178.

König, M., Winther, J.-G. & Isaksson, E. (2001) Measuring Snow and Glacier Ice Properties from Satellite. *Reviews of Geophysics*, **39**, 1–27.

Kopilevich, Y.I., Feygels, V.I., Tuell, G.H. & Surkov, A. (2005) Measurement of ocean water optical properties and seafloor reflectance with scanning hydrographic operational airborne LiDAR system (SHOALS): I. Theoretical Background. *Proceedings SPIE*, **Vol. 5885.**

Koponen, S., Pulliainen, J., Kallio, K. & Hallikainen, M. (2002) Lake water quality classification with airborne hyperspectral spectrometer and simulated MERIS data. *Remote Sensing of Environment*, **79**, 51–59.

Kostylev, V.E., Todd, B.J., Fader, G.B.J., Courtney, R.C., Cameron, G.D.M. & Pickrill, R.A. (2001) Benthic habitat mapping on the Scotian Shelf based on multibeam bathymetry, surficial geology and sea floor photographs. *Marine Ecology Progress Series*, **219**, 121–137.

Krabill, W., Hanna, E., Huybrechts, P., Abdalati, W., Cappelen, J., Csatho, B., Frederick, E., Manizade, S., Martin, C., Sonntag, J., Swift, R., Thomas, R. & Yungel, J. (2004) Greenland Ice Sheet: Increased coastal thinning, *Geophysical Research Letters*, **31**, L24402, doi:10.1029/2004GL021533.

Krabill, W.B., Wright, C.W., Swift, R.N., Frederick, E.B., Manizade, S.S., Yungel, J.K., Martin, C.F., Sonntag, J.G., Duffy, M., Hulslander, W. & Brock, J.C. (2000) Airborne laser mapping of Assateague National Seashore Beach. *Photogrammetric Engineering and Remote Sensing*, **66**, 65–71.

Kramer, H.J. (2002) *Observation of the Earth and its Environment*. Springer, Berlin.

Kraszewski, A. (1996). *Microwave aquametry: Electromagnetic wave interaction with water-containing materials*. IEEE Press, Piscataway, NJ.

Kravitz, B., Robock, A., Oman, L., Stenchikov, G. & Marquardt, A.B. (2009) Sulfuric acid deposition from stratospheric geoengineering with sulfate aerosols. *Journal of Geophysical Research*, **114**, D14109.

Kuffner, I.B., Brock, J.C., Grober-Dunsmore, R., Bonito, V.E., Hickey, T.D., & Wright, C.W. (2007) Relationships between reef fish communities and remotely sensed rugosity measurements in Biscayne National Park, Fla. *Environmental Biology of Fishes*, **78**, 71–82.

Kump, L.R. (2002) Reducing uncertainty about carbon dioxide as a climate driver. *Nature*, **419**, 188–190.

Kutser, T., Dekker, A.G. & Skirving, W. (2003) Modeling spectral discrimination of Great Barrier Reef benthic communities by remote sensing instruments. *Limnology and Oceanography*, **48**, 497–510.

Kutser, T., Vahtmäe, E. & Praks, E. (2009) A sun glint correction method for hyperspectral imagery containing areas with non-negligible water leaving NIR signal. *Remote Sensing of Environment*, **113**, 2267–2274.

Kwok, R., Cunningham, G.F. & Hibler III, W.D. (2003) Sub-daily sea ice motion and deformation from RADARSat observations. *Geophysical Research Letters*, **30**, 2218.

Kwok, R., Cunningham, G.F., Wensnahan, M., Rigor, I., Zwally, H.J. & Yi, D. (2009) Thinning and volume loss of the Arctic Ocean sea ice cover: 2003–2008. *Journal of Geophysical Research*, **114**, C07005.

Lachowski, H., Maus, P., Golden, M., Johnson, J., Landrum, V., Powell, J., Varner, V., Wirth, T., Gonzales, J. & Bain, S. (1995) *Guidelines for the Use of Digital Imagery for Vegetation Mapping*. US Department of Agriculture, Forest Service, Washington, DC.

Lamb, H.H. (1969) Volcanic dust in the atmosphere; with a chronology and assessment of its meteorological significance. *Philosophical Transactions for the Royal Society of London. Series A, Mathematical and Physical Sciences*, **266**, 425–533.

Lambeck, K. & Chappell, J. (2001) Sea level change during the Quaternary. *Science* **292**, 679–686.

Landsberg, J.J., Prince, S.D., Jarvis, P.G., McMurtrie, R.E., Luxmoore, R.J. & Medlyn, B.E. (1996) Energy conversion and use in forests: an analysis of forest production in terms of radiation utilization efficiency. In: Gholz, H.L., Nakane, K., Shimoda, H. (eds.) *The use of remote sensing in the modeling of forest productivity*, 273–298. Kluwer Academic Publishers, Dordrecht, The Netherlands.

Langdon, C., Takahashi, T., Sweeney, C., Chipman, D., Goddard, J., Marubini, F., Aceves, H., Barnett, H. & Atkinson, M.J. (2000) Effect of calcium carbonate saturation state on the calcification rate of an experimental coral reef. *Global Biogeochemical Cycles*, **14**, 639–654.

Larsen, C., Clark, I., Guntenspergen, G., Cahoon, D., Caruso, V., Hupp, C. & Yanosky, T. (2004) The Blackwater NWR Inundation Model. *USGS Open File Report*, Aug 18 2004, 04–1302.

Lathrop, R.G., Cole, M.B. & Showalter, R.D. (2000) Quantifying the habitat structure and spatial pattern of New Jersey (USA) salt marshes under different management regimes. *Wetlands Ecology and Management*, **8**, 163–172.

Latifovic, R., Trishchenko, A.P., Chen, J., Park, W.B., Khlopenkov, K.V., Fernandes, R., Pouliot, D., Ungureanu, C., Luo, Y., Wang, S., Davidson, A. & Cihlar, J. (2005) Generating historical AVHRR 1 km baseline satellite data over Canada suitable for climate change studies. *Canadian Journal of Remote Sensing*, **31**, 324–346.

Lautenbacher, C.C. (2005) One world, interconnected. *Sea Technology*, Sept 2005.

Lawford, R.G., (2008) Observing surface waters for global change applications. In: Chuvieco, E. (ed.) *Earth Observation of Global Change*. Springer-Verlag, Berlin.

le Maire, G., Francois, C. & Dufrene, E. (2004) Towards universal broad leaf chlorophyll indices using PROSPECT simulated database and hyperspectral reflectance measurements. *Remote Sensing of Environment*, **89**, 1–28.

Le Treut, H., Somerville, R., Cubasch, U., Ding, Y., Mauritzen, C., Mokssit, A., Peterson, T. & Prather, M. (2007) Historical overview of climate change. In: Solomon, S., Qin, D., Manning, M., Chen, Z., Marquis, M., Averyt, K.B., Tignor, M. & Miller, H.L. (eds.) *Climate Change 2007: The physical science basis. Contribution of Working Group I to the fourth assessment report of the Intergovernmental Panel on Climate Change*. Cambridge University Press, Cambridge UK & NY.

Leakey, R. & Lewin, R. (1995) *The sixth extinction: Biodiversity and its survival.* Doubleday, NY.

Leben, R.R., Born, G.H., Urban, T.J. & Schutz, B.E. (2003) A test of ocean mesoscale monitoring with ICESat altimetry. *American Geophysical Union, Fall Meeting 2003.*

Lee, Z., Carder, K.L., Mobley, C.D., Steward, R.G. & Patch, J.S. (1999) Hyperspectral remote sensing for shallow waters: 2. Deriving bottom depths and water properties by optimization. *Applied Optics,* **38**, 3831–3853.

Lee, Z., Carder, K.L., Chen, R.F. & Peacock, T.G. (2001) Properties of the water column and bottom derived from Airborne Visible Infrared Imaging Spectrometer (AVIRIS) data. *Journal of Geophysical Research,* **106**, 11639–11651.

Lefsky, M.A., Harding, D., Cohen, W.B., Parker, G. & Shugart, H.H. (1999) Surface LiDAR remote sensing of basal area and biomass in deciduous forests of eastern Maryland, USA. *Remote Sensing of Environment,* **67**, 83–98.

Lefsky, M.A., Cohen, W.B., Parker, G.G. & Harding, D.J. (2002) LiDAR remote sensing for ecosystem studies. *BioScience,* **52**, 19–30.

Legleiter, C.J. & Roberts, D.A. (2005) Effects of channel morphology and sensor spatial resolution on image-derived depth estimates. *Remote Sensing of Environment,* **95**, 231–247.

Legleiter, C.J. & Roberts, D.A. (2009) A forward image model for passive optical remote sensing of river bathymetry. *Remote Sensing of Environment,* **113**, 1025–1045.

Lehner, B., Verdin, K. & Jarvis, A. (2006) *HydrSHEDS technical documentation.* World Wildlife Fund, Washington, D.C. (Available at http://hydrosheds.cr.usgs.gov)

Lehner, B., Verdin, K. & Jarvis, A. (2008) New global hydrography derived from spaceborne elevation data. *Eos, Transactions, AGU,* **89**, 93–94.

Leica (2002) *ADS40 Airborne Digital Sensor.* Leica Geosystems, GIS and Mapping, LLC, Atlanta, GA.

Leitão, A.B. & Ahern, J. (2002) Applying landscape ecological concepts and metrics in sustainable landscape planning. *Landscape and Urban Planning,* **59**, 65–93.

Lenton, A., Codron, F., Bopp, L., Metzl, N., Cadule, P., Tagliabue, A. & Le Sommer, J. (2009) Stratospheric ozone depletion reduces ocean carbon uptake and enhances ocean acidification. *Geophysical Research Letters,* **36**, L12606.

Leprieur, C., Verstraete, M.M. & Pinty, B. (1994) Evaluation of the performance of various vegetation indices to retrieve vegetation cover from AVHRR data. *Remote Sensing Reviews,* **10**, 265–284.

Lesser, M.P. & Mobley, C.D. (2007) Bathymetry, water optical properties, and benthic classification of coral reefs using hyperspectral remote sensing imagery. *Coral Reefs,* doi: 10.1007/s00338-007-0271–5.

Levy, M.A. (1995) Is the environment a national security issue? *International Security,* **20**, 35–62.

Li, R. & Li, J. (2004) Satellite remote sensing technology for lake water clarity monitoring: An overview. *Environmental Informatics Archives,* **2**, 893–901.

Light, D.L. (2001) An airborne direct digital imaging system. *Photogrammetric Engineering and Remote Sensing,* 67, 1299–1305.

Lillesand, T.M. & Kiefer, R.W. (1994) *Remote Sensing and Image Interpretation,* 3rd edn. Wiley, NY.

Lillesand, T.M., Kiefer, R.W. & Chipman, J.W. (2004) *Remote Sensing and Image Interpretation,* 5th edn. Wiley, NY.

Lindsey, R. & Rothrock, D. (1993) The calculation of surface temperature and albedo of Arctic sea ice from AVHRR. *Annals of Glaciology,* **17**, 174–183.

Linehan, J.R. & Gross, M. (1998) Back to the future, back to basics: the social ecology of landscapes and the future of landscape planning. *Landscape and Urban Planning*, **42**, 207–223.

Lipfert, F.W. (1994) *Air Pollution and Community Health: A Critical Review and Data Sourcebook*. Van Nostrand Reinhold, NY.

Liu, G., Strong, A.E., Skirving, W. & Arzayus, L.F. (2005) Overview of NOAA coral reef watch program's near-real time satellite global coral bleaching monitoring activities. *Proceedings of the 10th International Coral Reef Symposium 2004*, **1**, 1783–1793. Okinawa, Japan.

Longhurst, A. (1998) *Ecological Geography of the Sea*. Academic Press, San Diego, CA.

Longhurst, A., Sathyendranath, S., Platt, T. & Caverhill, C. (1995) An estimate of global primary production in the ocean from satellite data. *Journal of Plankton Research*, **17**, 1245–1271.

Longley, P.A. (2002) Geographical information systems: Will developments in urban remote sensing and GIS lead to 'better' urban geography? *Progress in Human Geography*, **26**, 231–239.

Lorenz, E.N. (1969) Atmospheric predictability as revealed by naturally occurring analogues. *Journal of the Atmospheric Sciences*, **26**, 636–646.

Louchard, E.M., Reid, R.P., Stephens, F.C., Davis, C.O., Leathers, R.A. & Downes, T.V. (2003) Optical remote sensing of benthic habitats and bathymetry in coastal environments at Lee Stocking Island, Bahamas: a comparative spectral classification approach. *Limnology & Oceanography*, **48**, 511–521.

Lu, D. & Weng, Q. (2006) Use of impervious surface in urban land-use classification. *Remote Sensing of Environment*, **102**, 146–160.

Lu, D. & Weng, Q. (2007) A survey of image classification methods and techniques for improving classification performance. *International Journal of Remote Sensing*, **28**, 823–870.

Lunetta, R.S. & Elvidge, C.D. (1998) *Remote Sensing Change Detection: Environmental Monitoring Methods and Applications*. Ann Arbor Press, MI.

Lunetta, R.S. & Balogh, M.E. (1999) Application of multi-temporal Landsat 5 TM imagery for wetland identification. *Photogrammetric Engineering and Remote Sensing*, **65**, 1303–1310.

Luthcke, S.B., Zwally, H.J., Abdalati, W., Rowlands, D.D., Ray, R.D., Nerem, R.S., Lemoine, F.G., McCarthy, J.J. & Chinn, D.S. (2006) Recent Greenland Ice Mass Loss by Drainage System from Satellite Gravity Observations. *Science*, **314**, 1286–1289.

Lyon, J.G. & Hutchinson W.S. (1995) Application of a radiometric model for evaluation of water depths and verification of results with airborne scanner data. *Photogrammetric Engineering and Remote Sensing*, **61**, 161–166.

Lyon, J.G. & McCarthy, J. (1995) *Wetland and Environmental Applications of GIS*. Lewis Publishers, NY.

Lyon, J.G., Lunetta, R.S. & Williams, D.C. (1992) Airborne multispectral scanner data for evaluating bottom sediment types and water depths of the St. Marys River, MI. *Photogrammetric Engineering and Remote Sensing*, **58**, 951–956.

Lyon, J.G., Yuan, D., Lunetta, R.S. & Elvidge, C.D. (1998) A change detection experiment using vegetation indices. *Photogrammetric Engineering and Remote Sensing*, **64**, 143–150.

Lyzenga, D.R. (1981) Remote sensing of bottom reflectance and water attenuation parameters in shallow water using aircraft and Landsat data. *International Journal of Remote Sensing*, **2**, 71–82.

Lyzenga, D.R. (1991) Interaction of short surface and electromagnetic waves with ocean fronts. *Journal of Geophysical Research*, **93**, 10765–10772.

Lyzenga, D.R., Malinas, N.P. & Tanis, F.J. (2006) Multispectral bathymetry using a simple physically based algorithm. *IEEE Transactions on Geoscience and Remote Sensing*, **44**, 2251–2259.

Ma, Z. & Redmond, R.L. (1995) Tau coefficients for accuracy assessment of classification of remote sensing data. *Photogrammetric Engineering and Remote Sensing*, **61**, 435–439.

MacAyeal, D.R., Rignot, E. & Hulbe, C.L. (1998) Ice-shelf dynamics near the front of the Filchner-Ronne Ice shelf, Antarctica, revealed by SAR interferometry: model/interferogram comparison. *Journal of Glaciology*, **44**, 419–428.

Maeder, J., Narumalani, S., Rundquist, D., Perk, R., Schalles, J., Hutchins, K. & Keck, J. (2002) Classifying and mapping general coral-reef structure using IKONOS data. *Photogrammetric Engineering and Remote Sensing*, **68**, 1297–1305.

Mahowald, N.M., Baker, A.R., Bergametti, G., Brooks, N., Jickells, T.D., Duce, R.A., Kubilay, N., Prospero, J.M., & Tegen, I. (2005) The atmospheric global dust cycle and iron inputs to the ocean. *Global Biogeochemical Cycles*, **19**, GB4025, doi:10.1029/2004GB002402.

Maiden, M.E. & Greco, S. (1994) NASA's Pathfinder data set programme: land surface parameters. *International Journal of Remote Sensing*, **17**, 3333–3345.

Mainelli-Huber, M. (2000) On the role of the upper ocean in tropical cyclone intensity change. *M.S. thesis, Dept. of Meteorology and Physical Oceanography, University of Miami.*

Mallin, M.A. & Corbett, C.A. (2006) How hurricane attributes determine the extent of environmental effects: multiple hurricanes and different coastal systems. *Estuaries and Coasts*, **29**, 1046–1061.

Manzello, D.P., Kleypas, J.A., Budd, D.A., Eakin, M.C., Glynn, P.W. & Langdon, C. (2008) Poorly cemented coral reefs of the eastern tropical Pacific: Possible insights into reef development in a high-CO_2 world. *Proceedings of the National Academy of Sciences*, **105**, 10450–10455.

Marcus, W.A. & Fonstad, M.A. (2008) Optical remote mapping of rivers at sub-meter resolutions and watershed extents. *Earth Surface Processes and Landforms*, **33**, 4–24.

Margles, S.W., Peterson, R.B., Ervin, J. & Kaplin, B.A. (2010) Conservation without borders: Building communication and action across disciplinary boundaries for effective conservation. *Environmental Management*, **45**, 1–4.

Marlon, J.R., Bartlein, P.J., Carcaillet, C., Gavin, D.G., Harrison, S.P., Higuera, P.E., Joos, F., Power, M.J. & Prentice, I.C. (2008) Climate and human influences on global biomass burning over the past two millennia. *Nature Geoscience*, **1**, 697–702.

Martin, S. (2004) *An Introduction to Remote Sensing*. Cambridge University Press, Cambridge, UK.

Martin-Vide, J. (2004) Spatial distribution of a daily precipitation concentration index in peninsular Spain. *International Journal of Climatology*, **24**, 959–971.

Masek, J.G., Lindsay, F.E. & Goward, S.N. (2000) Dynamics of urban growth in the Washington DC metropolitan area, 1973–1996, from Landsat observations. *International Journal of Remote Sensing*, **21**, 3473–3486.

Massonnet, D. & Feigl, K. (1998) Radar interferometry and its application to changes in the Earth's surface. *Reviews of Geophysics*, **36**, 441–500.

Mather, J.R., Field, R.T. & Yoshioka, G.A. (1967) Storm damage hazard along the East Coast of the United States. *Journal of Applied Meteorology*, **6**, 20–30.

Matthews, H.D. & Caldeira, K. (2008) Stabilizing climate requires near-zero emissions. *Geophysical Research Letters*, **35**, L04705.

Maurer, F., Roberto, R. & Martini, R. (2007) Triassic stratigraphy, facies and evolution of the Arabian shelf in the northern United Arab Emirates. *International Journal of Earth Sciences*, doi: 10.1007/s00531–007–0194–y.

Mayaux, P., Eva, H., Brink, A., Achard, F. & Belwar, A. (2008) Remote sensing of land- cover and land-use dynamics. In: Chuvieco, E. (ed.): *Earth Observation of Global Change*. Springer-Verlag, Berlin.

Mazel, C.H., Strand, M.P., Lesser, M.P., Crosby, M.P., Coles, B. & Nevis, A.J. (2003) High-resolution determination of coral reef bottom cover from multispectral fluorescence laser line scan imagery. *Limnology and Oceanography*, **48**, 522–534.

McCoy, R. (2005) *Field Methods in Remote Sensing*. Guilford Press, NY.

McCubbin, D.R. & Delucchi, M.A. (1999) The health costs of motor vehicle related air pollution. *Journal of Transport Economics and Policy*, **33**, 253–286.

McFarland, M.J., Miller, R.N. & Neale, C.M.U. (1990) Land surface temperature derived from the SSM/I passive microwave brightness temperatures. *IEEE Transactions on Geoscience and Remote Sensing*, **28**, 839–845.

McKean, J., Nagel, D., Tonina, D., Bailey, P., Wright, C.W., Bohn, C. & Nayegandhi, A. (2009) Remote sensing of channels and riparian zones with a narrow-beam aquatic-terrestrial LiDAR, *Remote Sensing*, **1**, 1065–096.

McKenzie, R.L., Weinreis, C., Johnston, P.V., Liley, B., Shiona, H., Kotkamp, M., Smale, D., Takegawa, N. & Kondo Y. (2008) Effects of urban pollution on UV spectral irradiances. *Atmospheric Chemistry and Physics*, **8**, 5683–5697.

Meier, W.N., Stroeve, J. & Fetterer, F. (2007) Whither Arctic sea ice? A clear signal regionally, seasonally, and extending beyond the satellite record. *Annals of Glaciology*, **46**, 428–434, doi: 10.3189/172756407782871170.

Meine, C. (1998) The continent indissoluble. In: Knight, R.L. & Landres, P.B. (eds.) *Stewardship across boundaries*. Island Press, Washington, DC.

Mellin, C., Andréfouët, S., Kulbicki, M., Dalleau, M. & Vigliola, L. (2009) Remote sensing and fish–habitat relationships in coral reef ecosystems: Review and pathways for multi-scale hierarchical research. *Marine Pollution Bulletin*, **58**, 11–19.

Mellin, C., Andréfouët, S. & Ponton, D. (2007) Spatial predictability of juvenile fish species richness and abundance in a coral reef environment. *Coral Reefs*, doi: 10.1007/s00338-007-0281-3.

Milford Neck Wildlife Conservation Area: An exploratory analysis. Report prepared for Delaware Division of Fish and Wildlife, under contract AGR 199990726 and the Nature Conservancy under contract DEFO–0215000–01.

Miller, J.L. & Goodberlet, M. (2004) Development and applications of STARRS: a next generation airborne salinity imager. *International Journal of Remote Sensing*, **25**, 1319–1324.

Miller, J.L., Goodberlet, M.A. & Zaitzeff, J.B. (1998) Airborne salinity mapper makes debut in coastal zone. *AGU EOS Transactions*, **79**, 173–177.

Miller, P.I., Shutler, J.D., Moore, G.F. & Groom, S.B. (2006) SeaWiFS discrimination of harmful algal bloom evolution. *International Journal of Remote Sensing*, **27**, 2287–2301.

Minnisa, P., Ayersb, J.K., Palikondab, R. & Phanb, D. (2004) Contrails, Cirrus Trends, and Climate. *Journal of Climate*, **17**, 1671–1685.

Mishra, D., Narumalani, S., Rundquist, D. & Lawson, M. (2006) Benthic habitat mapping in tropical marine environments using QuickBird multispectral data. *Photogrammetric Engineering and Remote Sensing*, **72**, 1037–1048.

Mobley, C.D. (1994) *Light and water; Radiative transfer in natural waters*. Academic Press, London.

Mobley, C.D. & Sundman, L.K. (2001) *Hydrolight 4.2 user's guide.* Redmond, WA7 Sequoia Scientific.

Mobley, C.D., Sundman, L.K., Davis, C.O., Downes, T.V., Leathers, R.A., Montes, M.J., Bowles, J.H., Bissett, W.P., Kohler, D.D.R., Reid, R.P., Louchard, E.M. & Gleason, A. (2005) Interpretation of hyperspectral remote-sensing imagery via spectrum matching and look-up tables. *Applied Optics*, **44**, 3576–3592.

Molina, M.J. & Rowland, F.S. (1974) Stratospheric sink for chlorofluoro-methanes: chlorine atom catalyzed destruction of ozone. *Nature,* **249**, 810–14.

Monaco, M.E., Christensen, J.D. & Rohmann, S.O. (2001) Mapping and monitoring of US coral reef ecosystems. *Earth System Monitor*, **12**, 1–16.

Montzka, S.A., Butler, J.H., Elkins, J.W., Thompson, T.M., Clarke, A.D. & Lock, L.T. (1999) Present and future trends in the atmospheric burden of ozone-depleting halogens. *Nature*, **398**, 690–694.

Moore G.K. (1982) *Ground-water applications of remote sensing.* USGS Open File Report, 82–240. US Geological Survey, Reston, VA.

Moore, G.K. (1996) Using thermal-infrared imagery to delineate ground-water discharge –Discussion. *Ground Water*, **34**, 962–962.

Mora, O., Mallorqui, J.J., Duro, J. & Broquetas, A. (2001) Long-term subsidence monitoring of urban areas using differential interferometric SAR techniques. *International Geoscience and Remote Sensing Symposium (IGARSS)*, **3**, 1104–1106.

Morel, A., Prieur, L. (1977) Analysis of variations in ocean color. *Limnology and Oceanography*, **22**, 709–722.

Morris, J.T., Sundareshwar, P.V., Nietch, C.T., Kjerfve, B. & Cahoon, D.R. (2002). Responses of coastal wetlands to rising sea level. *Ecology*, **83**, 2869–2877.

Morse, W.C., Nielsen-Pincus, M., Force, J. & Wulfhorst, J. (2007) Bridges and barriers to developing and conducting interdisciplinary graduate-student team research. *Ecology and Society*, **12**, 8.

Morton, D.C., DeFries, R.S., Shimabukuro, Y.E., Anderson, L.O., Del Bon Esprito-Santo, F., Hansen, M. & Carroll, M. (2005) Rapid Assessment of Annual Deforestation in the Brazilian Amazon Using MODIS Data. *Earth Interactions*, **9**, 1–22.

Morton, R.A. & Miller, T.L. (2005) National assessment of shoreline change: Part 2. Historical shoreline change and associated coastal land loss along the US southeast Atlantic coast. *US Geological Survey Open-File Report 2005–1401.*

Moyer, R.P., Riegl, B., Banks, K. & Dodge, R.E. (2005) Assessing the accuracy of acoustic seabed classification for mapping coral reef environments in South Florida (Broward County, USA). *Revista de Biologia Tropical*, **53** Suppl. 1, 175–184.

Muller, E., Décamps, H. & Dobson, M.K. (1993) Contributions of space remote sensing to river studies. *Freshwater Biology*, **29**, 301–312.

Muller, E.M., Rogers, C.S., Spitzack, A.S. & VanWoesik, R. (2008) Bleaching increases likelihood of disease on *Acropora palmata* (Lamarck) in Hawksnest Bay, St John, US Virgin Islands. *Coral Reefs*, **27**, 191–195.

Muller R.A. & MacDonald, G.J. (1997) Glacial Cycles and Astronomical Forcing. *Science*, **277**, 215–218.

Mumby, P.J. & Edwards, A.J. (2002) Mapping marine environments with IKONOS imagery: enhanced spatial resolution can deliver greater thematic accuracy. *Remote Sensing of Environment*, **82**, 248–257.

Mumby, P.J. & Harborne, A.R. (1999) Development of a systematic classification scheme of marine habitats to facilitate regional management and mapping of Caribbean coral reefs. *Biological Conservation*, **88**, 155–163.

Mumby, P.J., Green, E.P., Edwards, A.J. & Clark, C.D. (1997) Coral reef habitat-mapping: how much detail can remote sensing provide? *Marine Biology*, **130**, 193–202.

Myint, S.W. (2003) Fractal approaches in texture analysis and classification of remotely sensed data: comparisons with spatial autocorrelation techniques and simple descriptive statistics, *International Journal of Remote Sensing*, **24**, 1925–1947.

Myint, S.W. (2006) A new framework for effective urban land use land cover classification: a wavelet approach. *GIScience and Remote Sensing*, **43**, 155–178.

Myint, S.W., Lam, N.S.N. & Tyler, J. (2002) An evaluation of four different wavelet decomposition procedures for spatial feature discrimination within and around urban areas. *Transactions in GIS*, **6**, 403–429.

Myint, S.W., Wentz, L. & Purkis, S.J. (2007) Employing spatial metrics in urban land-use/land-cover mapping: comparing the Getis and Geary Indices. *Photogrammetric Engineering & Remote Sensing*, **73**, 1403–1415.

N'Tchayi, M.G., Bertrand, J.J. & Nicholson, S.E. (1997) The diurnal and seasonal cycles of wind-borne dust over Africa north of the equator. *Journal of Applied Meteorology*, **36**, 868–882.

Namin, K.S., Risk, M.J., Hoeksema, B.W., Zohari, Z. & Rezai, H. (2010) Coral mortality and serpulid infestations associated with red tide, in the Persian Gulf. *Coral Reefs*, **29**, 509.

Narayan, U. & Lakshmi, V. (2006) High Resolution Change Estimation of Soil Moisture by Combination of TMI Brightness Temperature and PR Surface Cross Section. IEEE International Conference on Geoscience and Remote Sensing Symposium. *IGARSS 2006*, 2336–2337.

NASA (2000) *Exploring our home planet: Earth Science Enterprise Strategic Plan*. NASA Headquarters, Washington, D.C.

Nature Conservancy (1994) *Standardized National Vegetation Classification System*. The Nature Conservancy, Arlington, Virginia.

Nayegandhi, A. & Brock, J.C. (2008) Assessment of Coastal Vegetation Habitats using LiDAR. In: Yang X. (ed) *Lecture Notes in Geoinformation and Cartography – Remote Sensing and Geospatial Technologies for Coastal Ecosystem Assessment and Management*, 365-389. Springer Publication.

Nayegandhi, A., Brock, J.C. & Wright, C.W. (2009) Small-footprint, waveform-resolving LiDAR estimation of submerged and sub-canopy topography in coastal environments. *International Journal of Remote Sensing*, **30**, 861–878.

Netzband, M., Redman, C.L. & Stefanov, W.L. (2005) Challenges for applied remote sensing science in the urban environment. *The International Archives of the Photogrammetry, Remote Sensing, and Spatial Information Sciences*, 36(8/W27).

Neuenschwander, A.L., Magruder, L.A. & Tyler, M. (2009) Landcover classification of small-footprint, full-waveform LiDAR data. *Journal of Applied Remote Sensing*, doi: 10.1117/1.3229944.

Neumann, A.C. & Hearty, P.J. (1996) Rapid sea-level changes at the close of the last interglacial (substage 5e) recorded in Bahamian island geology. *Geology*, **24**, 775–778.

Nghiem, S. V., Chao, Y., Neumann, G., Li, P., Perovich, D.K., Street, T. & Clemente-Colón, P. (2006), Depletion of perennial sea ice in the East Arctic Ocean, *Geophysical Research Letters*, **33**, L17501, doi:10.1029/2006GL027198.

Nicholls, R.J. & Hoozemans, F.M.J. (2005) Global vulnerability analysis. In: Schwartz, M.L. (ed.) *Encyclopedia of Coastal Science*, 486–491. Springer, Dordrecht, The Netherlands.

Nicholls, R.J., Hoozemans, F.M.J. & Marchand, M. (1999) Increasing flood risk and wetland losses due to global sea-level rise: regional and global analyses. *Global Environmental Change*, **9**, 69–87.

Njoku, E.G. & Entekhabi, D. (1996) Passive microwave remote sensing of soil moisture. *Journal of Hydrology*, **184**, 101–129.

NOAA (1999) *Trends in US Coastal Regions, 1970–1998. Addendum to the Proceedings: Trends, and Future Challenges for US National Ocean and Coastal Policy*. NOAA, August 1999.

NOAA (2005) *Blackwater Wildlife Refuge Wetlands Restoration Project*. National Ocean Service, Tidewater Currents Report.

NOAA (2006) *Hurricanes: Unleashing Nature's Fury. A Preparedness Guide*, 1–24. NOAA National Weather Service, US Department of Commerce.

NOAA (2008) *Hurricane Katrina*. NOAA Homepage. http://www.katrina.noaa.gov/

NOAA/NHC (2008) *Hurricane preparedness: SLOSH model*. NOAA National Weather Service, National Hurricane Center.

NOAA/NWS/NHC (2008) *Hurricane Ike*. NOAA National Weather Service, National Hurricane Center.

Noble, I. (2008) The changing climate: What we know, what we will never know, and what we are learning. Environment Matters at the World Bank. Climate Change and Adaptation. *The World Bank*, 14–17.

Noe, G.B. & Zedler, J.B. (2001) Variable rainfall limits the germination of upper intertidal marsh plants in Southern California. *Estuaries*, **24**, 30–40.

Norstrom, A.V., Nystrom, M., Lokrantz, J. & Folke, C. (2009) Alternative states on coral reefs: beyond coral-macroalgal phase shifts. *Marine Ecology Progress Series*, **376**, 295–306.

Odum, E.P. (1993) *Ecology and Our Endangered Life-Support Systems*, 2nd edn. Sinauer Associates Inc., Sunderland, MA.

Oerlemans, J. (1982) Glacial cycles and ice-sheet modelling. *Climate Change*, **4**, 353–374.

Oke, T.R. (1982) The energetic basis of the urban heat island. *Quarterly Journal of the Royal Meteorological Society*, **108**, 1–24.

Orbimage (2003) *OrbView-3 Satellite and Ground Systems Specifications*. Orbimage Inc., VA.

Ormsby, J.P., Blanchard, B.J. & Blanchard, A.J. (1985) Detection of lowland flooding using active microwave systems. *Photogrammetric Engineering and Remote Sensing*, 51, 317–328.

Orr, J.C. (1993) Accord between ocean models predicting uptake of anthropogenic CO_2. *Water, Air, & Soil Pollution*, **70**, 465–481.

Orr, J.C., Maier-Reimer, E., Mikolajewicz, U., Monfray, P., Sarmiento, JL., Toggweiler, J.R., Taylor, N.K., Palmer, J., Gruber, N., Sabine, C.L., Le Quéré, C., Key, R.M. & Boutin, J. (2001) Estimates of anthropogenic carbon uptake from four three-dimensional global ocean models. *Global Biogeochemical Cycles*, **15**, 43–63.

Orth, R.J., Carruthers, T.J.B., Dennison, W.C., Duarte, C.M., Fourqurean, J.W., Heck Jr., K.L., Hughes, A.R., Kendrick, G.A., Kenworthy, W.J., Olyarnik, S. Short, F.T., Waycott, M. & Williams, S.L. (2006) A global crisis for seagrass ecosystems. *BioScience*, **56** (12), 987–996.

Oyama, Y., Matsushita, B., Fukushima, T., Matsushige, K. & Imai, A. (2009) Application of spectral decomposition algorithm for mapping water quality in a turbid lake (Lake Kasumigaura, Japan) from Landsat TM data. *ISPRS Journal of Photogrammetry and Remote Sensing*, **64**, 73–85.

Paduan, J.D. & Graber, H.C. (1997) Introduction to high-frequency radar: Reality and myth. *Oceanography*, **10**, 36–39.

Palandro, D.A., Andréfouët, S., Hu, C., Hallock, P., Muller-Karger, F.E., Dustan, P., Brock, J.C., Callahan, M.K., Kranenburg, C. & Beaver, C.R. (2008) Quantification of two decades of coral reef habitat decline in the Florida Keys National Marine Sanctuary using Landsat data (1984-2002). *Remote Sensing of Environment*, **112**, 3388–3399.

Parkinson, C.L. (2003) Aqua: an Earth-observing satellite mission to examine water and other climate variables. *IEEE Transactions on Geoscience and Remote Sensing*, **41**, 173–183.

Parkinson, C.L., Comiso, J.C., Zwally, H.J., Cavalieri, D.L., Gloersen, P. & Campbell, W.J. (1987) *Arctic Sea Ice, 1973–1976: Satellite Passive-Microwave Observations*. NASA SP-487, GPO, Washington, DC.

Parkinson, C.L., Cavalieri, D.J., Gloersen, P., Zwally, H.J. & Comiso, J.C. (1999) Variability of the Arctic sea ice cover 1978–1996. *Journal of Geophysical Research*, **104**, 837–856.

Pastol, Y., Le Roux, C. & Louvart, L. (2007) LITTO3D: a seamless digital terrain model. *The International Hydrographic Review*, **8**, 38–44.

Pe'eri, S. & Philpot, W. (2007) Increasing the existence of very shallow-water LiDAR measurements using the red-channel waveforms. *IEEE Transactions on Geoscience and Remote Sensing*, **45**, 1217–1223.

Peñuelas, J., Baret. F. & Filella, I. (1995) Semi-empirical indices to assess carotenoids/chlorophyll a ratio from leaf spectral reflectance. *Photosynthetica*, **31**, 221–230.

Perez, T., Wesson, J., & Burrage, D. (2006) Airborne remote sensing of the Plata plume using STARRS. *Sea Technology*, **Sept 2006**, 31–34.

Perry, C.T., Smithers, S.G. & Johnson, K.G. (2009) Long-term coral community records from Lugger Shoal on the terrigenous inner-shelf of the central Great Barrier Reef, Australia. *Coral Reefs*, **28**, 941–948.

Philpot, W. (2007) Estimating atmospheric transmission and surface reflectance from a glint-contaminated spectral image. *IEEE Transactions on Geoscience and Remote Sensing*, **45**, 448–457.

Philpot, W.D., Davis, C.O., Bissett, P., Mobley, C.D., Kohler, D.D., Lee, Z., Snyder, W.A., Steward, R.G., Agrawal, Y., Trowbridge, J., Gould, R. & Arnone, R. (2004) Bottom characterization from hyperspectral image data. *Oceanography*, **17**, 76–85.

Phinn, S.R., Stow, D.A. & Zedler, J.B. (1996) Monitoring wetland habitat restoration in Southern California using airborne multispectral video data. *Restoration Ecology*, **4**, 412–422.

Phinn, S.R., Menges, C., Hill, G.J.E. & Stanford, M. (2000) Optimizing remotely sensed solutions for monitoring, modelling and managing coastal environments. *Remote Sensing of Environment*, **73**, 117–132.

Phinn, S., Stanford, M., Scarth, P., Murray, A.T. & Shyy, P.T. (2002) Monitoring the composition of urban environments based on the vegetation-impervious surface-soil (VIS) model by subpixel analysis techniques. *International Journal of Remote Sensing*.

Phinn, S.R., Roelfsema, C.M. & Stumpf, R. (2010) Remote sensing: The promise and the reality. In: Dennison, W. (Ed.) *Coastal Assessment Handbook*, Ch. 15. University of Maryland.

Pierce, H. & Lang, S. (2005) TRMM – Tropical Rainfall Measurement Mission: Katrina Intensifies into a Powerful Hurricane, Strikes Northern Gulf Coast. http://trmm.gsfc. nasa.gov/ (accessed May 2, 2008).

Pinet, P.R. (2009) *Invitation to Oceanography*, 5th Edition. Jones & Bartlett, Sudbury, MA.

Piñol, J., Terradas, J. & Lloret, F. (1998) Climate warming, wildfire hazard, and wildfire occurrence in coastal eastern Spain. *Climatic Change*, **38**, 345–357.

Piotrowitz, S.R. (2006) Integrated ocean observing system development. *Sea Technology*, Jan. 2006, 43.

Pittman, S.J., Christensen, J.D., Caldow, C., Menza, C. & Monaco, M.E. (2007) Predictive mapping of fish species richness across shallow-water seascapes in the Caribbean. *Ecological Modelling*, **204**, 9–21.

Pope, A.C., Thun, M.J., Namboodir, M.M., Dockery, D.W., Evans, J.S., Spezer, F.E. & Heath, C.W., Jr. (1995) Particulate air pollution as a predictor of mortality in a prospective study for US adults. *American Journal of Respiratory and Critical Care Medicine*, **151**, 669–674.

Pope, R.M. & Fry, E.S. (1997) Absorption spectrum (380–700 nm) of pure water. II. Integrating cavity measurements. *Applied Optics*, **36**, 8710–8723.

Porter, D.E. (2006) RESAAP Final Report: NOAA/NERRS Remote Sensing Applications Assessment Project. *University of South Carolina*, SC.

Portnoy, J.W., Nowicki, B.L., Roman, C.T. & Urish, D.W. (1998) The discharge of nitrate-contaminated groundwater from developed shoreline to marsh-fringed estuary. *Water Resources Research*, **34**, 3095–3104.

Pounder, E.R., (1965) *Physics of Ice*, 151. Pergamon Press, Oxford.

Pratchett, M.S., Munday, M.S., Wilson, S.K., Graham, N.A.J., Cinner, J.E., Bellwood, D.R., Jones, G.P., Polunin, N.V.C. & McClanahan, T.R. (2008) Effects of climate-induced coral bleaching on coral-reef fishes: ecological and economic consequences. *Oceanography and Marine Biology: An Annual Review*, **46**, 251–296.

Preisendorfer, R.W. (1961) Generalised invariant imbedding relation. *Proceedings of the National Academy of Sciences*, **47**, 591–594.

Preisendorfer, R.W. (1976) *Hydrologic optics*. US Department of Commerce, National Oceanic and Atmospheric Administration, Environmental Research Laboratories, Pacific Marine Environmental Laboratory (CD-ROM).

Prevot, L., Champion, I. & Guyot, G. (1993) Estimating surface soil-moisture and leaf area index of a wheat canopy using a dual-frequency (C and X-bands) scaterometer. *Remote Sensing of Environment*, **46**, 331–339.

Prince, S.D. & Justice, C.O. (1991) Coarse resolution remote sensing of the Sahelian environment. *International Journal of Remote Sensing*, **12**, 1133–1421.

Prospero, J.M. & Carlson, T.N. (1972) Vertical and areal distribution of Saharan dust over the western equatorial North Atlantic ocean. *Journal of Geophysical Research*, **77**, 5255–5265.

Prospero, J.M. & Lamb, P.J. (2003) African Droughts and Dust Transport to the Caribbean: Climate Change Implications. *Science*, **302**, 1024–1027.

Prospero. J.M. & Nees, R.T. (1977) Dust concentration in the atmosphere of the equatorial North Atlantic Possible relationship to Sahelian drought. *Science*, **196**, 1196–1198.

Prospero, J.M. & Nees, R.T. (1986) Impact of the North African drought and El Niño on mineral dust in the Barbados trade winds. *Nature*, **320**, 735–738.

Prospero, J.M., Ginoux, P., Torres, O., Nicholson, S.E. & Gill, T.E. (2002) Environmental characterization of global sources of atmospheric soil dust identified with the NIMBUS 7 TOMS absorbing aerosol product. *Reviews of Geophysics*, **40**, 2.1–2.31.

Provencher, J. (2007). Stronger storms are bad news for coastal ecosystems. *Ocean News*, **7**, 2–4.

Purkis, S.J. (2005) A 'reef-up' approach to classifying coral habitats from IKONOS imagery. *IEEE Transactions on Geoscience and Remote Sensing*, **43**, 1375–1390.

Purkis, S.J. & Kohler, K.E. (2008) The role of topography in promoting fractal patchiness in a carbonate shelf landscape. *Coral Reefs*, **27**, 977–989.

Purkis, S.J. & Pasterkamp, R. (2004) Integrating *in situ* reef-top reflectance spectra with Landsat TM imagery to aid shallow-tropical benthic habitat mapping. *Coral Reefs*, **23**, 5–20.

Purkis, S.J., Kenter, J.A.M., Oikonomou, E.K. & Robinson, I.S. (2002) High-resolution ground verification, cluster analysis and optical model of reef substrate coverage on Landsat TM imagery (Red Sea, Egypt). *International Journal of Remote Sensing*, **23**, 1677–1698.

Purkis, S.J., Riegl, B. & Andréfouët, S. (2005) Remote sensing of geomorphology and facies patterns on a modern carbonate ramp (Arabian Gulf, Dubai, U.A.E.). *Journal of Sedimentary Research*, **75**, 861–876.

Purkis, S.J., Myint, S. & Riegl, B. (2006) Enhanced detection of the coral *Acropora cervicornis* from satellite imagery using a textural operator. *Remote Sensing of Environment*, **101**, 82–94.

Purkis, S.J., Kohler, K.E., Riegl, B.M. & Rohmann, S.O. (2007) The Statistics of Natural Shapes in Modern Coral Reef Landscapes. *Journal of Geology*, **115**, 493–508.

Purkis, S.J., Graham, N.A.J. & Riegl, B.M. (2008) Predictability of reef fish diversity and abundance using remote sensing data in Diego Garcia (Chagos Archipelago). *Coral Reefs*, **27**, 167–178.

Purkis, S.J., Rowlands, G.P., Riegl, B.M. & Renaud, P.G. (2010) The paradox of tropical karst morphology in the coral reefs of the arid Middle East. *Geology*, **38**, 227–230.

Qi, J., Chehbouni, A., Huete, A.R., Kerr, Y.H. & Sorooshian, S. (1994) A modified soil adjusted vegetation index. *Remote Sensing of Environment*, **48**, 119–126.

Qu, J.J., Hao, X. Kafatos, M. & Wang, L. (2006) Asian Dust Storm Monitoring Combining Terra and Aqua MODIS SRB Measurements. *IEEE Geoscience and Remote Sensing Letters*, **3**, 484–486.

Quattrochi, D.A. & Goodchild, M.F. (eds.) (1997) *Scale in Remote Sensing and GIS*. Lewis Publishers, NY.

Rack, W. & Rott, H. (2004) Pattern of retreat and disintegration of the Larsen B ice shelf, Antarctic Peninsula. *Annals of Glaciology*, **39**, 505–510.

Rahmstorf, S. (1997) Risk of sea-change in the Atlantic. *Nature*, **388**, 825–826.

Rahmstorf, S. & Ganopolski, A. (1999) Long-term global warming scenarios computed with an efficient coupled climate model. *Climate Change*, **43**, 353–367.

Rairoux, P., Schillinger, H., Niedermeier, S., Rodriguez, M., Ronneberger, F., Sauerbrey, R., Stein, B., Waite, D., Wedekind, C., Wille, H., Wöste, L. & Ziener, C. (2000) Remote sensing of the atmosphere using ultrashort laser pulses. *Applied Physics B Lasers and Optics*, **71**, 573–580.

Ramanathan, V., Cess, R.D., Harrison, E.F., Minnis, P., Barkstrom, B.R., Ahmad, E. & Hartmann, D. (1989) Cloud-radiative forcing and climate: results from the earth radiation budget experiment. *Science*, **243**, 57–63.

Ramaswamy V., Schwarzkopf, M.D., Randel, W.J., Santer, B.D., Soden, B.J. & Stenchikov, G.L. (2006) Anthropogenic and natural influences in the evolution of lower stratospheric cooling. *Science*, **311**, 1138–1141.

Ramon, D. & Santer, R. (2001) Operational Remote Sensing of Aerosols over Land to Account for Directional Effects. *Applied Optics*, **40**, 3060–3075.

Ramsey, E. (1995) Monitoring flooding in coastal wetlands by using radar imagery and ground-based measurements. *International Journal of Remote Sensing*, **16**, 2495–2502.

Randall, C.J. & Szmant, A.M. (2009) Elevated temperature reduces survivorship and settlement of the larvae of the Caribbean scleractinian coral, *Favia fragum* (Esper). *Coral Reefs*, **28**, 537–545.

Randall, D.A., Wood, R.A., Bony, S., Colman, R., Fichefet, T., Fyfe, J., Kattsov, V., Pitman, A., Shukla, J., Srinivasan, J., Stouffer, R.J., Sumi, A. & Taylor, K.E. (2007) Climate models and their evaluation. In: *Climate change 2007: The physical science basis*. Contribution of

Working Group I to the Fourth Assessment Report of the Intergovernmental Panel on Climate Change.

Randel, W.J. & Wu, F. (1999) Cooling of the arctic and Antarctic polar stratospheres due to ozone depletion. *Journal of Climate*, **12**, 1467–1479.

Rankey, E.C., Enos, P., Steffen, K. & Druke, D. (2004) Lack of impact of Hurricane Michelle on tidal flats, Andros Island, Bahamas: Integrated remote sensing and field observations: *Journal of Sedimentary Research*, **74**, 654–661.

Rasher, M.E. & Weaver, W. (1990). *Basic Photo Interpretation: A Comprehensive Approach to Interpretation of Vertical Aerial Photography for Natural Resource Applications.* US Department of Agriculture, Washington, DC.

Read, J.M., Clark, D.B., Venticinque, E.M. & Moreira, M.P. (2003) Application of merged 1-m and 4-m resolution satellite data to research and management in tropical forests. *Journal of Applied Ecology*, **40**, 592–600.

Reed, D.J. (2005) Wetlands. In: Schwartz, M.L. (ed.) *Encyclopedia of Coastal Science*, 1077-1081. Springer, Dordrecht, The Netherlands.

Remillard, M.M. & Welch, R.A. (1993) GIS technologies for aquatic macrophyte studies: Modelling applications. *Landscape Ecology*, **8**, 163–175.

Retalis, A., Cartalis, C. & Athanassiou, E. (1999) Assessment of the distribution of aerosols in the Athens with the use of Landsat Thematic Mapper data. *International Journal of Remote Sensing*, **20**, 939–945.

Retalis, A., Sifakis, N., Grosso, N., Paronis, D. & Sarigiannis, D. (2003) Aerosol optical thickness retrieval from AVHRR images over the Athens urban area. *Proceedings of the International Geoscience & Remote Sensing Symposium (IGARSS).* Toulouse, France.

Richards, S.A., Possingham, H.P. & Tizard, T. (1999) Optimal fire management for maintaining community diversity. *Ecological Applications*, **9**, 880–892.

Richardson, S.D. & Reynolds, J.M. (2000) An overview of glacial hazards in the Himalayas. *Quaternary International*, 65–66, 31–47.

Ridd, M.K. (1995) Exploring a V-I-S (vegetation-impervious surface-soil) model for urban ecosystem analysis through remote sensing: comparative anatomy for cities. *International Journal of Remote Sensing*, **16**, 2165–2185.

Ridd, M.K. & Liu, J. (1998) A comparison of four algorithms for change detection in an urban environment. *Remote Sensing of Environment*, **63**, 95–100.

Ridgwell, A. & Zeebeb, R.E. (2005) The role of the global carbonate cycle in the regulation and evolution of the Earth system. *Earth and Planetary Science Letters*, **234**, 299–315.

Riegl, B. & Purkis, S.J. (2005) Detection of shallow subtidal corals from IKONOS satellite and QTC View (50, 200 kHz) single-beam sonar data (Arabian Gulf; Dubai, UAE). *Remote Sensing of Environment*, **95**, 96–114.

Riegl, B., Moyer, R.P., Morris, L.J., Virnstein, R.W. & Purkis, S.J. (2005) Distribution and seasonal biomass of drift macroalgae in the Indian River Lagoon (Florida, USA) estimated with acoustic seafloor classification (QTCView, Echoplus). *Journal of Experimental Marine Biology and Ecology*, **326**, 89–104.

Riegl, B., Bruckner, A., Coles, S.L., Renaud, P. & Dodge, R.E. (2009) Coral reefs: threats and conservation in an era of global change. *The Year in Ecology and Conservation Biology, 2009: Annals of the New York Academy of* Sciences, **1162**, 136–186.

Riegl, B.M. & Piller, W. (2003) Possible refugia for reefs in times of environmental stress. *International Journal of Earth Sciences*, **92**, 520–531.

Riegl, B.M., Halfar, J., Purkis, S.J. & Godinez-Orta, L. (2007) Sedimentary facies of the Eastern Pacific's northernmost reef-like setting (Cabo Pulmo, Mexico). *Marine Geology*, **236**, 61–77.

Ries, J.B., Cohen, A.L. & McCorkle, D.C. (2009) Marine calcifiers exhibit mixed responses to CO_2-induced ocean acidification. *Geology*, **37**, 1131–1134.

Riggs, G.A., Hall, D.K. & Ackerman, S.A. (1999) Sea ice extent and classification with the Moderate Resolution Imaging Spectroradiometer Airborne Simulator (MAS). *Remote Sensing of Environment*, **68**, 152–163.

Rignot, E. (2002) Mass balance of East Antarctic glaciers and ice shelves from satellite data. *Annals of Glaciology*, **34**, 217–227.

Rignot, E., Padman, L., MacAyeal, D.R. & Schmeltz, M. (2000) Analysis of sub-ice-shelf tides in the Weddell Sea using SAR interferometry. *Journal of Geophysical Research*, **105**, 615–620.

Rignot, E., Bamber, J.L., van den Broeke, M.R., Davis, C., Li, Y., van de Berg, W.-J. & van Meijgaard, E. (2008) Recent Antarctic ice mass loss from radar interferometry and regional climate modelling. *Nature Geoscience*, **1**, 106–110.

Rindfuss, R.R. & Stern, P.C. (1998) Linking Remote Sensing and Social Science: The need and the challenges. In: *People and Pixels*, National Academy Press. Washington, DC.

Roberts, N.C. & Bradley, R.T. (1991) Stakeholder collaboration and innovation: a study of public policy initiatives at the state level. *Journal of Applied Behavioral Science*, **27**, 209–227.

Robinson, I.S. (2004) *Measuring the Ocean from Space: The Principles and Methods of Satellite Oceanography*. Springer-Verlag, Berlin.

Robinson, J. (2006) Conservation biology and real-world conservation. *Conservation Biology*, **20**, 658–669.

Rodell, M.J., Famiglietti, S., Chen, J., Seneviratne, S.I., Viterbo, P., Holl, S. & Wilson, C.R. (2004) Basin scale estimates of evapotranspiration using GRACE and other observations. *Geophysical Research Letters*, **31**, L20504.1–L20504.4

Rodó, X., Baert, E. & Comín, F.A. (1997) Variations in seasonal rainfall in Southern Europe during the present century: relationships with the North Atlantic Oscillation and the El Niño-Southern Oscillation. *Climate Dynamics*, **13**, 275–284.

Rohling, E.J., Grant, K., Bolshaw, M., Roberts, A.P., Siddall, M., Hemleben, C. & Kucera, M. (2009) Antarctic temperature and global sea level closely coupled over the past five glacial cycles. *Nature Geoscience*, **2**, 500–504.

Rohmann, S.O. & Monaco, M.E. (2005) *Mapping southern Florida's shallow-water coral ecosystems: An implementation plan*. NOAA Technical Memorandum NOS NCCOS 19 (online).

Rohmann, S.O., Hayes, J.J., Newhall, R.C., Monaco, M.E. & Grigg, R.W. (2005) The area of potential shallow-water tropical and subtropical coral ecosystems in the United States. *Coral Reefs*, **24**, 370–383.

Romanovsky, V., Oechel, W. C., Morison, J., Zhang, J. & Barry, R. G. (2000) Observational evidence of recent change in the northern high-latitude environment. *Climatic Change*, **46**, 159–207.

Rosenberg, E., Kushmaro, A., Kramarsky-Winter, E., Banin, E. & Yossi, L. (2009) The role of microorganisms in coral bleaching. *ISME Journal*, **3**, 139–146.

Rosenfeld, D., Rudich, Y. & Lahav, R. (2001) Desert dust suppressing precipitation: a possible desertification feedback loop. *Proceedings of the National Academy of Sciences of the United States of America*, **98**, 5975–5980.

Roth, A. (2003) TerraSAR-X: *A new perspective for scientific use of high resolution spaceborne SAR data*. 2nd GRSS/ISPRS Joint Workshop on Remote Sensing and Data Fusion over Urban Areas, URBAN 2003, 22–23.

Roth, M., Oke, T.R. & Emery, W.J. (1989) Satellite derived urban heat islands from three coastal cities and the utilization of such data in urban climatology. *International Journal of Remote Sensing*, **10**, 1699–1270, doi: 10.1080/01431168908904002.

Rowlands, G.P., Purkis, S.J. & Riegl, B.M. (2008) The 2005 coral-bleaching event Roatan (Honduras): Use of pseudo-invariant features (PIFs) in satellite assessments. *Journal of Spatial Science*, **53**, 99–112.

Royal Society (2005) Ocean acidification due to increasing atmospheric carbon dioxide. *Policy document 12/05 Royal Society, London*. The Clyvedon Press Ltd, Cardiff.

Ruddick, K.G. (2001). Optical remote sensing of chlorophyll-a in case 2 waters by use of an adaptive two-band algorithm with optimal error properties. *Applied Optics*, **40**, 3575–3585.

Rühlemann, C., Mulitza, S., Müller, P.J., Wefer, G. & Zahn, R. (1999) Warming of the tropical Atlantic Ocean and slowdown of thermohaline circulation during the last deglaciation. *Nature*, **402**, 511–514.

Rundquist, D., Murray, G. & Queen, L. (1985) Airborne thermal mapping of a flow-through lake in the Nebraska sandhills. *Water Resources Bulletin*, **21**, 989–994.

Rutherford, S. & D'Hondt, S. (2000) Early onset and tropical forcing of 100,000-year Pleistocene glacial cycles. *Nature*, **408**, 72–75.

Ryan, J.P., Yoder, J.A., Cornillon, P.C. & Barth, J.A. (1999) Chlorophyll enhancement and mixing associated with meanders of the shelf break front in the Mid-Atlantic Bight. *Journal of Geophysical Research*, **104**, 23479–23493.

Sabine, C.L., Feely, R.A., Gruber, N., Key, R.M., Lee, K., Bullister, J.L., Wanninkhof, R., Wong, C.S., Wallace, D.W.R., Tilbrook, B., Millero, F.J., Peng, T.H., Kozyr, A., Ono, T. & Rios, A.F. (2004) The oceanic sink for anthropogenic CO_2. *Science*, **305**, 367–371.

Sabins, F.F. (1978) *Remote Sensing: Principles and Interpretation*, 2nd edn. W.H. Freeman and Company, NY.

Sabins, F.F. (2007) *Remote Sensing: Principles and Interpretation*, 3rd edn. Waveland Press Inc., Long Grove, IL

Sadler, G.J., Barnsley, M.J. & Barr, S.L. (1991) Information extraction from remotely-sensed images for urban land analysis, *Proceedings of the Second European Conference on Geographical Information Systems (EGIS &llenis;91), Brussels, Belgium*, **April**, 955–964. EGIS Foundation, Utrecht.

Sale, P.F. (2008) Management of coral reefs: where have we gone wrong and what can we do about it? *Marine Pollution Bulletin*, **56**, 805–809.

Sallenger, A.H., Krabill, W.B., Brock, J.C., Swift, R.N., Jansen, M.., Manizade, S., Richmond, B., Hampto, M. & Eslinger, D. (1999) Airborne laser study quantifies El Niño-induced coastal change. *American Geophysical Union, EOS Transactions*, **80**, 89–93.

Sallenger, A.H., Stockdon, H.F., Fauver, L., Hansen, M., Thompson, D., Wright, C.W. & Lillycrop, J. (2006) Hurricanes 2004: an overview of their characteristics and coastal change. *Estuaries and Coasts*, **29**, 880–888.

Sarthou, G., Baker, A.R., Blain, S., Achterberg, E.P., Boye, M., Bowie, A.R., Croot, P., Laan, P., De Baar, H.J.W., Jickells, T.D. & Worsfold, P.J. (2003) Atmospheric iron deposition and sea-surface dissolved iron concentrations in the east Atlantic. *Deep Sea Research Part I*, **50**, 1339–1352.

Saunders, M.A. & Lea, A.S. (2008) Large contribution of sea surface warming to recent increase in Atlantic hurricane activity. *Nature*, **451**, 557–561.

Sawaya, K.E., Olmanson, L.G., Heinert, N.J., Brezonik, P.L. & Bauer, M.E. (2003) Extending satellite remote sensing to local scales: land and water resource monitoring using high-resolution imagery. *Remote Sensing of Environment*, **88**, 144–156.

Scambos, T.A., Bohlander, J.A., Shuman, C.A. & Skvarca, P. (2004) Glacier acceleration and thinning after ice shelf collapse in the Larsen B embayment, Antarctica. *Geophysical Research Letters*, **31**, L18402.1–L18402.4.

Schellnhuber, H.J., Crutzen, P.J., Clark, W.C., Claussen, M. & Held, H. (eds.) (2004) *Earth system analysis for sustainability*. MIT Press, Cambridge, MA.

Schmidt, K.S., Skidmore, A.K., Kloosterman, E.H., Van Oosten, H., Kumar, L. & Janssen, J.A.M. (2004) Mapping coastal vegetation using an expert system and hyperspectral imagery. *Photogrammetric Engineering and Remote Sensing*, **70**, 703–716.

Schneider, A. & Woodcock, C.E. (2008) Compact, dispersed, fragmented, extensive? A comparison of urban growth in 25 global cities using remotely sensed data, pattern metrics and census information. *Urban Studies*, **45**, 659–692.

Schneider, A., Seto, K.C. & Webster, D.R. (2005) Urban growth in Chengdu, Western China: linking remote sensing, urban planning and policy perspectives. *Environment and Planning B*, **32**, 323–345.

Schofield, O., Arnone, R.A., Bissett, W.P., Dickey, T.D., Davis, C.O., Finkel, Z., Oliver, M. & Moline, M.A. (2004) Watercolors in the coastal zone: what can we see? *Oceanography*, **17(2)**, 24–31.

Schofield, O., Kohut, J. & Glenn, S. (2008) Evolution of coastal observing networks. *Sea Technology*, **49(2)**, 31–36.

Schutz, B.E., Zwally, H.J., Shuman, C.A., Hancock, D. & DiMarzio, J.P. (2005) Overview of the ICESat mission. *Geophysical Research Letters*, **32**, L21S01.

Schuur, E.A., Vogel, J.G., Crummer, K.G., Lee, H., Sickman, J.O. & Osterkamp, T.E. (2009) The effect of permafrost thaw on old carbon release and net carbon exchange from tundra. *Nature*, **28**, 556–559.

Sellers, P.J. & Schimel, D. (1993) Remote sensing of the land biosphere and biochemistry in the EOS era: science priorities, methods and implementation – EOS land biosphere and biogeochemical panels. *Global and Planetary Change*, **7**, 279–297.

Serreze, M.C., Walsh, J.E., Chapin III, F.S., Osterkamp, T., Dyurgerov, M., Stocker, T.F. & Schmittner, A. (1997) Influence of CO_2 emission rates on the stability of the thermohaline circulation. *Nature*, **388**, 862–865.

Serreze, M.C., Walsh, J.E., Chapin, F.S., Osterkamp, T., Dyurgerov, M., Romanovsky V., Oechel, W.C., Morison, J., Zhang, T. & Barry, R.G. (2000) Observational evidence of recent change in the northern high-latitude environment. *Climatic Change*, **46**, 159–207.

Shan, J. & Sampath, A. (2007) Urban terrain and building extraction from airborne LiDAR data. In: Weng, Q., Quattrochi, D.A. (eds) *Urban remote sensing*, 412. CRC Press.

Shay, L.K., Goni, G.J. & Black, P.G. (2001) Effects of a Warm Oceanic Feature on Hurricane Opal. *Monthly Weather Review*, **14**, 1366–1383.

Sheppard, C.R.C. (2003) Predicting recurrences of mass coral bleaching in the Indian Ocean. *Nature*, **425**, 294–297.

Sheppard, C.R.C., Harris, A. & Sheppard, A.L.S. (2008) Archipelago-wide coral recovery patterns since 1998 in the Chagos Archipelago, central Indian Ocean. *Marine Ecology Progress Series*, **362**, 109–117.

Sherman, C.E., Glenn, C.R., Jones, A.T., Burnett W.C. & Schwarcz, H.P. (1993) New evidence for two highstands of the sea during the last interglacial, oxygen isotope substage 5e. *Geology*, **21**, 1079–1082.

Shinn, E.A., Smith, G.W., Prospero, J.M., Betzer, P., Hayes, M.L., Garrison, V. & Barber, R.T. (2000) African Dust and the Demise of Caribbean Coral Reefs. *Geophysical Research Letters*, **27**, 3029–3032.

SHOM-SPT. (1990) Les dossiers de spatio-préparation des campagnes hydrographiques aux Tuamotu-Gambier. In: *Proceedings of Pix'Iles 90: Int. Workshop on Remote Sensing and Insular Environments in the Pacific: integrated approaches, Nouméa-Tahiti*, 593–595.

Siddall, M., Stocke, T.F. & Clark, P.U. (2009) Constraints on future sea-level rise from past sea-level change. *Nature Geoscience*, **2**, 571–575.

Siegenthaler, U., Stocker, T.F., Monnin, E., Lüthi, D., Schwander, J., Stauffer, B., Raynaud, D., Barnola, J.-M., Fischer, H., Masson-Delmotte, V. & Jouzel, J. (2005) Stable Carbon Cycle–Climate Relationship During the Late Pleistocene. *Science*, **310**, 1313–1317.

Sifakis, N. & Deschamps, P.Y. (1992) Mapping of air pollution using SPOT satellite data. *Photogrammetric Engineering and Remote Sensing*, **58**, 1433–1437.

Sifakis, N., Soulakellis, N.A. & Paronis, D. (1998) Quantitative mapping of air pollution density using Earth observations: A new processing method and application to an urban area. *International Journal of Remote Sensing*, **19**, 3289–3300.

Sinclair, M. (1999) Laser hydrography – commercial survey operations. *Proceedings of US Hydrographic Conference*. Mobile, AL.

Singer, S.F. (1999) *Hot talk cold science: Global warming's unfinished debate,*2nd edn. The Independent Institute, Oakland CA.

Sitch, S., Cox, P.M., Collins, W.J. & Huntingford, C. (2007) Indirect radiative forcing of climate change through ozone effects on the land-carbon sink. *Nature*, **448**, 791–794.

Skole, D. & Tucker, C. (1993) Tropical Deforestation and Habitat Fragmentation in the Amazon: Satellite Data from 1978 to 1988. *Science*, **260**, 1905–1910.

Slonecker, E.T., Jennings, D.B. & Garofalo, D. (2001) Remote sensing of impervious surfaces: a review. *Remote Sensing Reviews*, **20**, 227–255.

Smale, D.A., Brown, K.M., Barnes, D.K.A., Fraser, K.P.P. & Clarke, A. (2008) Ice Scour Disturbance in Antarctic Waters. *Science*, 321, 371.

Smith, D.M. (1996) Extraction of winter total sea-ice concentration in the Greenland and Barents Seas from SSM/I data. *International Journal of Remote Sensing*, **17**, 2625–2646.

Smith, E., Digby, S., Vazquez, J., Tran, A. & Sumagaysay, R. (1996) Satellite-derived sea surface temperature data available from the NOAA/NASA Pathfinder program, *EOS Electronic Supplement*, http://www.AGU.org/eos_elec/95274e.html, April 2, 1996.

Smith, J.M., Quinn, T.M., Helmle, K.P. & Halley, R.B. (2006). Reproducibility of geochemical and climatic signals in the Atlantic coral *Montastraea faveolata*. *Paleoceanography*, **21**, PA1010, doi:10.1029/2005PA001187.

Smith, M.W. & Riseborough, D.W. (1996) Ground temperature monitoring and detection of climate change. *Permafrost and Periglacial Processes 7*, 301–310.

Smith, R.C. & Baker, K.S. (1981) Optical properties of the clearest natural waters (200–800 nm). *Applied Optics*, **20**, 177–184.

Smith, S.L. & Burgess, M.M. (1999) Mapping the sensitivity of Canadian permafrost to climate warming. *Interactions between the cryosphere, climate and greenhouse gases*. Proceedings of IUGG 99 Symposium HS2, Birmingham, July 1999. *IAHS Publication*, **256**, 71–80.

Sokolik, I.N., Toon, O.B. & Bergstrom, R.W. (1998) Modelling the radiative characteristics of airborne mineral aerosols at infrared wavelengths. *Journal of Geophysical Research*, **103**, 8813–8826.

Solomon, S. (1999) Stratospheric ozone depletion: a review of concepts and history. *Reviews of Geophysics*, **37**, 275–316.

Souza, C.M. & Roberts, D. (2005) Mapping forest degradation in the Amazon region with IKONOS images. *International Journal of Remote Sensing*, **26**, 425–429.

Space Imaging (2003) *IKONOS Imagery Products and Product Guide* (version 1.3). Space Imaging LLC., CO.

SPOT (2009) http://www.spotimage.fr/web/en/234-preprocessing-levels-and-location-accuracy.php.

Spreen, G., & Kaleschke, L. (2008) *AMSR-E ASI 6.25 km Sea-ice Concentration Data, V5.4.* Institute of Oceanography, University of Hamburg, Germany, digital media (ftp://ftp-projects.zmaw.de/seaice/).

Spreen, G., Kaleschke, L. & Heygster, G. (2008) Sea ice remote sensing using AMSR-E 89 GHz channels, *Journal of Geophysical Research*, doi:10.1029/2005JC003384.

Steinacher, M., Joos, F., Frölicher, T.L., Plattner, G.-K. & Doney, S.C. (2009) Imminent ocean acidification in the Arctic projected with the NCAR global coupled carbon cycle-climate model. *Biogeosciences*, **6**, 515–533.

Stephenson, D. & Sinclair, M. (2006) NOAA LiDAR data acquisition & processing report: Project OPR-I305-KRL-06. *NOAA Data Acquisition & Processing Report NOS OCS* (online).

Stern, H.L. & Moritz, R.E. (2002) Sea ice kinematics and surface properties from RADARSAT synthetic aperture radar during the SHEBA drift. *Journal of Geophysical Research*, **107**, 8028.

Stilla, U. & Soergel, U. (2007) Reconstruction of buildings in SAR imagery of urban areas. In: Weng, Q., Quattrochi, D.A. (eds) *Urban remote sensing*. CRC Press.

Stilla, U., Soergel, U., Thoennessen, U. & Brenner, A. (2005) Potential and limits for reconstruction of buildings from high resolution SAR data of urban areas. *Proceedings of 28th General Assembly of International Union Radio Science (URSI)*. New Delhi.

Stockdon, H.F., Sallenger, A.H., List, J.H. & Holman, R.A. (2002) Estimation of shoreline position and change using airborne topographic LiDAR data. *Journal of Coastal Research*, **18**, 502–513.

Stockdon, H.F., Doran, K.S. & Sallenger, A.H. (2009) Extraction of LiDAR-based dune-crest elevations for use in examining the vulnerability of beaches to inundation during hurricanes. *Journal of Coastal Research*, **53**, 59–65.

Stocker, T.F. & Schmittner, A. (1997) Influence of CO_2 emission rates on the stability of the thermohaline circulation. *Nature*, **388**, 862–865.

Storey, M., Duncan, R.A. & Swisher III, C.C. (2007) Paleocene-Eocene Thermal Maximum and the opening of the Northeast Atlantic. *Science*, **316**, 587–589.

Storlazzi, C.D., Logan, J.B. & Field, M.E. (2003) Quantitative morphology of a fringing reef tract from high-resolution laser bathymetry: Southern Molokai, Hawaii. *Geological Society of America Bulletin*, **115**, 1344–1355.

Story, M., Congalton, R. (1986) Accuracy Assessment: A User's Perspective. *Photogrammetric Engineering and Remote Sensing*, **52**, 397–399.

Stott, P.A., Huntingford, C., Jones, C.D. & Kettleborough, J.A. (2008) Observed climate change constrains the likelihood of extreme future global warming. *Tellus B*, **60**, 76–81.

Streten, N.A. (1973) Satellite observations of the summer decay of the Antarctic sea-ice. *Meteorology and Atmospheric Physics*, **22**, 119–134.

Streutker, D.R. (2002) A remote sensing study of the urban heat island effect of Houston, Texas. *International Journal of Remote Sensing*, **23**, 2595–2608.

Stroeve, J. C., Serreze. M.C., Fetterer, F., Arbetter, T., Meier, W., Maslanik, J. & Knowles, K. (2005) Tracking the Arctic's shrinking ice cover: Another extreme September minimum in 2004. *Geophysical Research Letters*, **32**.

Strozzi, T., Kääb, A. & Frauenfelder, R. (2004) Detecting and quantifying mountain permafrost creep from *in situ* inventory, space-borne radar interferometry and airborne digital photogrammetry. *International Journal of Remote Sensing*, **25**, 2919–2931.

Stumpf, R. P. (2001). Applications of satellite ocean color sensors for monitoring and predicting harmful algal blooms. *Human and Ecological Risk Assessment*, **7(5)**, 1363–1368.

Stumpf, R. P., Culver, M. E., Tester, P.A, Kirkpatrick, G. J., Pederson, B., Tomlinson, M. C., Truby, E., Ransibrahmanakul, V., Hughes, K. & Soracco, M. (2003). Use of satellite imagery and other data for monitoring Karenia brevis blooms in the Gulf of Mexico. *Harmful Algae*, **2**, 147–160.

Stumpf, R.P., Holderied, K. & Sinclair, M. (2003). Determination of water depth with high-resolution satellite imagery over variable bottom types. *Limnology and Oceanography*, **48**, 547–556.

Suhong, L., Xiang, Z., Peijuan, W. & Chun, Y (2004) A study on river and vegetation dynamical monitoring in a typical mine environment using Landsat TM. *Proceedings of the IEEE International Geoscience and Remote Sensing Symposium (IGARSS) 2004*, **7**, 4640–4642.

Sutton, P., Roberts, D., Elvidge, C.D. & Baugh, K. (2001) Census from heaven: An estimate of the global human population using night-time satellite imagery. *International Journal of Remote Sensing*, **22(16)**, 3061–3076.

Sutton, P.C., Taylor, M.J., Anderson, S. & Elvidge, C.D. (2007) Sociodemographic characterization of urban areas using nighttime imagery, Google Earth, Landsat, and 'social' ground truthing. In: Weng, Q. & Quattrochi, D.A. (eds). *Urban Remote Sensing*. CRC Press, Boca Raton, FL.

Swap, R., Ulanski, S., Cobbett, M. & Garstang, M. (1996) Temporal and spatial characteristics of Saharan dust outbreaks. *Journal of Geophysical Research*, **101**, 4205–4220.

Swart, P.K. & Grottoli, A. (2004) Proxy indicators of climate in coral skeletons: a perspective. *Coral Reefs*, **22**, 313–315.

Tanner, J.E. (1997) Interspecific competition reduces fitness in scleractinian corals. *Journal of Experimental Marine Biology and Ecology*, **214(1/2)**, 19–34.

Tanner, J.E., Hughes, T.P. & Connell, J.H. (1996) The role of history in community dynamics: A modelling approach. *Ecology*, **77**, 108–117.

Tanré, D., Kaufman, Y.J., Herman, M. & Mattoo, S. (1997) Remote sensing of aerosol over oceans from EOS-MODIS. *Journal of Geophysical Research*, **102**, 16971–16988.

Tanskanen, A., Määttä, A., Krotkov, N., Kaurola, J., Koskela, T., Karpetchko, A., Fioletov, V. & Bernhard, G. (2005) Validation of the OMI surface UV data. In: *AGU Fall Meeting*, American Geophysical Union.

Tegen, I. & Fung, I. (1995) Contribution to the mineral aerosol load from land surface modification. *Journal of Geophysical Research*, **18**, 707–726.

Tegen, I., Lacis, A.A., Fung, I. (1996) The influence on climate forcing of mineral aerosols from disturbed soils. *Nature*, **380**, 419–422.

Teillet, P.M., Gauthier, R.P., Chichagov, A. & Fedosejevs, G. (2002) Towards integrated Earth sensing: Advanced technologies for *in situ* sensing in the context of Earth observation. *Canadian Journal of Remote Sensing*, **28**, 713–718.

Tester, P. A., Stumpf, R. P., Vukovich, F. M., Fowler, P. K. & Turner, J. F. (1991) An expatriate red tide bloom: Transport, distribution and persistence. *Limnology and Oceanography*, **36**, 1053–1061.

Thieler, E.R. & Danforth, W.W. (1994) Historical shoreline mapping: improving techniques and reducing positioning errors. *Journal of Coastal Research*, **10**, 539–548.

Thomas, A.C. & Weatherbee, R.A. (2006) Satellite-measured temporal variability of the Columbia River plume. *Remote Sensing of Environment*, **100**, 167–178.

Thomas, N., Hendrix, C. & Congalton, R.G. (2003) A comparison of urban mapping methods using high-resolution digital imagery, *Photogrammetric Engineering & Remote Sensing*, **69**, 963–972.

Thomas, R., Frederick, E., Krabill, W., Manizade, S. & Martin, C. (2009) Recent changes on Greenland outlet glaciers. *Journal of Glaciology*, **55**, 147–162.

Thomas, V., Treitz, P., Mccaughey, J.H., Noland, T. & Rich, L. (2008) Canopy chlorophyll concentration estimation using hyperspectral and LiDAR data for a boreal mixedwood forest in northern Ontario, Canada. *International Journal of Remote Sensing*, **29**, 1029–1052.

Thompson, L.G., Mosley-Thompson, E., Davis, M.E., Henderson, K.A., Brecher, H.H., Zagorodnov, V.S., Mashiotta, T.A., Lin, P.-N., Mikhalenko, V.N., Hardy, D.R. & Beer, J. (2002) Kilimanjaro ice core records: Evidence of Holocene climate change in tropical Africa. *Science*, **298**, 589–593.

Thompson, L.G., Brecher, H.H., Mosley-Thompson, E., Hardy, D.R. & Mark, B.G. (2009) Glacier loss on Kilimanjaro continues unabated. *Proceedings of the National Academy of Sciences of the United States of America*, **106**, 19770–19775.

Thorncroft, C. & Hodges, K. (2001) African easterly wave variability and its relationship to Atlantic tropical cyclone activity. *Journal of Climate*, **14**, 1166–1179.

Tomás, R., Marquez, Y., Lopez-Sanchez, J.M., Delgado, J., Blanco, P., Mallorqui, J.J., Martinez, M., Herrera, G. & Mulas, J. (2005) Mapping ground subsidence induced by aquifer over-exploitation using advanced Differential SAR Interferometry: Vega Media of the Segura River (SE Spain) case study. *Remote Sensing of Environment*, **98**, 269–283.

Tomlinson, M.C., Stumpf, R.P., Ransibrahmanakul, V., Truby, E.W., Kirkpatrick, G.J., Pederson, B.A., Vargo, G.A. & Heil, C.A. (2004). Evaluation of the use of SeaWiFS imagery for detecting Karenia brevis harmful algal blooms in the eastern Gulf of Mexico. *Remote Sensing of Environment*, **91**, 293–303.

Torgersen, C.E., Faux, R.N., McIntosh, B.A., Poage, N.J. & Norton, D.J. (2001) Airborne thermal remote sensing for water temperature assessment in rivers and streams. *Remote Sensing of Environment*, **76**, 386–398.

Tortell, P.D., Payne, C.D., Li, Y., Trimborn, S., Rost, B., Smith, W.O., Riesselman, C., Dunbar, R.B., Sedwick, P. & DiTullio, G.R. (2008) CO_2 sensitivity of Southern Ocean phytoplankton. *Geophysical Research Letters*, **35**, L04605.

Towson (2009) http://chesapeake.towson.edu/data/all_image.asp.

Travis, D.J., Carleton, A.M. & Lauritsen, R.G. (2002) Jet aircraft contrails: Surface temperature variations during the aircraft groundings of September 11–13, 2001. *American Meteorological Society 10th Conference on Aviation, Range, and Aerospace Meteorology. May 14. Portland, OR.*

Trendberth, K.E. (2007) Warmer Oceans, Stronger Hurricanes. *Scientific American*, July 2007, 45–51.

Tselioudis, G., DelGenio, A.D., Kovari, W. & Yao, M.-S. (1998) Temperature dependence of low cloud optical thickness in the GISS GCM: Contributing mechanisms and climate implications. *Journal of Climate*, **11**, 3268–3281.

Tucker, C.J. & Sellers, P.J. (1986) Satellite remote sensing of primary production. *International Journal of Remote Sensing*, **7**, 1395–1416.

Tucker, C.J., Dregne, H.E. & Newcomb, W.W. (1991) Expansion and contraction of the Saharan Desert from 1980 to 1990. *Science*, **253**, 299–301.

Tuell, G.H. & Park, J.Y. (2004) Use of SHOALS Bottom Reflectance Images to Constrain the Inversion of a Hyperspectral Radiative Transfer Model Proceedings. In: Kammerman, G. (ed.) *Laser Radar and Technology Applications IX*, SPIE Vol. 5412, 185–193.

Tuell, G.H., Feygels, V.I., Kopilevich, Y.I., Cunningham, A.G., Weidemann, A.D., Mani, R., Podoba, V., Ramnath, V., Park, J.Y. & Aitken, J. (2005) Measurement of ocean water optical properties and seafloor reflectance with scanning hydrographic operational airborne LiDAR system (SHOALS): II. Practical results and comparison with independent data. *Proceedings SPIE*, Vol. 5885.

Tupin, F. & Roux, M. (2003) Detection of building outlines based on the fusion of SAR and optical features. *ISPRS Journal of Photogrammetry and Remote Sensing*, **58**, 71–82.

Turley, C.M., Roberts, J.M. & Guinotte, J.M. (2007) Corals in deep-water: will the unseen hand of ocean acidification destroy cold-water ecosystems? *Coral Reefs*, **26**, 445–448.

UNEP (2005) The Global Earth Observation System of Systems (GEOSS). *Global Change Newsletter*, **61**, March 2005.

UNFP (1999) *The state of the world population 1999*. United Nations Population Fund, United Nations Publications, NY. http://www.unfpa.org/swp/1999/index.htm.

US Department of Commerce, NOAA (2004) US Integrated Ocean Observing System. *Earth System Monitor*, **14**(2).

US Geological Survey (1997) NDVI Data Information and References. *Africa Data Dissemination Service*, **Sept. 9**, 1–3.

US Geological Survey (2004). *The Blackwater NWR Inundation Model*. Open File Report 04–1302, Version 1.0.

US Geological Survey (2007) *A leading source of land information for exploring our changing planet*. http://eros.usgs.gov.

Ustin, S.L., Roberts, D.A., Gamon, J.A., Asner, G.P. & Green, R.O. (2004) Using imaging spectroscopy to study ecosystem properties and processes. *Bioscience*, **54**, 523–534.

Van de Griend, A.A. & Owe, M. (1994) Microwave vegetation optical depth and inverse modelling of soil emissivity using Nimbus/SMMR satellite observations. *Meteorology and Atmospheric Physics*, **54**, 225–239.

Van der Werf, G.R., Randerson, J.T., Giglio, L., Collatz, G.J., Kasibhatla, P.S. & Arellano Jr., A.F. (2006) Interannual variability in global biomass burning emissions from 1997 to 2004. *Atmospheric Chemistry and Physics*, **6**, 3423–3441.

Varotsos, C.A., Ghosh, S.S., Chronopoulos, G.J., Katsikis, S.C. & Cracknell, A.P. (1998) Total ozone measurements over Athens: intercomparison between Dobson, TOMS (version 6) and SBUV measurements. *International Journal of Remote Sensing*, **19**, 3327–3333.

Vaughan, D.G. & Doake, C.S.M. (1996) Recent atmospheric warming and retreat of ice shelves on the Antarctic Peninsula. *Nature*, **379**, 328–331.

Velders, G.J.M., Fahey, D.W., Daniel, J.S., McFarland M. & Andersen, S.O. (2009) The large contribution of projected HFC emissions to future climate forcing. *Proceedings of the National Academy of Sciences of the United States of America*, **106**, 10949–10954.

Vellinga, M. & Wood, R.A. (2002) Global climatic impacts of a collapse of the Atlantic thermohaline circulation. *Climatic Change*, **54**, 251–267.

Veron, J.E.N. (2008) *A Reef in Time: The Great Barrier Reef from Beginning to End*. Harvard University Press, Cambridge, MA.

Vogelzang, J. (1997) Mapping submarine sand waves with multiband imaging radar 1. Model development and sensitivity analysis. *Journal of Geophysical Research*, **102**, 1163–1181.

Waggoner, A.P. & Weiss, R.E. (1985) The colour of the Denver haze. *Proceedings of the 73rd Annual Meeting of the Air Pollution Control Association*, paper 80–58.

Wald, L., Antipolis, S. & Baleynaud, J.-M. (1999) Observing air quality over the city of Nantes by means of Landsat thermal infrared data. *International Journal of Remote Sensing*, **5**, 947–959.

Walker, B.K., Riegl, B. & Dodge, R.E. (2008) Mapping coral reef habitats in southeast Florida using a combined technique approach. *Journal of Coastal Research*, **24**, 1138–1150.

Walsh, J. E., Chapman, W. L. & Shy, T. L. (1996) Recent decrease of sea-level pressure in the Central Arctic. *Journal of Climate*, **9**, 480–486.

Walton, C.C., Pichel, W.G., Sapper, J.F. & May, D.A. (1998) The development and operational application of nonlinear algorithms for the measurement of sea surface temperatures with the NOAA polar-orbiting satellites. *Journal of Geophysical Research*, **103**, 27999–28012.

Wang, C-K. & Philpot, W.D. (2007) Using airborne bathymetric LiDAR to detect bottom type variation in shallow waters. *Remote Sensing of Environment*, **106**, 123–135.

Wang, L., Sousa, W.P. & Gong, P. (2004) Integration of object-based and pixel-based classification for mapping mangroves with IKONOS imagery. *International Journal of Remote Sensing*, **25**, 5655–5668.

Wang, X. & Mauzerall, D.L. (2004) Characterizing distributions of surface ozone and its impact on grain production in China, Japan, and South Korea: 1900 and 2020. *Atmospheric Environment*, **38**, 4383–4402.

Wang, Y. (2010) *Remote Sensing of Coastal Environments*. CRC Press, Taylor and Francis Group, Boca Raton, FL.

Washington, W.M., Parkinson, C.L. (1986) *An Introduction to Three-Dimensional Climate Modeling*. University Science Books, Mill Valley, CA.

Weatherbee, O.P. (2000) Application of satellite remote sensing for monitoring and management of coastal wetland health. In: Gutierrez, J. (ed.) *Improving the Management of Coastal Ecosystems through Management Analysis and Remote Sensing/GIS Applications*. Sea Grant Report, University of Delaware.

Weatherhead, E.C. & Andersen, S.B. (2006) *The search for signs of recovery of the ozone layer. Nature*, **441**, 39–45.

Weeks, W. (1981) Sea ice: the potential of remote sensing. *Oceanus*, **24**, 39–48.

Wehner, M., Oliker, L. & Shalf, J. (2008) Towards ultra-high resolution models of climate and weather. *The International Journal of High Performance Computer Applications*, **22**, 149–165.

Weijers, J.W.H., Schouten, S., Sluijs, A., Brinkhuis, H. & Damste, J.S.S. (2007) Warm arctic continents during the Palaeocene-Eocene thermal maximum. *Earth and Planetary Science Letters*, **261**, 230–238.

Weng, Q. (2002) Land use change analysis in the Zhujiang Delta of China using satellite remote sensing, GIS and stochastic modelling. *Journal of Environmental Management*, **64**, 273–284.

Weng, Q. (2004) Modelling Urban Growth Effects on Surface Runoff with the Integration of Remote Sensing and GIS. *Journal Environmental Management*, **28**, 737–748.

Weng, Q. (2009) Thermal infrared remote sensing for urban climate and environmental studies: methods, applications, and trends. *ISPRS Journal of Photogrammetry and Remote Sensing*, **64**, 335–344.

Went, F.W. (1964) The nature of Aitken condensation nuclei in the atmosphere. *Proceedings of the National Academy of Sciences of the United States of America*, **51**, 1259–1267.

Wild, M. (2009) Global dimming and brightening: A review. *Journal of Geophysical Research*, **114**, D00D16.

Wilen, B.O. & Bates, M.K. (1995). The US Fish and Wildlife Service's National Wetlands Inventory Project. *Vegetatio*, **118**, 153–169.

Wilkinson, B.H. & Drummond, C.N. (2004) Facies mosaics across the Persian Gulf and around Antigua – stochastic and deterministic products of shallow-water sediment accumulation. *Journal of Sedimentary Research*, **74**, 513–526.

Wilkinson, C. (2006) Status of the reefs of the world: summary of threats and remedial action. In: Cote, I.M., Reynolds, J.D. (eds.) *Coral Reef Conservation*, 4–39. Zoological Society of London. Cambridge University Press.

Willis, Z. & Cohen, K. (2007) The way forward on IOOS. *Marine Technology Reporter*, **Oct. 2007**, 28–31.

Wilson, M.F.J., O'Connell, B., Brown, C., Guinan, J.C. & Grehan, A.J. (2007) Multiscale terrain analysis of multibeam bathymetry data for habitat mapping on the continental slope. *Marine Geodesy*, **30**, 3–35.

Wilson, S.K., Dolman, A.M., Cheal, A.J., Emslie, M.J., Pratchett, M.S. & Sweatman, H.P.A. (2009) Maintenance of fish diversity on disturbed coral reefs. *Coral Reefs*, **28**, 3–14

Wingham, D.J., Shepherd, A., Muir, A. & Marshall, G.J. (2006) Mass balance of the Antarctic ice sheet. *Philosophical Transactions of the Royal Society A*, **364**, 1627–1635.

Winterbottom, S.J. & Gilvear, D.J. (1997) Quantification of channel bed morphology in gravel-bed rivers using airborne multispectral imagery and aerial photography. *Regulated Rivers: Research and Management*, **13**, 489–499.

Woldai, T., Oppliger, G. & Taranik, J. (2009) Monitoring dewatering induced subsidence and fault reactivation using interferometric synthetic aperture radar. *International Journal of Remote Sensing*, **30**, 1503–1519.

Wood, R. (1999) *Reef evolution*, 414. Oxford University Press.

World Health Organization (WHO) (2003) *Health aspects of air pollution with particulate working group*. WHO Regional Office for Europe, Copenhagen.

Wray-Lake, L., Flanagan, C. & Osgood, D.W. (2010) Examining trends in adolescent environmental attitudes, beliefs and behaviors across three decades. *Environment and Behavior*, **42**, 61–85.

Wright, P. & Stow, R. (1999) Detecting mining subsidence from space. *International Journal of Remote Sensing*, **20**, 1183–1188.

Wulder, M.A., Franklin, S.E. (2003) *Remote sensing of forest environments: concepts and case studies*. Kluwer Academic Publishers: Boston, MA.

Xian, G. & Crane, M. (2005) Assessments of urban growth in the Tampa Bay watershed using remote sensing data. *Remote Sensing of Environment*, **97**, 203–215.

Yaffee, S.L. (1998) Cooperation: a strategy for achieving stewardship across boundaries. In: Knight, R.L., Landres, P.B. (eds) *Stewardship across boundaries*. Island Press, Washington, DC.

Yamano, H. & Tamura, M. (2004) Detection limits of coral reef bleaching by satellite remote sensing: simulation and data analysis. *Remote Sensing of Environment*, **90**, 86–103.

Yan, X.-H., Ho, C., Zheng, Q. & Klemas, V. (1993) Using satellite IR in studies of the variabilities of the Western Pacific Warm Pool. *Science*, **262**, 440–441.

Yang, X. (2009), *Remote Sensing and Geospatial Technologies for Coastal Ecosystem Assessment and Management*. Springer-Verlag, Berlin

Yao, M.-S. & Del Genio, A.D. (1999) Effects of cloud parameterization on the simulation of climate changes in the GISS GCM. *Journal of Climate*, **12**, 761–779.

Yirdawa, S.Z., Snelgroveb, K.R. & Agbomab, C.O. (2008) GRACE satellite observations of terrestrial moisture changes for drought characterization in the Canadian Prairie. *Journal of Hydrology*, **356**, 84–92.

Yoshioka, M., Mahowald, N.M., Conley, A.J., Collins, W.C., Fillmore, D.W., Zender, C.S. & Coleman, D.B. (2007) Impact of desert dust radiative forcing on Sahel precipitation: relative importance of dust compared to sea surface temperature variations, vegetation changes, and greenhouse gas warming. *Journal of Climate*, **20**, 1445–1467.

Younger, P.L. (2007) *Groundwater in the environment: an introduction*. Blackwell.

Yuan, F. & Bauer, M.E. (2007) Comparison of impervious surface area and normalized difference vegetation index as indicators of surface urban heat island effects in Landsat imagery. *Remote Sensing of Environment*, **106**, 375–386.

Yuan, Y., Smith, R.M. & Limp, W.F. (1997) Remodeling census population with spatial information from Landsat TM Imagery. *Computers, Environment and Urban Systems*, **21(3/4)**, 245–258.

Zainal, A.J.M., Dalby, D.H. & Robinson, I.S. (1993) Monitoring of marine ecological changes on the east coast of Bahrain with Landsat TM. *Photogrammetric Engineering and Remote Sensing*, **59**, 415–421.

Zaneveld, J.R.V. & Boss, E. (2003) The influence of bottom morphology on reflectance: theory and two-dimensional geometry model. *Limnology and Oceanography*, **48**, 374–379.

Zaneveld, J.R.V, Boss, E. & Moore, C.M. (2001) A diver operated optical and physical profiling system. *Journal of Atmospheric and Oceanic Technology*, **18**, 1421–1427.

Zhang, T. & Armstrong, R. (2005) *Arctic Soil Freeze/Thaw Status from SMMR and SSM/I,* Version 2. Boulder, CO: National Snow and Ice Data Center/World Data Center for Glaciology. Digital media.

Zhang, T., Barry, R.G., Knowles, K., Heginbottom, J.A. & Brown, J. (1999) Statistics and characteristics of permafrost and ground ice distribution in the Northern Hemisphere. *Polar Geography*, **23**, 147–169.

Zhang, T., Barry, R.G. & Armstrong, R.L. (2004) Applications of satellite remote sensing techniques to frozen ground studies. *Polar Geography*, **3**, 163–196.

Zhang, X., Friedl, M.A., Schaaf, C.B., Strahler, A.H. & Schneider, A. (2004) The footprint of urban climates on vegetation phenology. *Geophysical Research Letters*, **31**, L12209, doi:10.1029/2004GL020137.

Zhao, Z., Klemas, V., Zheng, Q. & Yan, X-H. (2004) Remote sensing evidence for baroclinic tide origin of internal solitary waves in the northeastern South China Sea. *Geophysical Research Letters*, **31**, L06302.

Zheng, G. & Moskal, L.M. (2009) Retrieving Leaf Area Index (LAI) using remote sensing: theories, methods and sensors. *Sensors*, **9**, 2719–2745.

Zhu, X.R., Prospero, J.M. & Millero, F.J. (1997) Diel variability of soluble Fe (II) and soluble total Fe in North African dust in the trade winds at Barbados. *Journal of Geophysical Research*, **102**, 297–305.

Ziemke, J.R., Chandra, S., Duncan, B.N., Froidevaux, L., Bhartia, P.K., Levelt, P.F. & Waters, J.W. (2006) Tropospheric ozone determined from Aura OMI and MLS: evaluation of measurements and comparison with the global modelling initiative's chemical transport model. *Journal of Geophysical Research*, 111, D19303, doi:10.1029/2006JD007089

Zilioli, E., Brivio, P.A. & Gomarasca, M.A. (1994) A correlation between optical properties from satellite data and some indicators of eutrophication in Lake Garda (Italy). *Science of the Total Environment*, **158**, 127–133.

Zimov, S.A., Schuur, E.A.G. & Chapin, F.S. (2006) Permafrost and the global carbon budget. *Science*, **312**, 1612–1613.

Zumpichiati, W., Domingos, C. & Dias, E. (2005) The use of satellite derived upper ocean heat content to the study of climate variability in the South Atlantic. *Revista Brasileira De Cartografia*, **57/02**, 2005.

Zwally, H.J., Comiso, J.C., Parkinson, C.L., Campbell, W.J., Carsey, F.D. & Gloersen, P. (1983) *Antarctic Sea Ice, 1973–1976: Satellite Passive-Microwave Observations.* NASA SP-459, p. 206. GPO, Washington, DC.

Index

Remote Sensing and Global Environmental Change, First Edition. Samuel Purkis and Victor Klemas.
© 2011 Samuel Purkis and Victor Klemas. Published 2011 by Blackwell Publishing Ltd.